FROM SEED GERMINATION TO YOUNG PLANTS

ECOLOGY, GROWTH AND ENVIRONMENTAL INFLUENCES

BOTANICAL RESEARCH AND PRACTICES

Additional books in this series can be found on Nova's website
under the Series tab.

Additional e-books in this series can be found on Nova's website
under the e-book tab.

ENVIRONMENTAL HEALTH - PHYSICAL, CHEMICAL AND BIOLOGICAL FACTORS

Additional books in this series can be found on Nova's website
under the Series tab.

Additional e-books in this series can be found on Nova's website
under the e-book tab.

FROM SEED GERMINATION TO YOUNG PLANTS

ECOLOGY, GROWTH AND ENVIRONMENTAL INFLUENCES

CARLOS ALBERTO BUSSO
EDITOR

New York

For permission to use material from this book please contact us:
Telephone 631-231-7269; Fax 631-231-8175
Web Site: http://www.novapublishers.com

NOTICE TO THE READER

Additional color graphics may be available in the e-book version of this book.

Library of Congress Cataloging-in-Publication Data

From seed germination to young plants : ecology, growth and environmental influences / editor: Carlos Alberto Busso.
       p. cm.
Includes index.
ISBN: 978-1-62618-653-8 (hardcover)
1. Germination. 2. Seeds--Ecology. 3. Seeds--Growth. I. Busso, Carlos Alberto.
QK740.F76 2013
575.6'8--dc23
                                    2013011834

Published by Nova Science Publishers, Inc. † New York

Contents

About the Editor

Carlos Alberto Busso is Professor of Ecology at the Departamento de Agronomía, Universidad Nacional del Sur (UNSur), and Principal Researcher of the Consejo Nacional de Investigaciones Científicas y Tecnológicas de la República Argentina (CONICET), Buenos Aires, Argentina. He received his Agronomy Engineer and MSc degrees at UNSur in 1978 and 1983, respectively, and his PhD degree in Range Ecology at Utah State University, Logan, Utah, USA. He started his work at UNSur as Teaching Assistant in Ecology in 1979, and continued with no interruption until today in the same discipline. He is currently teaching the undergraduate-level courses on General Ecology and Autoecology of Rangeland Plants, and the postgraduate-level course on How to Write and Publish Scientific Articles. He conducted Postdoctoral Studies during 1995/96 at Texas A&M University, College Station, Texas, USA. He was a Faculty member, Research Assistant Professor, at the Department of Agriculture, Biotechnology and Natural Resources at the University of Nevada, Reno, Nevada, USA, during 2003/04. During 2010, he was invited as a Visiting Professor at the Institute of Ecology, Chinese Academy of Sciences (CAS), Guangzhou, China. In 2012, C.A. Busso obtained the Third World Academy of Sciences Associatedship Award, which allowed him to visit the Institute of Applied Ecology, CAS, Shenyang, China during April and May 2013 and 2014. He is Editor-in-Chief of *Phyton, International Journal of Experimental Botany* since 2005. He has published more than 90 research articles, including books, and over 200 communications to Scientific Meetings (www.rangeecologybusso.com.ar; www.uns.edu.ar), over a wide spectrum of subjects in ecology.

Preface

Plants go through various developmental morphology stages throughout their growing cycle. Among all these developmental morphology stages, germination and seedling establishment are the most sensitive to any disturbance. Once established, the plant must successfully interact with the biotic (disturbances of different origin: anthropogenic, fire, grazing, logging) and abiotic (e.g., drought, extreme temperaturas, salinity, fire, nutrient defficiency; heavy metal toxicity) environmental factors to persist. Persistence of plant species at any place contribute to determine the community structure and ecosystem functioning at that place. Therefore, it is essential to understand the response of plants to changing environmental conditions at different scales of their development, and at various temporal and spatial scales of plant growth and development, within a period of climate change and global warming.

This book deals with the ecology and growth at, and environmental influences on, different developmental morphology stages at various study scales: from molecules to young plants. It contains **18 Chapters** written by leading experts in their respective fields of knowledge and expertise. 'From seed to seedling: An ecophysiological point of view', by *M. Á. Ruiz, A. D. Golberg and M. L. Molas*, constitutes the **1st Chapter** of this book. It is an overview of the fate of seeds in the soil, and the biotic and abiotic factors affecting the processes of germination, emergence, and seedling recruitment and establishment. The next four chapters deal mostly with basic research: they are as follows: A contribution to a broader understanding on the translocation and utilization of storage compounds during germination and greening of *Cucumis sativus* seedlings, with respect to the formation of aromatic compounds, is explained in **Chapter 2** by *P. Siekel, J. Stano, K. Mičieta and M. Koreňova*. Light-induced exaggeration of the hypocotyl hook – Its developmental basis and significance, is the subject of **Chapter 3** by *Ch. Chichijo and T. Hashimoto*. These authors studied some physiological and abiotic factors that affect the light-induced hook exageration (LIHE). Species either exhibiting or not LIHE presented different morphophysiological characteristics. This chapter is illustrated with various, nice movies. The hypothesis on the existence of an ''oxidative window'' for germination is discussed by *H. Causin, G. Roqueiro and S. Maldonado* in **Chapter 4**. This window defines (1) that a critical free radical level must not be overcome, which would otherwise affect normal germination and eventually prevent it, and (2) a reactive oxygen species (ROS) threshold level below which root growth (and/or normal development) cannot occur. Within this oxidative window, specific ROS may play various roles (e.g., defense against pathogens, modification of cell wall properties, cell signaling). The next **Chapter 5** deals with the role of root border cells in the formation of a root-microenvironment system in 1- to 3-day-old wheat seedlings by *A.L. Bozhkov, Y. A. Kuznetsova, N.G. Menzyanova and M.K. Kovaleva.* These authors indicate the quantitative and qualitative composition of the root microenvironment, which depended on the root growth rate. Also, they demonstrated that the border cell number increase in the root microenvironment was not accompanied by an increased content of carbohydrates and proteins in the root exudates. Besides reporting antibacterial activity for such root exudates, they discussed the possible mechanisms of autoregulation in the root-microenvironment system. The remaining thirteen chapters are more related with applied research in various types of ecosystems. In the **6ᵗʰ Chapter,** *S.M. Smith* provides valuable information that can be used to predict the potential for passive (i.e., natural) or active (i.e., managed) re-vegetation of dieback areas in the salt marshes of the Cape Cod National Seashore in Massachusetts. The important role of the availability of safe sites for the establishment of new individuals and plant species regeneration is the theme of the **7ᵗʰ chapter** by *A. Loydi and G. Peter.* In **Chapter 8**, factors that affect germination of the cypselas of *Hyalis argentea* (Asteraceae) are addressed by *J. L. Camina, E. Tourn, A. Andrada and C. Pellegrini.* Results showed that *H. argentea*, which often grows successfully on sand dunes, was insensitive to light. This represents an advantage to seeds of this species which are either totally covered or exposed during the movement of dunes due to winds. Greater seed germination during short (i.e., autumn) than long (i.e., spring) days may allow seedlings of this species to develop a sufficiently mature and deep root system that might allow a greater competitive and survival ability during the summer. The physiological processes involved in the adaptation to water stress on seedlings of three native grasses of semiarid Argentina are detailed in the **9ᵗʰ Chapter** by *R.E. Brevedan, M.N. Fioretti, Sandra S. Baioni and C. Cabeza.* In **Chapter 10**, *Y. A. Torres, C. A. Busso, O.A. Montenegro, H. D. Giorgetti, G. D. Rodríguez and L. S. Ithurrart* compared the performance of young plants of native and introduced, warm-season rangeland grasses in arid Argentina. This was because of the need of introducing water-stress tolerant grasses in this region that characterizes by insufficient palatable forages to domestic livestock during the warm-season. They found that various morphophysiological traits, but plant establishment, were greater or similar, but not lower, in the introduced than in the most abundant native genotype. As a result, these authors emphazised that future research is needed to substantially improve plant establishment of the introduced, study genotypes, which have the advantage that conventional drilling can be used for its seeding. In **the 11ᵗʰ Chapter**, *P.M. López Bernal, M.F. Urretavizcaya and G.E. Defossé* make an excelent review of the most important environmental and biotic factors that influence seedling dynamics of three conspicuous species grown in a forest-steppe gradient in Patagonia, Argentina. The effects of water stress and temperature on the germination, and water stress on the early seedling growth under controlled conditions were assessed on the grass *Digitaria eriantha* to determine its potential for revegetation in the arid parts of Argentina; this was the theme of **Chapter 12** by *R. E. Brevedan, C. A. Busso, M. N. Fioretti, M. B. Toribio, S. S. Baioni, Y. A. Torres, O. A. Fernández, H. D. Giorgetti, D. Bentivegna, L. Entío, L.S. Ithurrart, O. A. Montenegro, M. M. Mujica, G. D. Rodríguez and G. Tucat.* In **Chapter 13**, *N. J. Carnevale, C. Alzugaray and R. M. Freire* made clear that deforestation followed by agriculture in the Humid Chaco Region, Argentina, has caused top-soil salinization of large areas. They studied the germination and early seedling responses of seven tree and two shrub, native species to increasing concentrations of sodium chloride (and polyethylen glycol). Results showed that all study species behaved as glycophytes, or slightly halophytes. Whatsoever, some of them showed a possible adaptation to salinity increases. The environmental and physiological requirements to optimize germination and growth of some native species of the High Mediterranean Mountains in Sierra Nevada, Spain, are explained in **Chapter 14** by *F. A. Serrano-Bernardo, J.J. De la Torre-Betts, M. Beltrán-Hermoso, K. Garcete and J. L. Rosúa-Campos.* Results from this study might contribute to restore degraded lands, and reestablish the vegetation cover, in an area where skiing is common. The fact that (1) management of native mycorrhizal strains is more appropriate than using introduced fungal strains to improve growth of Malagasy endemic tree seedlings, and (2) use of sun-tolerant shrubs, which can have a positive effect on soil mycorrhizal communities, is of great importance to the plantation program of endemic trees or to the forest ecosystem regeneration in Madagascar is the subject matter of **Chapter 15** by *H. Ramanankierana, R. Baohanta, J. Thioulouse, Y. Prin, H. Randriambanona, E. Baudoin, N. Rakotoarimanga, A. Galiana, E. Rajaonarimamy, M. Lebrun and R. Duponnois.* In **the next Chapter 16**, *T. Massad* elaborated a delicious review article that shows the intricacy of the herbivory-tropical forest seedling interactions: the importance of the abiotic environment-herbivory interactions in modyfing plant investments relatively more into storage or defense compounds, or growth; the connection between herbivory and plant diversity, and the importance of knowledge on plant-herbivore dynamics to successfully address reforestation. *Kojiro Suzuki* discusses the importance of the leaf litter layer (i.e., forest floor) for the developmental morphology of spring ephemeral species in the deciduous, broad-leaved forests of Satoyama, Japan in **Chapter 17**. Finally, in **Chapter 18**, *N.S. Aggangan, M.R.R. Edradan, G.B. Alvarado, P.C. Macana, L.K.S. Noel and D.R.R. Edradan* showed that *Acacia mangium* inoculated with native mycorrhizal fungi was the most prominent plant species for contributing to the rehabilitation of marginal and mine tailing sites at some locations in the Philippines.

The study of such broad coverage in this book would have not been possible without the assistance from many notable experts. I thank all individuals who contributed their knowledge and experience as either articles or reviews in this book. Finally, I thank Dra. Cecilia Pellegrini for incorporating the editorial corrections which I made to all chapters in this book, and Nova Science Publishers, Inc. for the cover photo.

Carlos Alberto Busso

Acknowledgments

The study of such broad coverage in this book would have not been possible without the assistance from many notable experts. I thank all individuals who contributed their knowledge and experience as either articles or reviews in this book. Finally, I thank Dra. Cecilia Pellegrini for incorporating the corrections from my editing to all manuscripts in this book, and Nova Science Publishers, Inc. for the cover photo.

Invitation Review Chapter: The Basic Ecological and Physiological Processes

In: From Seed Germination to Young Plants
Editor: Carlos Alberto Busso

ISBN: 978-1-62618-653-8
© 2013 Nova Science Publishers, Inc.

Chapter 1

From Seed to Seedling: An Ecophysiological Point of View

*María Ruiz[1], Alberto Golberg[*2] and María Lía Molas[2]*

[1]Estación Experimental Agropecuaria Anguil, INTA, Ruta
Anguil, La Pampa, Argentina
[2]Facultad de Agronomía, Universidad Nacional de La Pampa,
Santa Rosa, La Pampa, Argentina

Abstract

The initial phase of the plant life is a major event in their life cycle. The present chapter addresses this subject from an ecophysiological viewpoint. We will first examine the germination process, and then we will discuss seed dormancy, taking into account that this process has major consequences for the plant's future individual fitness. The soil seed bank, including quiescent and dormant seeds, is the third topic. The various environmental constraints that diminish or inhibit the ability of seed germination are discussed in the section dedicated to abiotic stress. Thereafter, we will examine some ecological topics such as the effect of fire on germination, and the recruitment of seeds. Finally, some aspects related to seedling establishment will be examined.

Introduction

The process that starts with the seed maturation and ends with the seedling emergence is a highly complex event that can last for years if seeds have some mechanism of dormancy, or several months if an environmental factor is limiting. In the last case, the seed is in a quiescence state until appearance of that factor –light, water, temperature, oxygen concentration- to meet the requirement. Hence, from the ecophysiological viewpoint, processes preceding the emergence become important. This chapter will examine the various

---

[*] Corresponding author: E-mail address: golberg@cpenet.com.ar.

complex processes that occur when a seed comes in contact with surrounding water, thereby triggering germination until seedling establishment.

Germination

In seed plants -Spermatophytes, from the Greek *sperma*= seed; *fiton*= plant- seed germination is a process of enormous relevance. In fact, this process starts the life cycle and places the future plant under appropriate environmental conditions that will ensure its ecological success. Many seeds of non-cultivated species, and some cultivated as forage, have sophisticated mechanisms to monitor the environment where they must survive and reproduce. Plants, unlike most animals, must live in the place where they began their existence, that is, where germination occurred.

Germination can be defined as the set of events triggered by water absorption by the quiescent seed, which concludes with the elongation of the embryonic axis [1, 2]. Thus, the visible sign that germination has been completed is the radicle going through the structures surrounding the embryo, which is called visible germination. The seed contains an embryo, which represents a miniature plant. The embryo is structurally and physiologically equipped to regenerate one individual. Thus, the seed must be well provided with reserves to sustain the growing embryo until the seedling become independent from seed reserves and develops into an autotrophic organism [3].

The first physiological process related to germination is the absorption of water and, consequently, the seed imbibition. Once this event is complete, the seed that was in a state of quiescence quickly resets its metabolic activity, and the respiratory activity becomes important. After this restoration, respiratory rate decreases momentarily until the root penetrates the surrounding structures and, finally, a new peak of respiration takes place [1]. These events are associated with the beginning of pentose cycle, glycolysis and the Krebs cycle, as well as the mobilization of seed reserves towards the embryo [4].

The emergence of the radicle through the structures surrounding the embryo is the event that determines the end of germination and marks the beginning of seedling growth [3]. The elongation of the radicle is a process associated with increased cellular turgor and yielding of the cell walls from the embryonic axis, located between the root cap and the base of hypocotyls [5]. According to Bewley [3] there are three possible causes for the radicle growth. In first term, the osmotic potential of the radicle during germination might be more negative than the apoplast and surrounding cells, therefore the flow of water towards the root cells produce an increase in turgidity and expansion [6]. A reduction in the osmotic potential would occur by accumulation of solutes derived from the hydrolysis of polymers located inside the radicle. However, this explanation presents a drawback, since no change has been verified in the osmotic potential of the radicle cells during germination [6].

The second possibility is linked to the extensibility of cells from the radicle, which would allow elongation. However in this case there is also a weakness, since no particular differences are observed in the cell wall structure of radicle cells and other surrounding cells. The cell wall loosening may result from the cleavage and rejoining of xyloglucans that bind to the cellulose microfibrils [7] as well as the action of expansions [5]. Again, there is no evidence to support this hypothesis since cell wall proteins that loose cell walls have not been observed in the radicle. Moreover, it is known that both expansins and xyloglucan endotransglycosylase (XET complex) activities are increased by the action of indole acetic acid, (IAA), but this hormone resulted ineffective in promoting germination [3].

The third possibility is the weakening of tissues around the root apex, thereby allowing the radicle elongation [3]. This might be true if the cell turgor of radicle cells is sufficient to permit cell expansion, considering that the surrounding tissue exerts no resistance to root elongation. This last possibility seems most likely, since measurements of the mechanical strength of the structures surrounding the root cap have been measured in the endosperm of pepper seeds. In coincidence with the emergence of the radicle, a decrease in the resistance of the cell wall -attributable to the action of hydrolases as hemicellulases- was observed [8].

Once growth of the radicle is initiated within the seed, it must overcome the resistance that opposes the seed coat. In some cases there is a breakdown of this structure because of the force of the growing radicle or an increasing volume due to imbibitions of the endosperm. In other cases there are formations in the tissue, resembling valves, allowing the passage of the radicle [9]. Finally, operculum is mentioned as a structure through which the seed opens like a box, enabling the emergence of the radical [4].

Hormones play a major role throughout the germination stage. It has been cited that the interaction of phytohormones such as abscisic acid (ABA), gibberellins (GA), ethylene (ET), brassinosteroids (BR), auxin (IAA) and cytokinins (CK) are involved in different processes. Abscisic acid is a positive regulator of dormancy, while GA promotes germination releasing the effects of ABA. Ethylene and BR also promote germination and counteract the effects of ABA [2].

Among the main abiotic factors affecting germination we can mention temperature, water, light, oxygen and carbon dioxide. Temperature has an effect on the germination rate through three independent physiological processes: firstly, seeds that remain in a quiescent state, both in the soil seed bank or through a conservation device, undergo a continuous deterioration of its viability and eventually die, unless they find favorable conditions for germination. The rate of degradation depends mainly on the humidity and temperature. The Q_{10} of the viability loss of orthodox seed increases by about 2 to 10°C, and to 10 to 70°C. The loss of quiescence depends on temperature and other factors such as hydration, presence of oxygen, and light in photoblastic species. There are species that responds to thermoperiod for germination instead of continuous temperature. Once the seeds lose their quiescence, the germination rate shows a linear and positive response ranging from a base temperature (i.e., a temperature below which no germination occurs) to an optimum temperature (i.e., the temperature of maximum germination). After the optimum has been reached, germination rate becomes negative [10].

The relationship between seeds and water is vital; from the point of view of adaptation to the environment, the seed must be able to integrate their history and sense the local environment in order to determine the chances of seedling survival. In this sense, water availability is one of the most critical factors in determining the survival of seeds [11]. When the water potential increases during seed imbibition the absorption gradient drops gradually to reach a plateau. This plateau is considered germination *sensu stricto*, and ends with the emergence of the radical [12].

The need of a certain light wavelength to seed germination in the lettuce cultivar Grand Rapids was described many years ago by Flint & McAllister in 1936. Subsequently, Borthwick et al. [13] determined the spectrum of the photoreaction, further showing that this was reversible; several years later the pigment involved in this process was discovered and termed phytochrome [14]. After such discovery, seeds of many species demonstrated to germinate only under certain photoperiods [14, 15].

Oxygen and carbon dioxide are the most important gases acting in germination. Oxygen is relevant for germination in a wide range of crops, such as rice, wheat, corn, soybean, sunflower, sorghum, pea, radish, lettuce, turnip, cabbage and flax. It was observed that maximum rates of respiration and germination were obtained at oxygen partial pressures close to its concentration in the air. Decreases in the gas partial pressure produced a gradual decrease in germination rates. According to the response to oxygen concentration, species could be divided into two groups: I, consisting of lettuce, sunflower, radish, turnip, cabbage, flax and soy, whose germination is inhibited at an oxygen partial pressure of 2 kPa; II, including rice, wheat, maize, sorghum and peas where the germination rate decreases gradually as the gas partial pressure also decreases, but they are unable to germinate at pressures as low as 0.1 kPa. The difference between the two groups could be attributed to different sensitivities of breathing oxygen [16].

The effect of CO_2 on germination is negative, unlike oxygen. However, its deleterious action on germination has been overestimated according to Grable & Danielson [17]. For instance, germination was not affected by CO_2 concentrations higher than those found in the soil in soybeans and corn. When the gas concentration was high enough to inhibit germination, the effect was reversed as soon as the seeds were exposed to air. Similar results were found in peas, cabbage, barley, beans and *Brassica* by Kidd [18]. This author observed that germination was delayed or inhibited by CO_2 partial pressures above the atmospheric concentration, but the effect was fully reversible when seeds were placed in an atmosphere having a standard gas concentration.

Seed Dormancy and Seed Bank

Dormancy

Once the seed has fully developed, changes that will lead to the establishment of a resting state are initiated [19]. This resting state, called quiescence, is accompanied by a decreased metabolism, basically due to the lack of water. Instead, we speak of dormancy when the seed does not germinate despite being under optimal conditions in terms of temperature and humidity [3]. Thus, we can define dormancy as follows: *Dormancy is an internal condition of the seed that impedes its germination under adequate hydric, thermal and gaseous conditions.* This means that once the impedance has been removed, seed germination would proceed under a wide range of environmental conditions [20].

The dormancy is an adaptive strategy for survival under unfavorable environmental conditions. The establishment of dormancy is genetically controlled through endogenous physiological mechanisms, which, in turn, interact with environmental factors. These include fluctuations in temperature and humidity, microclimatic variations (e.g. light spectral quality and thermoperiod) as well as site-specific characteristics at which plants are adapted to. Climatic variations, in addition to plant hormonal and nutritional conditions, have a large influence on the establishment of seed dormancy during its development. Hence, depending on the season and location, seeds of different harvests from a single species might differ in their degree of dormancy [21, 22].

Seed dormancy status can vary in a continuous scale between some point where dormancy is maximum and some point where it is minimum [23]. The degree or level of dormancy of a seed population establishes the width of the range of environmental conditions that will allow germination. A low dormancy level is characterized by a wide range of environmental conditions enabling seed germination, while seeds presenting a high dormancy level show a narrow range of environmental conditions enabling seed germination [24].

Two types of seed dormancy can also be distinguished on the basis of the timing of dormancy onset: a) innate or primary, b) induced or secondary.

Innate or primary dormancy occurs when the embryo ceases to grow and the seed is still in the plant. This constraint continues until the endogenous impediment comes to an end and the seeds are able to germinate under appropriate environmental conditions [3]. More likely, the main causes of this impediment are the presence of chemical germination inhibitors or the embryo immaturity. The duration of innate dormancy is highly variable depending on the species; in some cases can even vary between seeds from the same individual [25-27].

Regarding this type of dormancy, we performed studies on native species of the genus *Bromus,* and two *Panicum* species of importance as forage in the semi-arid pampas of Argentina: *Panicum virgatum* and *Panicum coloratum* [25-27]. In *P. virgatum* it was suggested that the floral parts surrounding the seed could be blocking the release of inhibitors, as the presence of ABA was detected. Also, the water penetrates the seed, hence there is not a problem of coat impermeability [28]. On the other hand, just harvested seeds of *P. coloratum* exhibit dormancy, which last about 6 months after storage at room temperature [29]. According to these authors, scarification with sulfuric acid (chemical scarification) could remove an inhibitor or weaken the bracts so that the inhibitor can be released. However, it has been suggested that different levels and dormancy mechanisms are acting in concert; thus, the response to different treatments changes as the post-harvest period elapse [29].

In *P. coloratum* cv. Green, the duration of dormancy in seeds from different lots, and the effect of storage temperature on dormancy were evaluated [26]. We examined two samples that were harvested in the Province of La Pampa, Argentina. Regarding the duration of dormancy, the sample that was manually harvested lost its dormancy after five to seven months from storage, reaching more than 90% germination of viable seeds after that period. Instead, when a harvester was used, seeds presented a high germination capacity after a month post-harvest, possibly due to a mechanical scarification during threshing. In both samples, low storage temperatures (7 and -20°C) kept dormancy for seven months [26].

Similar evaluations were conducted in 11 cultivars of *P. virgatum* produced in the Province of La Pampa, Argentina [27], where cultivars responded in a different manner. Cultivars Pahfinder, Cave-in-Rock, Trailblazer, Alamo, Kanlow and Greenville showed an increased ability to germinate up to 210 days post-harvest, which may be due to the loss of dormancy. On the other hand, Alamo, Caddo and Pizzo decreased their germination capability, while Blackwell and Summer did not show a definite pattern. Similarly to *P. coloratum* [26], *P. virgatum* prolonged seed dormancy under lower storage temperatures (7 and -20°C) compared with results at room temperature.

Induced or secondary dormancy occurs when seeds are able to germinate, but they do not do it because of an adverse environment (e.g., high CO_2 concentrations, high temperature, anoxia). These unfavorable environments can produce physiological changes in the seeds that are reversible. In these cases, the seeds fall into a state of secondary dormancy, and they cannot germinate despite they are still alive.

The release from primary dormancy followed by a subsequent entrance into secondary dormancy, whenever conditions are given for this entrance, may lead to dormancy cycling. Evidence for dormancy cycling has been obtained for seeds of many weed species [20].

Dormancy might be originated by seed internal barriers or external impediments. On this basis, two types of seed dormancy have been recognized: coat-imposed dormancy and embryo dormancy. *Coat-imposed dormancy* is imposed on the embryo by the seed coat and other enclosing tissues (e.g., endosperm, pericarp). The second dormancy type, the *embryo dormancy*, refers to a dormancy that is inherent to the embryo, and is not due to any influence of the seed coat or other surrounding tissues. Embryo dormancy is due to the presence of inhibitors (e.g., ABA), the absence of growth promoters (e.g., gibberellic acid-GA] and/or embryo inmaturity [30].

Many plants produce seeds with a hard external coat which is an obstacle to water or gases; even the micropyle is provided with a barrier that prevents water penetration to the embryo. This feature is common in several plant families, particularly in the Fabaceae, Bombacaceae and Malvaceae. These seeds are impermeable to water and gases and require scarification before planting [31].

In forest soils, the seed coat gradually becomes permeable by weathering, microbial degradation, or the effect of temperature fluctuations; finally, the seeds germinate slowly. Often, transit through the digestive tract of animals is a major factor for disrupting the coat-imposed dormancy. In the Phytogeographical Province of the Espinal, Argentina, the typical example is caldén (*Prosopis caldenia* Burk), a tree whose pods are eaten by cattle because of its high nutritional value [32].

Another example of mammalian herbivores playing a role in the endozoochorous dispersal of *Prosopis* seeds in the arid zones of Argentina was studied by Campos et al. [33]. They observed that the introduction of exotic mammals can change relevant parameters of the plant reproductive ecology. They quantified seeds of *Prosopis flexuosa*, *P. chilensis* and *P. torquata* contained in the feces of the following mammals: native mara (*Dolichotis patagonum*), guanaco (*Lama guanicoe*), exotic donkey (*Equus asinus*) and hare (*Lepus europaeus*). The mortality and germination percentage and speed of seeds borne in the feces was compared with those collected from the trees. *Prosopis torquata* and *P. flexuosa* seeds were found in the dung of guanaco, hare and mara, whereas only *P. flexuosa* seeds appeared in the donkey's feces. *Prosopis chilensis* seeds were only found in guanaco's feces. There was a notable relative abundance of seeds in small herbivore's feces in comparison to large herbivores's. The passage through the digestive tract of animals modified seed germination capacity and speed, with great variability between animal species. Guanacos had greater effects than donkeys on mortality, germination percentage and germination speed. Differences on seed responses were much smaller among smaller-size herbivores [33].

Environmental Control of Dormancy

In most cases, environmental variables such as soil temperature and humidity play a central role in breaking dormancy, and define the pattern of emergence [34]. There are two different kinds of environmental factors affecting dormancy, those that govern changes in the degree of dormancy of a seed population –i.e. temperature and its interactions with hydric conditions-, and those that remove the ultimate constrains for seed germination once the degree of dormancy is low enough –i.e. light, fluctuating temperatures, nitrate concentration [20].

In adapted species, dormancy is either released or alleviated during the season preceding the period with favorable conditions for seedling development and plant growth, and secondary dormancy is induced in a period preceding the season with environmental conditions unsuitable for plant survival [20].

Temperature

In summer and winter annual species, the high seed dormancy immediately after dispersal is shown by the fact that germination does not occur at any temperature. When the seed population is released from dormancy, the thermal range that allows germination expands. In summer annuals, this expansion occurs through a progressive decrease in the minimum temperature for germination and, in winter annuals, through a progressive increase in the maximum temperature for germination. Re-induction of dormancy results in a narrowing of the permissive thermal range through an increase in the minimum temperature for germination in summer annuals, and a decrease in the maximum temperature in winter annuals. In both cases, germination occurs in the field when soil temperatures enter the permissive range [20, 35].

Thermoperiod

Inside the forest, the temperature in the soil surface remains relatively constant during the day and night, while in open spaces it can change to 10°C per day. In temperate and cool regions, the seasonality of temperature is critical for germination. Some species produce seeds whose germination is favored by alternating temperatures either daily or seasonal. Seeds that respond to this environmental change have various sensing mechanisms (e.g., the presence of an impermeable coat that becomes permeable once heated). Another case is the existence of an endogenous chemical mechanism that only can activate germination when temperature fluctuations occur. It is thought that the fluctuation of temperatures allows the activation of certain enzymes and makes permeable some membranes, which finally brings the onset of germination [36].

The ecological significance of this requirement has been related to the possibility of detecting canopy gaps and depth of burial [37-39]. It could also act as an effective mechanism to distribute germination over a longer time period [40, 41]. This is the case of seeds from several weed species, such as Johnsongrass (*Sorghum halepense* L.), one of the most widespread in Argentina [39]. This grass responds to fluctuating temperatures, which is directly linked to the depth of burial. The loss of dormancy in Johnsongrass occurs after exposure to fluctuating temperatures [40].

Light

An ecological interpretation of the requirements of light and alternating temperatures to complete dormancy release in many wild species has been related to the possibility of detecting canopy gaps, light flashes during tillage operations, and depth of burial under field situations [20, 42].

The main wavelengths that influence on germination are red (R, 660 nm), far red (FR, 730 nm) and blue light (400 and 500 nanometers), although the impact is far less clear in the last case. In areas covered with vegetation the light reaching the soil is poor in R and rich in FR. However, in open places both wavelengths come in equal proportion, because the light is not filtered through the canopy. The mechanism of light uptake by the phytochrome is mainly sensitive to the quality of light, allowing seeds to detect the proportion of R:FR reaching the ground. Thus, seeds from sunny plants can remain dormant when they are dispersed to places that are densely covered with vegetation since these conditions are not favorable for establishment [43, 44].

This type of dormancy regulated by light is very frequent on seeds that remain dormant in the dark or buried in the ground until they are exposed because of agricultural practices. For example, seeds of *Datura ferox* (jimson weed; a weed of summer crops) possess a very high sensitivity to light and depend on vegetation cover or depth of burial. Thus they are affected by agricultural practices [45].

The regulation of germination produced by temperature fluctuations, light quality or a combination of both factors is common in colonizing plants. Under field conditions, light and alternating temperature are the most important factors interrupting dormancy of buried seeds. However, there are many other factors that can break dormancy under specific field conditions (e.g., nitrate, ethylene, carbon dioxide) [24].

In temperate zones, seeds of many species respond to photoperiod, germinating just under either long days or special photoperiods, thus preventing germination during unfavorable times [14].

Fire

High temperatures caused by fire can break the seed teguments. This type of dormancy is particularly common in Fabaceae, Malvaceae, and other plants that germinate plenty in soils of recently burned forests or other communities (e.g., California chaparral, pine forests and some tropical savannas). Teguments may also change their structure after being exposed to direct sunlight for prolonged periods. For example, the Ericaceae family includes some of the characteristic shrubs species from the European Atlantic region with such requirements [46]. This author has studied the germination response to fire of different shrub species in relation to different temperature regime and smoke. Seeds of *Calluna vulgaris, Erica ciliaris* and *E. tetralix* increase their germination with temperatures reaching 150°C. The smoke effect is clearly observed in *C. vulgaris, E. erigena, E. scoparia, E. umbellata* and *E. vagans*, which increase germination [46].

Soil Seed Bank

Seed bank is the reserve of viable seeds in the plants -aerial seed bank-, buried in the soil or mulch in the vegetable waste -soil seed bank- [10]. The soil seed bank (SSB) is the regeneration potential of plant communities [47], and is also an important component of plant dynamics, constituting a survival strategy of the species over time [48, 49]. The SSB can play a key role in the recovery of areas submitted to drastic disturbance processes. Thus, it is necessary to implement management and conservation practices of seed banks, for the maintenance of the floristic diversity and the social and ecological sustainability of ecosystems [50]. Due to the intense anthropogenic degradation, there is an increased interest in studying the SSB in order to develop predictive models of plant succession [51, 52]. For instance, it should be discussed the inclusion of genetically modified organisms in agricultural lands. This is because of the transgenic canola (*Brassica napus* L.), that has the potential to be established as a new glyphosate resistant weed able to remain in the SSB up to 10 years [53].

Formation of Seed Banks in the Soil

The formation of a SSB is determined by the horizontal and vertical movement of seeds. It starts with the dispersal and primarily ends with the death or germination of seeds. Successful dispersal depends on factors strictly related to seed: production time and amount, way of transport, dispersal distance and seed dispersal index [54]. Seed horizontal transfer is due to the action of animals, rain, and runoff of rounded shape seeds. An example of such dispersal in the semiarid region of Argentina is the truck clover (*Medicago minima* L. Bartal), a naturalized legume whose pods containing the seeds, have small hooks that attach to the sheep wool contributing to dispersion [55, 56].

The vertical movement occurs because the seed penetrates into the soil through channels of biological origin, either by the action of animal or plant roots. Also, small seeds may penetrate the soil due to rain or settle in fissures caused by drought. There is a close relationship between seed size and the feasibility of soil penetration. In addition, tillage can bury the seeds regardless of their size. There are other types of seeds with particular features such as long, hygroscopic awns that contribute to burial. An example of this trait can be observed in many species of *Nassella* and *Piptochaetium,* from the semiarid, central region of Argentina, commonly known as "flechettes" by its similarity to arrows [55, 57]. Hence, seed penetration in the soil is a crucial fact because the burial is associated with the persistence of seeds.

In addition to dispersal strategies, another mechanism contributing to the survival and establishment of seed banks is the distribution of germination over time. Thus, all the seeds produced and dispersed in a given year do not necessarily germinate the following year since many of these seeds remain dormant [3, 57].

Types of Soil Seed Bank

Different authors propose to classify the SSB based on characteristics that contribute to the seed bank persistence, such as viability and longevity. According to seed viability, Thompson and Grime [37] classified the SSB into two categories: transitory SSB, where no seeds remain viable for more than a year, and persistent SSB, that include those seeds viable for more than one year. The same authors compared the germination of seeds present in the seed bank at a morphophysiological level. Based on the arising information, they extended the classification to four categories: *type I* consists of annual and perennial grasses from dry or disturbed habitats; *type II*, is represented by annual and perennial species that colonize areas in the early spring; *type III*, includes species that germinate in the fall, but maintain a small persistent seed bank, and *type IV*, consisting of annual or perennial herbs and shrubs with a large number of persistent seeds [50].

Walck et al. [58] divided the SSB into three types: a) transient: species which persist in the soil less than a year, usually a few months, b) short-term persistent, with seeds that persist in the soil between 1 and 5 years; c) long-term persistent, with seeds that persist in the soil for at least 5 years. Only this last type, termed "permanent" [59], is the contributor to the regeneration of degraded plant communities [50].

Role of Dormancy on Soil Seed Bank

Many species persist in the soil for years or decades as dormant seeds [60]. For most species, the role of dormancy in the persistence of the SSB is to regulate the time of the year in which they will respond to environmental stimuli, triggering or inhibition of germination [61]. Thus, dormancy is important in the fitness of the seed bank. For instance, Mayor et al. [62] suggest that the existence of dormancy in *Stipa tenuis* (syn.: *Nassella* tenuis) and *Piptochaetium napostaense* - dominant perennial forage grasses in the Caldenal of central Argentina- may contribute to explain that these species are dominant grasses, considering the occurrence of periodic disturbances (e.g., drought, fire).

Experiments with *B. napus* suggested that secondary dormancy acts as a mechanism of persistence in the seed bank. These authors observed that persistence of *B. napus* may be reduced by the increasing low dormancy genotypes and avoiding seed burial for one year after seed bank establishment [52, 63].

Longevity and SSB

The ability of seeds to prevent germination and still remain viable for different periods of time in the communities where they belong is known as *ecological longevity*. In all cases, their persistence as viable seeds depends on avoiding germination under adverse circumstances [60]. The survival of seeds and size of the SSB depend on the conditions of burial and habitat; seed age and density, and predation which can affect the inflow of seeds in the soil [64]. The crucial step in the formation of a persistent seed bank is seed burial. If seeds remain on the soil surface they probable will germinate or will be subject to predation; instead, both situations become unlikely after burial. In the Caldenal –central Argentina-, the relatively high number of damaged seeds suggests that predation may be an important factor in the seed bank dynamics [62].

It has also been suggested that the seed chemical composition can influence both its dispersion and ecological longevity. For instance, the presence of lipids makes seeds lighter, but also more sensitive to damage by lipid peroxidation [65]. Similarly, production of secondary metabolites such as tannins (i) can exert a repellent effect against predators, and (ii) make the seed covers impermeable to water and oxygen, thus contributing to their persistence in the SSB [66, 67].

Effects of Grazing on the Soil Seed Bank

Grazing may (i) affect the density and composition of seed banks through its influence on plant survival and seed production [68]; (ii) modify the persistence of seeds in the bank [69], and (iii) alter the relationship between the seed bank and the abundance of established species [70-72].

Results on the effect of grazing on the formation of seed banks in grassland are contradictory and a single pattern can not be established [73]. The effect of grazing on the standing vegetation and the germinable soil seed bank was studied in natural mountain grasslands in the Córdoba Province, Argentine. The germinable seed bank was analyzed at two depths (0-5 and 5-10 cm) in two different sites: (a) sites exposed to moderate to intense grazing, and (b) sites excluded from herbivores during the last 10 years. Grazing did not produce significant changes in the seed bank richness and diversity. Average seed bank density did not change after grazing. The only exceptions were the annual grass *Muhlenbergia peruviana*, where seed density of increased, and *Deyeuxia hieronymi*, where seed density decreased in grazed plots [73].

In La Pampa Province, Argentina, it was found that grazing affected the seed bank. The vegetation cover was highest in patches dominated by non-forage species, while diversity and richness were more important in patches dominated by co-dominant species and species producing lower amounts of forage [74].

Effect of Fire on the SSB

In semiarid regions, the dynamics of a plant community is partially conditioned by the occurrence of sporadic fires because they might affect individual plant survival, and subsequently, alter the SSB with consequences in the species composition of the community [75]. For instance, in the Caldenal, Argentina, fire is an important disturbance that, together with grazing and precipitation, shapes the community structure [76-78].

Such environments are periodically subjected to prescribed burns. Prescribed burning is a controlled use of fire by qualified personnel, in a specific area under selected weather conditions. This practice, regulated by law, has the purpose to accomplish well defined management objectives confining the fire to the treated area in a security framework [78]. Fernandez et al. [79] found that after a fire, in areas dominated by forage species, propagules of non-forage species increased. On the other hand, in areas dominated by non-forage species, controlled fire increased seeds of forage and non-forage species. At a site within the Phytogeographical Province of the Monte, a semiarid region of Argentina, no differences were found in the SSB between forbs and grasses in areas that endured fires of different severities [80]. Other studies have shown increased species diversity and decreased seed viability of the SSB in the upper layers [81]. Some herbaceous species increased after fire, indicating that fire promoted germination of some species [82, 83].

Finally, the SSB should be understood as the set of viable seeds contained in the soil and lying in the soil surface, which determine the current floristic resource and the regenerative potential of a plant community. The relationships between persistence, dispersal, seed size and predation have been the subject of numerous investigations. Hence, it is crucial to know the potential effects of climatic and anthropogenic disturbances, such as grazing and the use of fire, on the SSB and the plant communities for contributing to the sustainable use of natural resources.

Germination and Environmental Stress

It is a complex task to rank the different stages of the plant life cycle, assigning a relative importance to each of them. However, from the perspective of the ecological success of a population, germination is a critical stage. In the seeds of species that do not have a history of domestication, the genetic heritage has a significant amount of information related to environmental monitoring. The more restrictive the environment in which the seedling will have to grow, the more sophisticated the monitoring and alarm system developed by the seed.

Is it in the seed, in the first place, where evolution has placed the mission of survival of a population in a given ecosystem?

The stages of seed germination and emergency must overcome several environmental restrictions and should also anticipate future environmental conditions to prevent plant death. Seeds of species termed "ephemeral", which grow in arid regions where rains are occasional and torrential, have developed very sophisticated mechanisms which allow the plant to allocate its life cycle in the most favorable period for their survival.

During germination, seeds must face several environmental stresses that are different in nature. Such is the case of low water availability in contrast to an excessive accumulation of water in the soil, with consequences on the concentration of oxygen and carbon dioxide. Similarly, the contrasting effect of high and low temperatures and the action of high salt concentrations in the soil solution are often present in natural environments.

Water Stress

From an ecophysiological point of view, the ability of seeds to survive at very low levels of soil moisture for long periods, and then germinate and grow when the environment becomes favorable, has been a key factor in the adaptation system of plant communities. These capacities are essential for restoring plant communities and allowing their spread to regions that have adverse ecological features [6]. Water availability is, indeed, the most critical factor for having a successful seedling establishment [11].

During the physiological germination, radicle emergence from the seed surface occurs only if the soil water potential exceeds a critical value. This value varies in relation to the seed composition and the relative proportion of embryonic axes and storage tissues [6].

In natural ecosystems, germination and emergency depend on rainfall. Seeds have different mechanisms of dormancy-related processes that can delay germination until the soil water content reaches a certain level [84]. Qi and Redmann [85] compared the germination and seedling survival ability of species belonging to the C_3 or C_4 photosynthetic groups under water limitation. They concluded that both the germination percentage and rate of all species decreased gradually in the range of water potentials from 0.0 to -1.5 MPa. These authors found that seedlings able to survive drought were more vigorous in C_3 than C_4 species, contrary to what is known about the adaptive advantages of C_4 compared to C_3 species in terms of water availability on seedling survival. Another interesting finding in this study was that seedlings survival under water stress is not directly related to the ability of seeds to germinate under stress.

Garwood [86] studied the plant community in a tropical forest with a well-marked dry season. He noted that 75% of the species under study germinated with the first rains of the season. From this stage, a small percentage of species showed dispersal before the rain season, but remained dormant until the next rain season; at the same time almost half of the species showed no dormancy mechanisms and germinated in the same rain season.

Friedman et al. [87] studied seeds of *Anastatica hierochuntica* (Rose of Jericho), a resurrection species native to the deserts of Arabia and Palestine. Seeds were treated with cycles of drying and rewatering (imbibition-drought-imbibition) after which they retained the ability to germinate, although extended periods of drought sharply reduced recovery.

A very interesting aspect, recently observed by Hubbard et al. [88] is the effect of endophytic fungi in increasing tolerance to heat and water stress. These authors showed that the association of wheat caryopses with different strains of endophytic fungi strongly increased the germination percentage and fresh weight of seedlings.

Flooding

Though water limitation is an important restriction for germination, its opposite -the saturation profile- is usually a significant constraint to that process. In this case, the stress factor causing inhibition is the lack of oxygen in the soil profile, as water fills the soil's empty spaces and produces anaerobic conditions. Stress caused by water logging is also known as anoxia, this term emphasizes the absence of oxygen. In relation to germination, the absence of oxygen immediately causes a marked change in the energy balance as non-dormant seeds of most plant species require a certain oxygen concentration to germinate. From this point of view, the species can be divided into two groups: Group I-seeds that can germinate in conditions of pressures up to 2% O_2, represented by species whose seeds contain lipid reserves such as lettuce, sunflower, radish, turnip, cauliflower, soybean; Group II- seeds that can germinate at concentrations of 1% O_2, of the starchy type, such as rice, wheat, corn, sorghum, peas [89].

Except for rice and weeds associated with this crop, germination of most species is inhibited due to their failure in producing sustained energy in the absence of O_2 by fermentation processes. Consequently, in many native species in wetlands, germination is delayed until seeds can sense that the danger of flooding has disappeared. Such is the case of *Phragmites australis* whose germination is inhibited while the seeds are covered by 5 cm of water [90].

It has also been observed in many species that seeds can survive for long periods of anoxia, although most of the cultivated species do not germinate under anoxic conditions. Once germination has occurred, the seedling is much more sensitive than the seed to O_2 deficiency, which is in agreement with the susceptibility to anoxia showed by established plants [91].

Heat Stress

The interaction of heat and water on seed germination was studied by Gummerson [34]. He proposed the concept of hydrothermal time (i.e., temperature x water content) to measure the effects of these two major factors. This interaction is also evident in the case of stressful temperatures, either low or high. This is because the studies related to heat stress usually include the water stress component.

Extreme temperatures generally exert a dual role: they might either prevent germination or break dormancy [92]. The inhibition of germination by high temperature is widespread in species from arid and semiarid regions with winter rains. This action could be an adaptive action to prevent germination during the late spring rains or after exceptional summer rains. High temperatures can also reduce the dormancy imposed by hard coats through increasing the permeability to oxygen and water [92]. Xiao-Xia et al. [93] have observed in

Halocnemum strobilaceum -a halophyte from saline regions of northern Africa, southern Europe and western Asia- that seed exposure to high temperatures -around 35°C- do not inhibit germination; however, seedlings appeared deformed or died soon after germination. Seeds of *Zygophyllum dumosum* -a shrub that grows in the deserts of Israel- can germinate without drawbacks in the range of 10 to 25°C, but germination is strongly inhibited at temperatures between 30-35°C [94].

In lettuce cultivars "Grand Rapids" and "Great Lakes", it has been observed that supraoptimal temperatures (around 35°C) can induce dormancy, which, in turn, can be removed by the joint action of CO_2 and ethylene [95].

Salinity

The stressful effect of NaCl on plants involves two aspects that can be studied separately. Firstly, the decrease in osmotic potential and the concomitant difficulty in absorbing water from the soil, produce a similar effect as that caused by salinity and drought. Secondly, the effect of toxicity in the plant cell, produced primarily by Na^+ is another aspect for study [96].

Plants are divided into two groups, regarding their response to salinity: I-*halophytes*, which can grow in saline soils, and II- *glicophytes*, which grow in non-saline soils or fresh water bodies.

Although halophytes can germinate in water without NaCl, glicophytes differ from those in the following aspects: (i) the seeds of halophyte species can germinate in saline substrates while those of glicophytes cannot; (ii) halophytes seeds can be viable even if they are under sea water for long periods of time [97]. In a glycophyte species such as tomato, high salt (150 mM NaCl and 15 mM $CaCl_2$) in the substrate significantly delayed germination and reduced the germination rate [98]. In safflower, a crop moderately tolerant to salinity, soil conductivities higher than 5.1 ds.m^{-1} decreased germination, retarded the emergence and reduced yield in cultivars tested under field conditions and salinity [99].

An interesting finding was observed in the experiment carried out by Waheed et al. [100] with pigeon pea (*Cajanus cajanus*). They found that there was no positive correlation between the degree of salinity tolerance in the early stages of plant development (germination, emergence and seedling growth) and the adult plant. These results are in agreement to what was observed by other authors [101-103] in the sense that tolerance to salinity changes during plant ontogeny. Therefore, a crop performance projection made from results obtained during germination is not reliable. However, tolerance to salinity during germination is critical for plants that have to grow in soils with high content of NaCl [104].

Given the interest shown among researchers in relation to the mechanisms of salt tolerance of halophytes, there are numerous works related to this species group. Many studies have shown that halophytes exhibit a similar tolerance response when subjected to salt stress. In these studies, germination is delayed and reduced, while many seeds remain dormant due to low osmotic potential produced by the saline substrate [105]. Rubio-Casal et al. [105] studied the behavior of two pioneer halophytes: *Salicornia macrostachyum* and *Arthrocnemum ramossisima* growing in saline depressions of the western Mediterranean. Experiences showed that the germination of the two species was delayed when the NaCl concentration reached the 3%. Recovery experiments showed that exposure to high salt concentrations did not harm the seeds, and that unlike *Salicornia*, *Arthrocnemum* germination was stimulated by salt concentrations of around 3%. Thus, the authors concluded that *Arthrocnemum* might be better adapted than *Salicornia* to extreme environmental conditions [105].

Halocnemum strobilaceum is a species that belongs to the family of the Chenopodiaceae, and colonizes saline ecosystems of North Africa, Mediterranean Europe and East Asia. Germination percentages decreased in this species with increasing salt concentrations from 0.1 to 0.75 M; there was radicle growth at low salt concentrations but it decreased as salt was increased, being completely inhibited at 2 M NaCl. The results obtained allowed to conclude that *Halocnemum* is highly tolerant to salinity; also the physiological responses differ between ecotypes that grow in the Mediterranean salt steppes in Spain from those that colonize the salt deserts of northwestern China [93].

Song et al. [106] studied the adaptive strategies of (i) an euhalophyte (*Suaeda physophora*), (ii) a xerohalophytes (*Haloxylon ammodendron*) and (iii) a xeric species (*Haloxylon persicum*). Seeds of the two halophytes showed a better germination recovery after being treated with a lower osmotic potential of NaCl (-3.13 MPa) and thereafter transferred to distilled water. Also, a higher partitioning of NaCl was observed in the teguments of these species compared to that in the xerophytic *H. persicum*. The authors concluded that the seed morphological structure and adaptation to salinity may explain the geographic distribution of *S. physophora* and *H. ammodendron* in saline regions.

Besides salt stress, seeds of many species of saline ecosystems remain part of their cycle under hypoxic conditions, and are subject to fluctuating temperatures due to their underwater condition. Wetson et al. [107] conducted an interesting study to investigate how these factors interact in a halophyte species such as *Suaeda maritima*. The results showed that it was necessary a pretreatment with seawater at 5°C to obtain a high germination percentage; moreover, the more extended that pretreatment during seed dormancy, the greater the percentage of germination. No decrease of germination was observed when seeds were subjected to very low levels of O_2. The experience allowed concluding that under natural conditions -high salt concentration and anoxic conditions- seeds remain dormant, and they will germinate when temperatures reach 15°C and there is a decrease in the soil salt concentration.

Effect of Fire on Germination

There has been a great influence on plant communities and ecosystems since *Homo sapiens* learned to make and handle fire. Primitive men produced big fires to facilitate hunting, and then fire was used for clearing land for crops and animal husbandry with the advent of farming in the so-called Neolithic Revolution. Even today, fire is used as a tool mainly for pasture management. In plant multiplication via sexual reproduction, seeds acquire a decisive role. Thus, in this section we will examine the influence of fire on the germination of species of different plant communities such as arid forests, thorn scrubs, fynbos ecosystems or the California Chaparral (USA) and northwestern Mexico [61]. About the importance of fire in northwestern Mexico, Keeley [108] studied the germination response of trees, shrubs, dwarf shrubs and vines. He observed that a portion of the seed bank remained dormant until germination was activated by the action of fire. This was specially the case for chaparral shrubs and trees which germinated quickly when seeds were moistened after fire.

An ecosystem where fire has a major role in maintaining the balance between herbaceous and woody communities is the savannah. In Australia, the savannah is subjected to the action of fire at intervals of 1-3 years. Experiences in this ecosystem by Setterfield [109] were aimed at determining whether the availability of seeds of two species of woody savanna -*Eucalyptus miniata* and *Acacia oncinocarpa*- was affected by fire frequency, and its implications for the seedling recruitment of these species. The fire regime reduced seedling emergence of these species likely by reducing the canopy, which produces an unfavorable microclimate for germination and seedling emergence. The reduction of the woody stand made possible the establishment of herbaceous species, raising the competition between herbaceous and woody species. These facts modify the floristic structure of the savannah, leading to the introduction of fire-resistant species, especially those whose reproduction is vegetative.

In many Australian ecosystems characterized by a Mediterranean climate, fire is a major cause of disturbance, where many plants die because of the high temperatures. In these ecosystems subjected to the action of recurrent fires, non-tolerant species are replaced by tolerant ones. Seeds of many species of the Fabaceae family remain dormant in the ground because they possess hard coats, impermeable to water and gas. They require the action of elevated temperatures for tegument scarification, thereby permitting the flow of water and oxygen into the embryo. Auld and Connell [110] studied 35 species of Fabaceae from eastern Australia and showed that germination was promoted by fires in the range 80-100°C for 120 minutes, while fires of 40-60°C had no effect.

Fire plays a key role in the distribution, organization and evolution of Mediterranean ecosystems. Numerous populations of conifers from this ecosystem can only release their seeds contained in thecones by the action of fire, thus allowing natural regeneration of the species [111, 112]. Fire has also a great importance in breaking the dormancy of coniferous species whose seeds have hard teguments such as *Pinus brutia* [113].

Also related to the Mediterranean region, it is very interesting the case of *Cistus incanatos* ssp *creticus* and *Cistus salvifolius*, shrubby species from xeric regions. These species have heterospermy; it means that the seed population has a portion of seeds with soft teguments while those with hard teguments must be scarified or be subjected to the action of fire. Heterospermy, hence, gives adaptive advantages to these species in the ecological context of the region, with a clear dry season in summer. In both species, a small proportion of the seed pool (with soft coats) germinates quickly every year, while the largest proportion (with hard seed coats) remains in the soil until the heat produced by fire softens the seed coat and enables germination. The very low germination rate of seeds with hard seed coats prevents their premature germination, allowing for their fitness in the Mediterranean atmosphere [114].

Germination and Recruitment

From the ecological point of view, the Oxford Dictionary defines recruitment as *"the increase in a natural population as progeny grow and new members arrive"*. According to McAlpine and Jesson [115] recruitment is a multiphased process involving seed dispersal, germination, seedling establishment and subsequent survival. Overall, researchers engaged in the study of recruitment have given an important role to the dispersion process, assuming that this is essential for seedling establishment. In this regard, Nathan and Muller-Landau [116] stated that the dispersal of seeds is one of the key processes in determining the spatial structure of plant populations. Patterns of seed dispersion vary over time and among species, populations and individuals in relation to the parental distance. In addition, microsite availability for propagule establishment is critical. Various studies have emphasized the role of the underlying mechanisms behind these patterns, and their implications for recruitment success.

McAlpine and Jesson [115] studied the dispersion, germination and seedling recruitment of *Berberis darwinii*, an invasive woody species from New Zealand, observing that seed dispersal by birds was essential for the establishment of seedlings. Seedling survival was very low under the canopy of the parent trees, while transport by birds to favorable microsites for germination and seedling growth was essential for recruitment of the species.

There is abundant literature that has studied the action of ruminants, mainly cattle, on the dispersion and release of dormancy due to hard teguments, especially in woody species belonging to the Fabaceae family. In this case, the passage of seeds by the animal's digestive tract serves to weaken the teguments, which prevent the flow of oxygen and water into the embryo. Moreover, presence of seeds in animal droppings provides a favorable environment for germination; also, the animal walking contributes to dispersal [32, 117-121].

Regarding seed dispersal by animals, it is interesting the case of *Antiaris toxicaria*, a Ghanaian forest woody species, where dispersal agents are primates that move the seeds away from the parent trees giving a survival advantage to the seedlings [122].

Wind is an important dispersal agent and, therefore, essentially involved in the recruitment process [123]. Studies were conducted to test the hypothesis that in open habitats winds are stronger than in forests [123]. Thus, trees create significant changes in the structure of the turbulent winds, and such differences determine significant changes in the patterns of seed distribution. The results showed that in closed spaces, where the wind effects were minor, the seed density was very important in the vicinity of adult trees compared to that in open spaces. The highest seed density within the forest could reach a level at which predators arrive to meet their needs for food, allowing greater survival of seedlings and favoring selection toward reduced dispersal distances in comparison to open spaces.

Galinato and Van der Valk [124] studied the processes of germination and recruitment of species growing in a swamp in Manitoba, Canada (laurentianus Fern Aster, *Atriplex patula* L., *Chenopodium rubrum* L., *Jubalum hordeum* L., *Scochloa festucacea* (Wild) Link *Phragmites australis* (Cav) Trin and *Typha glauca* GODR) when the swamp was free of superficial water. They concluded that the germination percentages of all species were higher when seeds were on the surface, whereas germination decreased sharply when seeds were covered by 1 cm of sand. Seedlings from large seeds (e.g. *Hordeum* and *Scolochoa*) could emerge from a depth of 5 cm while those from small seeds (e.g. *Typha*) could only emerge from a depth of 1 cm. These differences had important implications for the recruitment and composition of plant populations of the swamp.

Establishment of Seedling

Seedling establishment in a given habitat is the synthesis of the processes outlined above. The environmental conditions that take place during regeneration of individuals are crucial in determining the distribution and abundance of plant populations [125]. In a plant community, composition and diversity are both influenced by differences among species in their regeneration strategy and safe-site requirements [126]. Seeds reaching the soil surface may remain there or be buried. Germination might occur immediately after seed dispersion or seeds may persist in the soil for a period of time. Seeds that do not geminate remain dormant or in a state of quiescence until the environmental conditions are appropriate for germination. A fraction of the seed stock might die; death will occur at any time along the way, either immediately after dispersion or during seedsettlement in the soil seed bank or during germination and seedling growth. Among the reasons causing death, we can mention abiotic – water and heat stress, salinity, anoxia- and biotic factors -interference from neighboring plants, microbial diseases, predation and senescence [127].

Grubb [128] emphasized the role of gaps in creating environments for the establishment of a new individuals: "*When a plant dies creates a gap and a new individual can take its place*". This new individual could belong to the same species or not. The heterogeneity of the microenvironment (physical-chemical and biotic) will determine if the replacement will be a plant of the same species or a different one. There is little doubt that, in most plant communities, gaps have different quality. In that context, the proportion of species able to germinate varies from one gap to another. Moreover, the gaps do not differ only in their environmental quality but also in their timing. Overall, germination time of different species in a community is spread over the year. In this way, a species A is more likely to succeed in a gap between such-and-such dates, without going through a catastrophic event. However, if this occurs, other species B may have greater success if germination occurs after. For instance, a herb whose germination occurs in autumn in temperate grasslands can lead to adult individuals in a certain gap if there is no frost after germination, but if frost occurs, species of a spring cycle will be more likely to settle.

In relation to the importance of the gap in the establishment of individuals, Schütz [129] studied the effect of ending the management of plant communities in the grasslands of central Europe. He observed an increase in the (i) fallow land and (ii) accumulation of biomass and litter, resulting in a reduction of suitable sites for the establishment of new species.

Invasion of exotic species on agricultural lands gives a good opportunity to study establishment mechanisms of invasive species in nature, commonly regarded as weeds. Martinez-Ghersa et al. [130] studied the dormancy mechanisms and germination response of several invasive species of agricultural lands in the Pampas, Argentina: *Amaranthus retroflexus, Datura ferox, Echinochloa cross-galli* and *Avena fatua*. According to these authors, the process of invasion by a new weed species starts with an introductory phase where propagules reach a site away from their original ecosystem as a result of a dispersion mechanism. This first phase is followed by a second one termed "colonization", where propagules are established and they reproduce and increase stand density of individuals of this new species, forming a colony. In the third phase, the new species is incorporated into the pre-existing community. Williamson [131] considered that the pressure of propagules (i.e. mass effect) is a major factor in the invasion process because it increases the possibility that the species finds a suitable habitat to avoid environmental and biotic adversities such as parasites and pathogens. Finally, the results obtained by Martinez-Ghersa et al. [130] using simulation models showed that interactions between the patterns of rainfall and soil types modified the fitness of the species mentioned above.

Weis [132] studied the germination and seedling growth of *Mirabilis hirsuta,* a species that grows only in disturbed areas due to foraging for badgers, in Iowa. It was shown that the seed weight was a good predictor of the speed of germination and vigor, since the seedling growth rate was proportional to the size of the seed.

The mechanism of seed dispersal is fundamental to the species whose ability to compete with surrounding individuals is weak. The efficiency of dispersion mechanisms can change every year in relation to wind speed and direction, the abundance of animals that act as vectors of seed dispersal, the incidence of floods, among others [133]. In Patagonia (Argentina), Fernandez et al. [134] observed that seeds with big wings were able to colonize suitable gaps since they could be carried out by strong winds blowing from the west. Also in Patagonia, Bertiller [135, 136] studied the effect of grazing management combined with topography on different stages of the seed bank and vegetation establishment. The results showed that grazing did not reduce the germination from the seed bank of perennial grasses in uplands compared with lowlands, where grazing pressure was considerably higher. The reduction of grazing towards the late spring and summer increased the germination from the seed bank and the establishment of perennial grass species in both the uplands and lowlands. From the standpoint of grazing management, the author stated that the exclusion of sheep towards the end of spring increased the germination from the seed bank and seedling establishment.

In the northeast of Patagonia, a much more humid area compared to that of the previous study, the forested ecosystem is dominated by *Austrocedrus chilensis* as a major tree species. In this environment Kitzberger et al. [137] studied seedling establishment of this species revealing the importance of shrubs that acted as nurse plants, promoting the regeneration of *Austrocedrus.* The authors noted that nurse plants protected the tree because they provided shade, thus avoiding the direct effects from sunlight.

References

[1]   Bewley, J. D. & Black, M. (1994). *Seed Physiology of Development and Germination,* 2nd Ed., Plenum Press, New York.

[2]   Larcher, W. (1995). *Physiological Plant Ecology,* 3rd Ed., Springer Verlag, New York.

[3]   Bewley, J. D. (1997). *Plant Cell, 9,* 1055.

[4]   Boesewinkel, F. D. & Bouman, F. (1995). In: *J. Kigel and G. Galili* (Eds.), Marcel Dekker Inc. New York.

[5]   Cosgrove, D. J. (1997). *Plant Cell, 9,* 1031.

[6]   Bradford, K. J. & Kigel, J. (1995). In: *J. Kigel and G. Galili* (Eds.), Marcel Dekker Inc. New York.

[7]   Wu, Y., Spollen, W. G., Sharp, R. E., Hetherington, P. R. & Fry, S. C. (1994). *Plant Physiol., 106,* 607.

[8]   Watkins, J. T. & Cantliffe, D. J. (1983). *Plant Physiol., 72,* 146.

[9]   Bregman, R. & Bouman, F. (1983). *Bot. J. Linn. Soc.*, *86*, 357.

[10]   Roberts, E. H. (1988). *Symp. Exp. Biol.*, *42*, 109.

[11]   Elberse, W. T. & Breman, H. (1990). *Oecologia*, *85*, 32

[12]   Bewley, J. D. & Black, M. (1978). *Physiology and Biochemistry of Seed*. Vol. *1*. Springer-Verlag, Berlin.

[13]   Borthwick, H. A., Hendricks, S. B., Parker, M. W., Toole, E. H. & Toole, V. K. (1952). *Proc. Nat. Acad. Sci.*, *38*, 662.

[14]   Heide, O. M. (2008). *Sci. Hort.*, *115*, 309.

[15]   Nwoke, F. I. O. (1982). *Ann. Bot.*, *49*, 23.

[16]   Al-Ani, A., Bruzau, F., Raymond, P., Saint-Gez, V., Leblanc, J. M. & Pradet, A. (1985). *Plant Physiol.*, *79*, 885.

[17]   R Grable, A. & Danielson, R. E. (1965). *J. Soil Sci. Soc. Am.*, *29*, 12.

[18]   Kidd, F. (2012). *Proc. Royal Soc.*, *87*, 408.

[19]   Ruiz, M. A., Pérez, M. A., Argüello, J. A. & Babinec, F. J. (2003). *RIA*, *32*, 3.

[20]   Benech-Arnold, R. L., Sánchez, R. A., Forcella, F., Kruka, B. C. & Ghersa, C. M. (2000). *Field Crops Res.*, *67*, 105.

[21]   Cohn, M. A. (1996). *Seed Sci. Res.*, *6*, 147.

[22]   A Bennett, M. & Evans, A. F. (2011). *Seed dormancy mechanisms in vegetable crops species*. Available: www.seedconsortium.org.

[23]   Batlla, D. & Benech-Arnold, R. L. (2004). *Seed Sci. Res.*, *14*, 277.

[24]   Batlla, D. & Benech-Arnold, R. L. (2010). *Plant Mol. Biol.*, *73*, 3.

[25]   A Ruiz, M., Covas, G. F. & F. J. Babinec, G. F. (1997). *RIA*, *28*, 37.

[26]   Checovich, M. & Ruiz, M. A. (2011). In: *XXXIII Jorn. Arg. Bot., Bol. SAB*, *46* (Supl.), 273.

[27]   Checovich, M. & Ruiz, M. A. (2011). *Análisis de semillas*, *5*, 58.

[28]   Murphy. B. D. (2008). Electronic Theses, South Dakota State University.

[29]   Tischler, C. R. & Young, B. A. (1983). *Crop Sci.*, *23*, 789.

[30]   Taiz, L. & Zeiger, E. (2006). *Plant Physiology*, 4th Ed.

[31]   McKee, G. W., Peiffer, R. A. & Mohsenin, N. N. (1977). *Agron. J.*, *69*, 53.

[32]   Peinetti, R., Cabezas, C., Pereyra, M. & Martínez, O. (1992). *Turrialba*, *42*, 415.

[33]   Campos, C. M., Peco, V. E. B., Campos, V. E., Malo, J. E., Giannoni, S. M. & Suárez, F. (2008). *Seed Sci. Res.*, *18*, 91.

[34]   Gummerson, R. J. (1986). *J. Exp. Bot. 37*, 729.

[35]   Vegis, A. (1964). *Ann. Rev. Plant Physiol.*, *15*, 185.

[36]   Azcon Bieto, J. & Talon, M. (2008). *Fundamentos de Fisiología Vegetal*, 2° Ed., Interamericana McGraw Hill, Madrid.

[37]   Thompson, K. & Grime, J. (1983). *J. Appl. Ecol.* 20, 141.

[38]   Benech-Arnold, R. L., Ghersa, C. M., Sánchez, R. A. & García Fernández, A. E. (1988). *Funct. Ecol.*, *2*, 311.

[39]   Ghersa, C. M., Benech Arnold, R. L. & Martínez Ghersa, M. A. (1992). *Funct. Ecol, 6*, 460.

[40]   Benech Arnold, R. L., Ghersa, C. M., Sánchez, R. A. & Insausti, P. (1990). *Weed Res.*, *30*, 81.

[41]   Benech Arnold, R. L., Ghersa, C. M., Sánchez, R.A. & Insausti, P. (1990). *Weed Res.*, *30*, 91.

[42] Casal, J. J. & Sánchez, R. A. (1998). *Seed Sci. Res.*, *8*, 317.

[43] Pearson, T. H. R., Burslem, D. F. R. P., Musllins, C. E. & Dalling, J. W. (2003). *Funct. Ecol.*, *17*, 394.

[44] Batlla, D. B., Kruck, B. C. & Benech Arnold, R. L. (2004). In: R. L. Benech Arnold and R. A. Sánchez (Eds.), Food Product Press, Oxford.

[45] Botto, J. F., Sánchez, R. A. &. Casal, J. J. (1998). *Seed Sci. Res.*, *8*, 423.

[46] Díaz Vizcaino, E., Lasanta, M. & Morey, M. (1989). *Stvdia Oecol.*, VI, *41*.

[47] Henderson, C. B., Petersen, K. E. & Redak, R. A. (1988). *J. Ecol.*, *76*, 717

[48] Maia, F. C., Medeiros, R. B., Pillar, V. P. & Focht, T. (2004). *Rev. Brás. Sem.*, *26*, 126.

[49] Maia, F. C., Medeiros, R. B., Pillar, V. P., Focht, T., Chollet, D. M. S. & Olmedo, M. O. M. (2003). *Iheringia (Sér. Bot.)*, *58*, 61.

[50] De Souza Maia, M. , Maia, F. C. & Pérez, M. A. (2006). *Agriscientia*, *23*, 33.

[51] Batlla, D. & Benech Arnold, R. L. (2003). *Plant Mol. Biol.*, *73*, 3.

[52] Medeiros, R. B. (1989). In: *Memórias Guarapuava*, UFPR. p 61.

[53] D'Hertefeldt, T., Jorgensen, R. B., Pettersson, L. B. (2008). *Biol. Lett.*, *4*, 314.

[54] Glenn-Lewin, R. K., Peet, R. K. & Veblen (Eds.), T. T. (1992). *Plant Succession: Theory and Prediction*, Chapman and Hall

[55] Cano, E. (1988). *Pastizales naturales de La Pampa. Descripción de las especies más importantes*. Ed. AACREA - Prov. La Pampa.

[56] Mayor, M. D., Boo, R. M., Pelaez, D. V. & Elia, O. R. (1999). *Phyton, Int. J. Exp. Bot.*, *64*, 141.

[57] Rúgolo de Agrasar, Z. E., Steibel, P. E. & Troiani, H. O. (2005). *Manual Ilustrado de las gramíneas de la provincia de La Pampa*. Ed. UNLPam.

[58] Walck, J. L., Baskina, J. M., Baskin, C. C. & Hidayati, S. N. (2005). *Seed Sci. Res.*, *15*, 189.

[59] Bakker, J. P., Bakker, E. S., Rosen, E., Vergueij, G. L. & Bekker, R. M. (1996). *J. Veg. Sci.*, *7*, 165.

[60] Thompson, K. (2000). *Seeds: The Ecology of Regeneration in Plant Communities*. 2° Ed., Ed. M. Fenner, Wallingford, UK: CAB International.

[61] Fenner, M. (1995). *Seed Development and Germination*. J. Kigel and G. Galili (Eds.), Marcel Dekker Inc. New York.

[62] Mayor, M. D., Bóo, R. M., Peláez, D. V. & Elía. O. R. (2003). *J. Arid Envir.*, *53*(4), 467.

[63] Gulden, R. H. (2003). University of Saskatchewan Library. Electronic Theses.

[64] M Alexander, H. & Schrag, A. M. (2003). *J. Ecol.* 91, 987.

[65] Lokesha, R., Hedge, S. G., Uma Shaanker, R. & Ganeshaiah, K. N. (1992). *Am. Nat.*, *140*, 520.

[66] Egley, G. H., Paul Jr., R. N., Duke, S. O. & Vaugham, K. C. (1985). *Plant Cell Environ.*, *8*, 253

[67] Hendry, G. A. F., Thompson, K., Moss, C. J., Edwards, E. & Thorpe, P. C. (1994). *Funct. Ecol.*, *8*, 658.

[68] Oesterheld, M. & Sala. O. E. (1990). *J. Veg. Sci.*, *1*, 353.

[69] Bakker, J. P. (1989). *Nature Management by Grazing and Cutting*, Kluwer Academic Publ., Dordrecht.

[70] Bertiller, M. B. (1996). *Seed Sci. Res.*, *8*, 39.

[71] Bertiller, M. B. (1992). *J. Veg. Sci.*, *3*, 47.

[72] Milberg, P. (1995). *Oikos*, *72*, 3.

[73] Márquez, S., Funes, G., Cabido, M. & Pucheta, E. (2002). *Rev. Chil. Hist. Nat.*, *75*, 327.

[74] Morici, E., Doménech-García, V., Gómez-Castro, G., Kin, A., Saenz, A. & Rabotnikof, C. (2009). *Agrociencia*, *43*, 529.

[75] Estelrich, H. D., Fernández; B., Morici, E. F. & Chirino, C. C. (2005). *Rev. Fac. Agr. UNLPam.*, *16*, 23.

[76] Boo, R. M. (1990). *Rev. Fac. Agr. UNLPam.*, *5*, 63.

[77] Distel, R. A. & Boo, R. M. (1995). In: *Fifth Int. Rangeland Cong.*, pp. 117 Salt Lake City, Utah.

[78] Weber, M. & Taylor, S. (1992). *For. Chron.*, *68*, 324.

[79] Fernández, B., Morici; E., Esterlich, H. & Chirino, C. (2001). In: *1° Congr .Nac. Man. Past. Nat.*, 65 Santa Fe.

[80] Ernst, R., Suárez, C., Chirino, C., Sosa, A., Kin, A. & Morici, E. (2007). In: *IV Congr. Nac. Man. Past. Nat. - I Congr. Mercosur Man. Past. Nat.*, San Luis.

[81] Blodgett, H., Hart, G. & Stanislaw, M. (2000). *Tillers*, *2*, 31.

[82] Riesco, A., Cesa, P., Estelrich, D. & Morici, E. (2004). In: *II Reunión Binac. Ecol.* 410, Mendoza.

[83] Estelrich, D., Fernández, B., Chirino, C. C. & Morici, E. F. (2004). In: *II Reunión Binac. Ecol.* 202, Mendoza.

[84] Frasier, G. W. & Evans, R. A. (1987). (Eds.), In: *Proc. Symp.Seed and Seedbed Ecol. of Rangeland Plants*, Tucson.

[85] Qi, M. Q. & Redman, R. E. (1993). *J. Arid Environ.*, *24*, 277.

[86] Garwood, N. C. (1983). *Ecol. Monogr.*, *53*, 159.

[87] Friedman, J., Stein, Z. & Rushin, E. (1981). *Oecologia*, *51*, 400.

[88] Hubbard, M., Germida, J. & Vujanovic, V. (2012). *Botany*, *90*, 137.

[89] Corbineau, F. & Côme, D. (1995). In: J. Kigel and G. Galili (Eds.), Marcel Dekker Inc., New York.

[90] Crawford, R. M. (1987). *Can. J. Bot.*, *81*, 1224.

[91] Vantoal, T. T., Saglio, P., Ricard, B. & Pradet, A. (1995). *Plant Cell Environ.*, *18*, 937.

[92] Kigel, J. (1995). In: J. Kigel and G. Galili (Eds.), Marcel Dekker Inc., New York.

[93] Xiao-Xia, Q., Zhen-Ying, H., Baskin, J. M. & Baskin, C. (2008). *Ann. Bot.*, *101*, 293.

[94] Agami, M. (1986). *Physiol. Plant.*, *67*, 305.

[95] Keys, R. D., Smith; O. E., Kumamoto, J. & Lyon, J. L. (1975). *Plant Physiol.*, *56*, 826.

[96] Munns, R. (2002). *Plant Cell Environ.*, *25*, 239.

[97] Ungar, I. A. (1995). In: J., Kigel, & G., Galili (Eds.), Marcel Dekker Inc., New York.

[98] Kaveh, H., Nemati, H., Farsi, M. & Vatandoost Jartroodeh. S. (2011). *J. Biol. Env. Sci.*, *15*, 159.

[99] Kaya, M. D. & Ipex, A. (2003). *Turk. J. Agr. For.*, *27*, 221.

[100] Waheed, A., Hafiz; I. S., Qadir, G. M., Mahmood, T. & Ashraf, M. (2006). *Pak. J. Bot.*, *38*, 1103.

[101] Kingsbury, R. W. & Epstein, E. (1984). *Crop Sci.*, *24*, 310.

[102] Shannon, M. C. & Grieve, C. M. (1999). *Sci. Hort.*, *78*, 5.

[103] Ashraf, M. & Harris, P. J. C. (2004). *Plant Sci.*, *166*, 3.

[104] Marañón, T., García, V. & Troncoso, A. (1989). *Plant Soil*, *119*, 223.

[105] Rubio-Casal, A. E., Castillo, J. M., Luque, C. J. & Figueroa, M. E. (2003). *J. Arid Environ.*, *53*, 145.

[106] Song, J., Feng, G., Tian, C. H. & Zhang, F. (2005). *Ann. Bot.*, *96*, 399.

[107] Wetson, A. M., Cassaniti, C. & Flowers, T. J. (2008). *Ann. Bot.*, *101*, 1319.

[108] Keeley, J. E. (1987). *Ecology*, *68*, 434.

[109] Setterfield, S. A. (2002). *J. Appl. Ecol.*, *39*, 949.

[110] Auld, T. D. & O´Connell, M. A. (1991). *Austr. J. Ecol.*, *16*, 53.

[111] Walter, H. (1973). *Vegetation of the Earth: In Relation to Climate and the Ecophysiological Conditions,* The English Universities Press Ltd., London.

[112] Castro, J. F., Bento, J. & Rego, F., Goldammer, J. G. & Jenkins, M. J. (1990). (Eds.), *Fire in Ecosystem Dinamics.* SPB Academic Publishing.

[113] Thanos, G. A., Marcou, S., Christodoukakis, P. & Yannitsaros, A. (1989). *Acta Oecol.-Oecol. Plant.*, *10*, 79.

[114] Thanos, C. A. & Georghiou, K. (1988). *Plant Cell Environ.*, *11*, 841.

[115] McAlpine, K. G. & Jesson, K. L. (2007). *Plant Ecol.*, *119*, 129.

[116] Nathan, R. & Muller-Landau, C. (2000). *Tree*, *15*, 278.

[117] Ortega Baes, P., de Viana, M. L., Larenas, G. & Saravia, M. (2001). *Rev. Biol. Trop.*, *49*, 25.

[118] Campos, C. M., Campos, V. E., Mongeaud, A., Borghi, C. E., De Los Ríos, C. & Giannoni, S. M. (2011). *Rev. Chil. Hist. Nat.*, *84*, 289.

[119] Malo, J. E. & Suárez, F. (1995). *Bot. J. Linn. Soc.*, *118*, 139.

[120] Peinetti, R., Pereyra, M., Kin, A. & Sosa, A. (1993). *J. Range Manage.*, *46*, 483

[121] Eilberg, B. A. (1973). *Ecología*, *1*, 56.

[122] Kankam, B. O. (2001). Thesis of Graduate Studies, Kwame Nkrumah Univ. Sci. Tech, Ghana.

[123] Nathan, R., Horn, H. S., Chave, J. & Levin, S. A. (2001). In: D. J. Levey, W. R. Silva and M. Galetti (Eds.), CAB.

[124] Galinato, M. L. & Van der Valk, A. G. (1986). *Aquatic Bot.*, *26*, 89.

[125] Gross, K. L. & Werner, P. A. (1982). *Ecology*, *63*, 921.

[126] Grime, J. P. & Hillier, H. S. (1992). In: M. Fenner (Ed.), Seeds: The Ecology of Regeneration in Plant Communities, CAB International, Wallingford.

[127] Clark, D. L. & Wilson. M. V. (2003). *Am. J. Bot.*, *90*, 730.

[128] Grubb, P. J. (1977). *Biol. Rev.*, *52*, 107.

[129] Schütz, W. (2000). *Ökol. Natur. 9*, 73.

[130] Martinez Ghersa, M. A., Ghersa, C. M., Benech Arnold, R. L., Donough, R. M. & Sanchez, R. A. (2000). *Plant Species Biol.*, *15*, 127.

[131] Williamson, M. H. (1996). *Biological Invasions*, Chapman & Hall, London.

[132] Weis, M. I. (1982) *Can J. Bot.*, *60*, 1868.

[133] Skellam, J. G. (1951). *Biometrika*, *38*, 196.

[134] Fernández, R. J., Golluscio, R. A., Bisigato, A. J. & Soriano, A. (2002). *Ecography*, *25*, 336.

[135] Bertiller, M. B. (1996). *Environ. Manage.*, *20*, 123.

[136] Bertiller, M. B. (1996). *Seed Sci. Res.*, *8*, 39.

[137] Kitzberer, T., Steinaker, D. F. & Veblen, T. T. (2000). *Ecology*, *81*, 1914.

Basic Research

In: From Seed Germination to Young Plants
Editor: Carlos Alberto Busso

ISBN: 978-1-62618-653-8
© 2013 Nova Science Publishers, Inc.

Chapter 2

Greening and the Influence of Light on the Enzyme Activity of *Cucumis sativus* Seedlings

Peter Siekel[1]*, Ján Stano[2], Karol Mičieta[3] and Marcela Koreňová[4]

[1]Department of Microbiology, Molecular Biology and Biotechnology, Food Research Institute, Priemyselná, Bratislava
[2]Department of Cell and Molecular Biology of Drugs, Faculty of Pharmacy, Comenius University, Odbojárov, Bratislava, Slovak Republic
[3]Department of Botany, Faculty of Natural Sciences, ComeniusUniversity, Révová, Bratislava
[4]Garden of Medicinal Plants, Faculty of Pharmacy, ComeniusUniversity, Odbojárov, Bratislava, Slovak Republic

Abstract

Cotyledons are not only saturated with nutrients, but they are also the first photosynthetic leaves; this makes them useful for greening studies. This paper contributes to a broader understanding on translocation and utilization of storage compounds during germination and greening of *Cucumis sativus* seedlings with respect to the formation of aromatic compounds. The activities of the shikimate NADP oxidoreductase and the glucose-6-phosphate NADP oxidoreductase were determined. Also, the contents of chlorophyll$_a$ and chlorophyll$_b$ were measured in the cotyledons. Light stimulated the activity of the shikimate NADP oxidoreductase (SOR, EC.1.1.1.25), and inhibited that of the glucose-6-phosphate NADP oxidoreductase (G-6-PDH, EC.1.1.1.49). The highest content of protochlorophyll, and most intensive forming of chlorophyll$_a$ and chlorophyll$_b$ were found in five-day-old gherkin seedlings.

* Corresponding author: E-mail address: siekel@vup.sk.

Introduction

Light affects many aspects of plant development from seed germination. In the absence of light, seedlings grow heterotrophically, using the seed's resources to eventually reach light. Etiolated seedlings are characterized by a long hypocotyl (primary stem), an apical hook, and unopened cotyledons (embryonic leaves). This may allow the seedlings to grow through a soil layer and emerge into the light. When the light is reached, photoreceptors interact with other signal transduction elements, which lead to further molecular and morphological responses. Then the biosynthesis of aromatic amino acids proceeds by way of the shikimic acid in microorganisms and plants. Shikimic acid was discovered in the Asian fruit of aniseed *Illicium anisatum* [1]. It is a key metabolite of a pathway leading to the formation of chorismate. Thereafter, the name of this pathway was derived from it [2]. The steps in this pathway lead from erythrose-4-phosphate via shikimate to chorismate, which is a common precursor of all the aromatic amino acids and other important compounds such as phenolic acids, flavonoids and alkaloids. Shikimate pathway is not a constituent of the animal metabolism. This is a good reason to use shikimate inhibitors in the design of novel antimicrobial agents [3-4].

The shikimate acid pathway proceeds in seven catalytic steps and combines the carbohydrate metabolism with the synthesis of aromatic amino acids. It begins with condensation of phosphoenol pyruvate (PEP) and erythrose-4-phosphate (E4P), and ends with the synthesis of chorismate [2, 5]. Chorismate is an intermediate in the synthesis of L-tryptophan, L-tyrosine, L-phenylalanine, and p-aminotenzoic acid [3, 6].

A stage initiating the shikimic acid pathway is the condensation of the PEP with E4P, which results in the formation of 3-deoxy-D-arabino-heptulosonate-7-phosphate. PEP is an intermediate originating from glycolysis, whereas E4P is an intermediate of penthase phosphate pathway [3]. The enzyme that catalyses the first step of the shikimate pathway is the 3-deoxy-arabino-heptulosonate-7-phosphate synthase (DAHP synthase, EC 4.1.2.15). The second stage of this pathway is 3-dehydroquinone synthase (DHQ synthase EC 4.6.1.3) [7]. The third enzyme, dehydroquinone dehydratase (EC 4.2.1.10), activates the formation of an intermediate referred to as dehydroshikimate (DHS). The DHS reduced to the shikimic acid (SA), and that stage of the shikimate pathway, are catalyzed by the shikimate dehydrogenase (EC 1.1.1.25). The next stage is the phosphorylation of SA followed by the condensation with the second molecule of PEP, thus producing 5-enolpyruvylshikimate (EPSP). The following intermediate compound loses its phosphoryl group and is subjected to the reduction to chorismate. The final step of the shikimic pathway is the synthesis of chorismate (EC 4.2.3.5) that is used further in the synthesis of aromatic amino acids and p-aminobenzoic acid.

The shikimate acid pathway leads to the synthesis of aromatic compounds and thus its regulation is strictly connected with the need of the regulation of the synthesis of aromatic amino acids that are a critical substrate for protein formation [3-4].

The seeds of the family Cucurbitaceae contain a high level of reserve proteins (35%), in addition to a large amount of reserve lipids (50%). During germination, the reserve lipids are transformed into carbohydrates, which are utilized for the growth and development of the seedlings [8-9]. The gherkin cotyledons are not only filled up with nutrients, but also they are the first photosynthesizing leaves to study leaf greening [6, 10].

Pentose phosphate pathway plays an important role in the biogenesis of aromatic compounds [3, 11]. These compounds are synthetized from erythrose-4-phosphate and phosphoenol pyruvate which are supplied by the pentose phosphate cycle and the glycolic breakdown of glucose, respectively.

Glucose-6-phosphate dehydrogenase G-6-PDH, D-glucose-6-phosphate: $NADP^+$ 1-oxidoreductase; (EC 1.1.1.49) is the key enzyme of the pentose phosphate pathway that catalyzes the conversion of glucose-6-phosphate to 6-phosphogluconolactone in presence of $NADP^+$ [12-13]. G-6-PDH is not only a key enzyme in the plant pentose pathway, but is an enzyme of vital importance in human medicine. This is because of its role in various haemolytic disorders, and at the same time has the potential of regulating for various biosynthetic pathways. NADPH is also active in the protection of the cell against oxidative agents by transferring its reductive power to glutathione disulphide via glutathione disulphide reductase [14]. NADPH plays an important role within reductive biosynthesis of fatty acids, isoprenoids, aromatic amino acids, etc.

Aromatic amino acids, proteins, lignins and polyphenols are synthetized not only in chloroplasts but also in leucoplasts, independently of light, indicating a different supply for a demand of ATP and NADPH than in chloroplasts [15-16].

This paper examines the (i) change of content of protochlorophyll during germination, (ii) amount of chlorophyll during greening, (iii) subcellular localization of SOR and glucose-6-phosphatedehydrogenase (G-6-PDH), and (iv) influence of light upon the activity of both enzymes.

Materials and Methods

Plant Cultivation

The seeds of gherkin (*Cucumis sativus* L. cv. Pálava) germinated in the dark at 25°C in Petri dishes on a polyurethane pad covered with a nylon cloth and soaked with distilled water [6].

Enzyme Preparation

Fifty pairs of etiolated or green cotyledons were homogenized in a precooled mortar with pestle and sand, in 20 ml of 0.1 mol.l^{-1} phosphate buffer, pH 7.4, with 2 mmol.l^{-1} cysteine. The next step was filtration and subsequent centrifugation at 120 000 m.s^{-2} during 15 min. The supernatant was used for assay of SOR and G-6-PDH activities. All procedures were done at 4°C.

SOR assay
The reaction mixture contained 0.2 ml of enzyme preparation, 0.5 μmol.l^{-1} NADP, 1 μmol.l^{-1} shikimic acid and 0.1 mol.l^{-1} Tris-HCl; buffer pH 8.8 was added to the total volume (3 ml).

G-6-PDH assay

The reaction mixture contained 0.2 ml of enzyme preparation, 0.9 μmol.l^{-1} NADP, 20 μmol.l^{-1} MgCl$_2$, 20 μmol.l^{-1} glucose-6-phosphate and 0.1 mol.l^{-1} Tris-HCl; buffer pH 8.8 was added to the total volume (3 ml).

Reaction mixtures were incubated at 25°C for 5 min and measured each 30 s. Concentration of NADP was enumerated using the molar extinction coefficient NADPH=6.22.cm^2.μmol^{-1}. Enzyme activities are given in katals. Protein content was determined according to the method of Doumas and co-workers [17]. Bovine serum albumin was used as a standard protein.

Influence of Light upon the Enzyme Activity

Five-day-old etiolated seedlings or extirpated cotyledons with hooks were used. These materials were illuminated (in Petri dishes) for 0.5, 1, 2, 3.5, 5 and 7 hours by light of 50 lm intensity [6, 10]. After illumination, the cotyledons were cleared of hooks and used for enzyme assay.

Determination of Chlorophyll Content

The contents of chlorophyll$_a$, chlorophyll$_b$ and protochlorophyll were measured in the diethyl etheric extracts of etiolated cotyledons, or in cotyledons illuminated as mentioned above. Fifty pairs of cotyledons were homogenized in 15 ml acetone, placed on the filter and washed by 30 ml ether and 10 ml acetone. Acetone from the filtrate was removed by washing two times by 1000 ml of distilled water. Diethyl ether was added to the filtrate to a total volume of 50 ml and, after drying by anhydrous sodium sulphate, the extract was measured at 624, 644 and 663 nm [6].

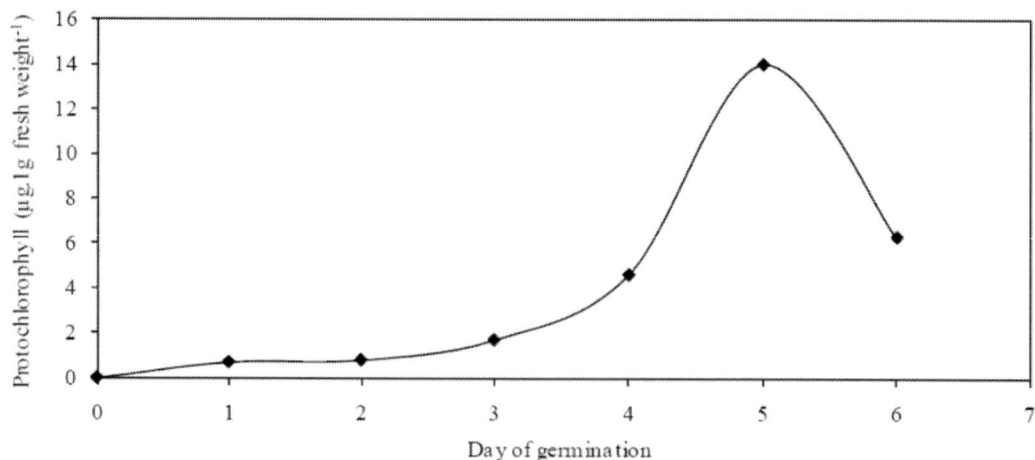

Figure 1. Changes of protochlorophyll (1 μg.1g fresh weight^{-1}).

Content of chlorophyll$_a$, chlorophyll$_b$ and protochlorophyll in cotyledons was calculated by the method of Koski [18] modified by Stano and others [6, 19].

The separation of cell organelles by fractional centrifugation was done by a modified method [20]. Five-day-old cotyledons (50 g) were homogenized in a rotating mixer using a homogenizing medium consisting of 0.5 mol.l^{-1} sucrose, 0.2 mol.l^{-1} Tris-HCl buffer pH 8, and 2 mmol.l^{-1} cysteine. The homogenate was filtered through the nylon cloth and the filtrate was subjected to fractionation by centrifugation.

The debris was removed at 10 000 m.s^{-2}. Chloroplasts were separated at 160 000 m.s^{-2} and mitochondria at 1 200 000 m.s^{-2}. Organelles were then disrupted by sonication (20 kHz), mechanical homogenization and osmotic shock. Homogenates were centrifuged (15 000 m.s^{-2} for 15 min) and the activity of both enzymes was assayed in the supernatant.

Results

Gherkin seeds were grown on the polyurethane pad soaked with distilled water and covered with a nylon cloth in Petri dishes at 25°C, 75% humidity. Seedlings were grown in the dark for 5 days; samples were taken every 24 h intervals. Changes of protochlorophyll content in the cotyledons of etiolated seedlings during germination are shown in Figure 1. After root protrusion through the testa (first three days of germination), the content of protochlorophyll increased slowly. After three days, there was an intensive formation of protochlorophyll which ended up on the fifth day, followed by the suppression of its biosynthesis. Figures 2 and 3 show the effect of light on the biosynthesis of chlorophyll$_a$ and chlorophyll$_b$ in 1- to 6-day-old etiolated cotyledons after 6 hours of illumination. The most intensive formation of both chlorophylls in (i) illuminated, extirpated cotyledons and (ii) cotyledons of intact seedlings occurred in 5-day-old cotyledons. Light stimulated the activity of SOR (Figure 4), contrary to the activity of G-6-PDH that was moderately inhibited (Figure 5).

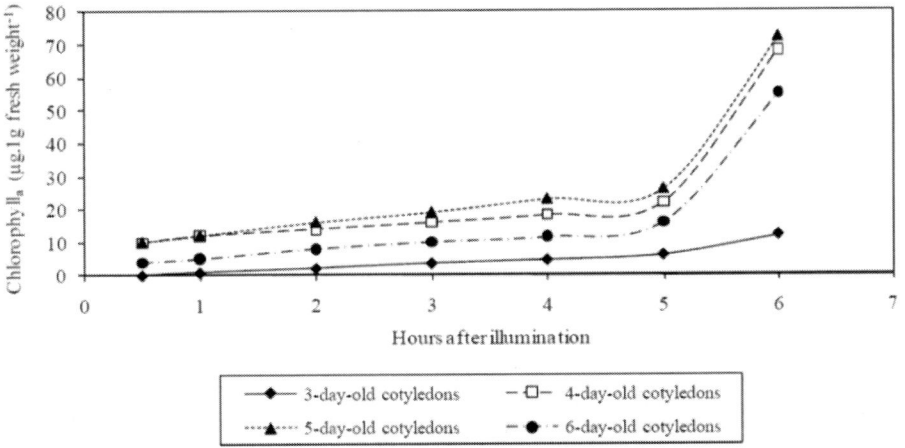

Figure 2. Changes in chlorophyll$_a$ content in cotyledons during greening (μg chlorophyll$_a$ .1g fresh weight^{-1}) with time from illumination.

Table 1. Subcellular localization of SOR and G-6-PDH

Cell fraction	% of activity			
	SOR		G-6-PDH	
	Dark	Light	Dark	Light
Etioplast	60	-	2	1
Chloroplast	-	65	-	-
Mitochondria	5	5	5	5
Microsomal fraction	-	-	7	7
Soluble fraction	35	30	86	87

The distribution activity of both enzymes in different cell fractions is summarized in Table 1. The major activity of G-6-PDH was located in the soluble fraction, while that of SOR was in the etioplasts, after chloroplasts were illuminated.

Discussion

Unusually high contents of fats and proteins, and a lack of starch as a reserve substance in the cotyledons of gherkins, and the fact that after utilization of reserve substances the cotyledons transform into the first photosynthetic leaves, caused that they are suitable to study the influence of light on the enzyme activity and morphogenesis [21-26]. Other plant models were used to study the influence of light on the morphogenesis and natural compound biotransformation [16, 23, 27-29].

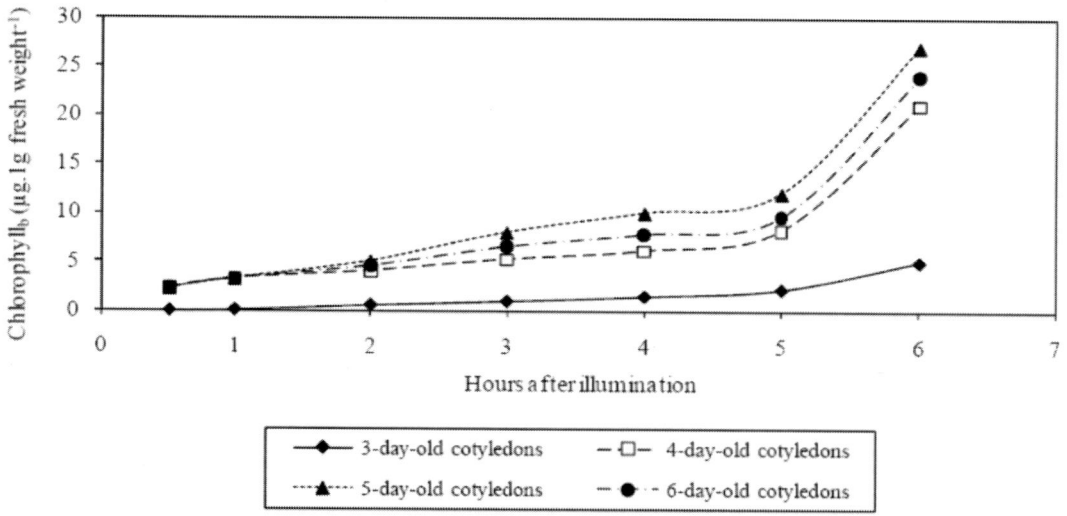

Figure 3. Changes in chlorophyll$_b$ content in cotyledons during greening (μg chlorophyll$_b$.1g fresh weight^{-1}) with time from illumination.

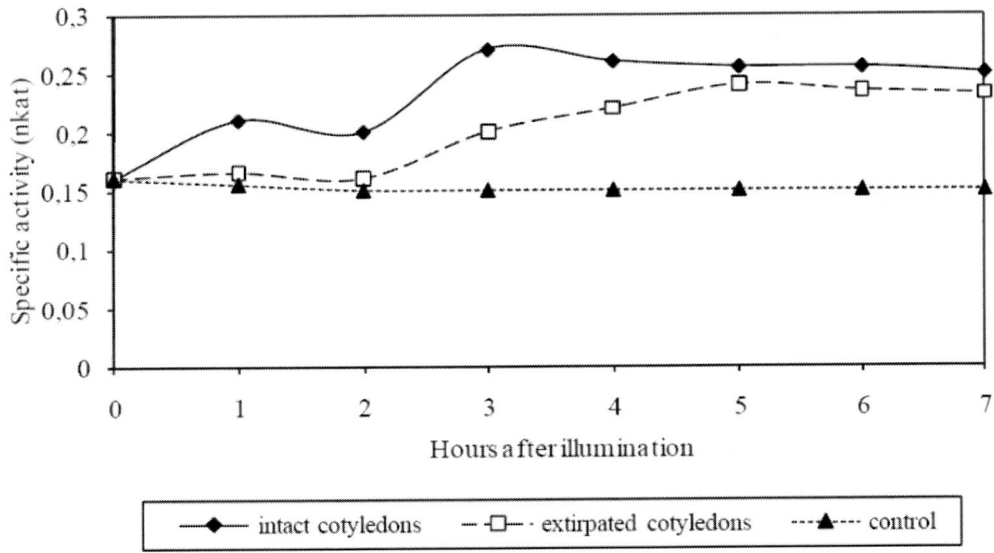

Figure 4. Influence of light upon the SOR activity; ♦ intact cotyledons; □ extirped cotyledons; ▲ control without light.

Biosynthesis of chloroplasts is a multistage process leading to fully differentiated and functionally mature plastids. During natural scotomorphogenesis, when the initial seedling growth occurs beneath the soil surface during germination in the dark, protoplastids differentiate into etioplasts. Etiolating is a natural process during the initial plant ontogenesis that might happen during the normal plant development [30]. Prolamellar bodies contain protochlorophyll, a precursor of chlorophyll, some enzymes and a few proteins [31]. The function of these bodies is not fully understood [32]. This complex is responsible for a regular, parocrystalinne structure [33] which protects the proteolysis of the chlorophyll synthase [32, 34] upon illumination, when young etiolated seedlings "switch" to photomorphogenesis, and etioplasts differentiate into chloroplasts [35]. Structural transformation and conversion from protochlorophyll to chlorophyll, and synthesis of proteins involved in photosynthesis, especially thylakoid proteins, gradually increased upon illumination [36], as was obvious from our experiment in this study.

During differentiation of chloroplasts, light perception and subsequent expression of nuclear plastid genes, biosynthesis of lipids and pigments, and chloroplast development take place [37-38].

Our results confirmed that changes in the protochlorophyll content during germination and greening correlated to chloroplast biogenesis, and to its relation to the structure and function of it. The majority of SOR was located in the chloroplasts, and thus the positive influence of light was obvious. The greening of etiolated gherkin cotyledons confirmed this assumption. These findings on light dependence of SOR are in agreement with data on bacteria and fungi [3]. The only difference is that the enzymes of the shikimate pathway were detected in the bacterial cytoplasm, while in plants they were found in plastids.

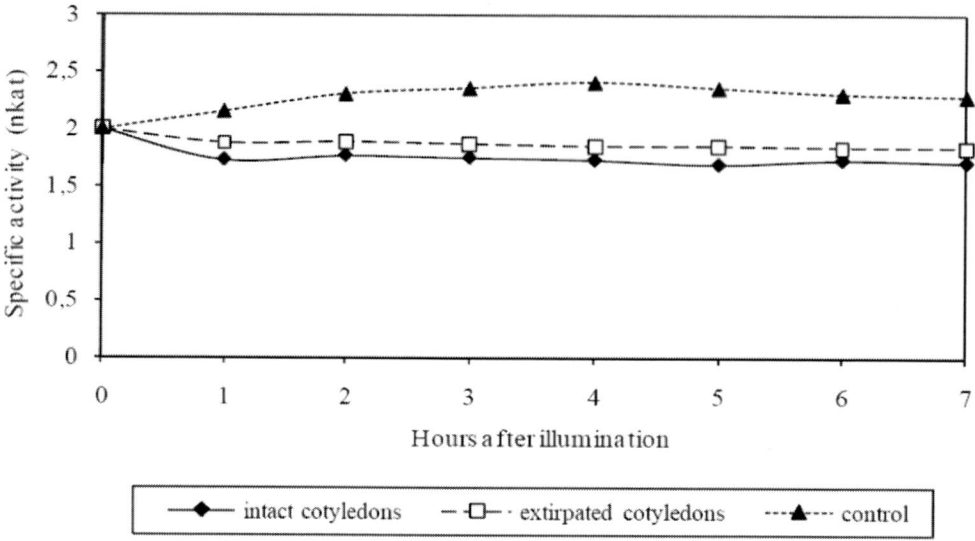

Figure 5. Influence of light upon the G-6-PDH activity; ♦ intact cotyledons; □ extirped cotyledons; ▲ control without light.

G-6-PDH has potential as a regulator for the availability of reduced NADPH, required for the reductive biosynthesis of fatty acids, isoprenoids and aromatic amino acids in the dark, and for nitrogen assimilation in heterotrophic tissues [39]. It is also an important source of NADPH on non-photosynthetic tissues [40] and fat-producing tissues, such as pollen and seeds. G-6-PDH is widely distributed, and it was isolated from microorganisms, plants and various mammalian tissues [41]. In higher plants, different G-6-PDH isoforms have been reported in cytosol, plastid stroma and peroxisomes [42]. Chloroplastic G-6-PDH has been reported to be under the control of photosynthetic redox modulation [43]. Chloroplast G-6-PDH is inactivated in the presence of light via reduced thioredoxin [44-45]. Cell growth can be stimulated with increased G-6-PDH activity [46]. G-6-PDH reported so far has a single submit type with a native form of dimer, tetramer or hexamer; however, only the dimeric and tetrameric forms were found to be enzymatically active [47]. A dark activation and light inactivation cycles have been reported as a way of activity regulation in isolated spinach chloroplasts [48]. All chloroplasts G-6-PDH homologues are inactivated by redox modification of the ferredoxin-thioredoxin system [44]. G-6-PDH is kept in a reduced state under light conditions, and becomes active only in its oxidized state in the dark.

The products of the shikimate and pentose phosphate pathways are very important for the biosynthesis of fatty acids, isoprenoids, aromatic amino acid and many aromatic compounds. Both enzymes can be regulated by these compounds, as well as other enzymes [14, 49].

Conclusion

The highest specific activity of SOR and the most intensive formation of chlorophyll in 5-day-gherkin cotyledons is why they were used as a model organism for studying the effects of light during greening on the activity of SOR and G-6-PDH. Our results showed the positive effects of light on SOR activity and its negative impact on G-6-PDH activity. The study of intracellular location of SOR pointed out that most of it is located in etioplasts. When light was applied most of its activity was present in the chloroplasts (60-65%), while the remaining was in the soluble fraction, with as little as 5% in the mitochondria.

The majority of G-6-PDH was located in the soluble fraction, while the rest in the mitochondria (5%), microsomal fraction (5%), etioplasts and chloroplasts (1-2%).

Acknowledgments

This publication was supported by Research & Development Operational Programme of the ERDF N° 26240120024 "The Centre of Excellence for Contaminants and Microorganisms in Food" and by the VEGA grant No. 1/0380/13.

References

[1] Jiang, S. & Singh, G. (1998). *Tetrahedron, 54*, 4697.
[2] Herrmann, K. M. & Weaver, I. (1999). *Ann. Rev. Plant Physiol. Plant Molec. Biol., 50*, 473.
[3] Gientka, I. & Duszkiewicz-Reinhard, W. (2009). *Pol. J. Food Nutr. Sci., 59*, 113
[4] Nagy, M., Grančai, D. & Mučaji, P. (2011). *Pharmacognosy. Biogenesis of natural compounds* Osveta, Martin, Slovakia.
[5] Stryer, L. (1997). *Biochemia*, Wydawnictwo Naukowe PWN, Warszawa, Poland.
[6] Stano, J., Mičieta, K., Neubert, K., Barth, A. & Koreňová, M. (2010). In: Proc. Analyse and synthese of Drugs, pp. 116, Harmonia.
[7] Bender, S. L., Widlanski, T. & Knowles, J. R. (1998). *Biochemistry, 28*, 7560.
[8] Becker, W. M., Leaver, C. H. J., Weir, E. H. & Reizman, H. (1978). *Plant Physiol., 62*, 542.
[9] Stano, J., Kovács, P., Pšenák, M., Barth, A. & Beneš, K. (1989). *Biológia, 44*, 321.
[10] Hardy, S. I., Castelfranco, P. A. & Rebeiz, C. A. (1970). *Plant Physiol., 46*, 705
[11] Leuschner, C. & Schultz, G. (1991). *Bot. Acta, 104*, 240.
[12] Warburg, O. & Christian, H. (1931). *Biochemie Zeitschrift, 238*, 131.
[13] Warburg, O. & Christian, H. (1931). *Biochemie Zeitschrift, 242*, 206.
[14] Singh, S., Anand, A. & Srivastava, P. K. (2012). *Int. J. Plant Physiol. Biochem., 4*, 1.
[15] Schmidt, C. L., Danneel, H. J., Schultz, G. & Buchanan, B. B. (1990). *Plant Physiol., 93*, 758.
[16] Sánchez-Mundo, M. L., Bautista-Muñoz, C. & Jaramillo-Flores, M. E. (2010). *Process Biochem., 45*, 1156.
[17] Doumas, B. T., Bayse, D. D., Carter, R. J., Peters, T. & Schaffer, R. (1981). *Clinic. Chem., 2*, 1642.
[18] Koski, V. L. (1950). *Arch. Biochem. Biophys., 28*, 339.
[19] Dinç, E., Ceppi, M. G., Tóth, S. Z., Bottka, S. & Schansker, G. (2012). *Biochim. Biophys. Acta, 1817*, 770.

[20] Bezáková, L., Mikuš, M., Šmogrovičová, H., Kovács, P. & Pšenák, M. (1996). *Biol. Plant.*, *38*, 377.

[21] French, C. H. J. & Smith, H. (1975). *Phytochemistry*, *14*, 963.

[22] Billet, E. E., Grayer-Berkmeijer, R. J., Johnson, C. B. & Harborne, J. B. *Phytochemistry*, (1981). 20, 1259.

[23] Johnson, C. H., Attridge, T. & Smith, H. (1975). *Biochim. Biophys. Acta*, *385*, 11.

[24] Attridge, T. H. & Smith, H. (1973). *Phytochemistry*, *12*, 1569.

[25] Iredale, S. E. & Smith, H. (1974). *Phytochemistry*, *13*, 575.

[26] Higashi, K. (1988). *Mutat. Res./Fundam. Molec. Mechan. Mutagen.*, *197*, 273.

[27] Reynolds, J. D., Kimbrough, T. D. & Weekley, L. B. (1985). *Biochem. Physiol. Pflanzen*, *180*, 345.

[28] Kim, S. H., Kronstad, J. W. & Ellis, B. E. (2001). *Phytochemistry*, *58*, 849.

[29] Yamada, M. K., Hasegawa, T. & Shigemori, H. (2008). *Phytochemistry*, *69*, 2781.

[30] Solymosi, K. & Schoefs, B. (2010). *Photosynt. Res.*, *105*, 143.

[31] Pogson, B. J. & Albrecht, V. (2011). *Plant Physiol.*, *155*, 1545.

[32] Blomqvist, L. A., Ryberg, M. & Sundqvist, C. (2008). *Photosynt. Res.*, *96*, 37

[33] Blomqvist, L. A., Ryberg, M. & Sundqvist, C. (2006). *Physiol. Plant.*, *128*, 368

[34] Abdelkader, A. F., Aronsson, H. & Sundqvist, C. (2007). *Physiol. Plant.*, *130*, 157.

[35] Adam, Z., Charuvi, D., Tsabari, O., Knopf, R. R. & Reich, Z. (2011). *Plant Molec. Biol.*, *76*, 221.

[36] Von Zychlinsky, A., Kleffman, T., Krishnamurthy, N., Sjolander, K., Baginsky, S. & Gruissem, W. (2005). *Molec. Cel. Proteom.*, *4*, 1072.

[37] Lopez-Juez, E. (2007). *J. Exp. Bot.*, *58*, 11.

[38] Waters, M. T. & Langdale, J. A. (2009). *EMBO J.*, *28*, 2861.

[39] Hauschild, R. & von Schaewen, A. (2003). *Plant Physiol.*, *133*, 47.

[40] Wakao, S., Andre, C. & Benning, C. (2008). *Plant Physiol.*, *146*, 277.

[41] Cardi, M., Chibani, K., Cafasso, D., Rouhier, N., Jacquot, J. P. & Esposito, S. (2011). *J. Exp. Bot.*, *62*, 4013.

[42] Corpas, F. J., Barroso, J. B., Sandalio, L. M., Distefano, S., Palma, J. M., Lupianez, J. A. & Del Rio, L. A. (1998). *Biochem. J.*, *330*, 777.

[43] Fickenscher, K. & Scheibe, R. (1986). *Arch. Biochem. Biophys.*, *247*, 393

[44] Buchanan, B. B. (1991). *Arch. Biochem. Biophys.*, *288*, 1.

[45] Lendzian, K. & Ziegler, H. (1970). *Planta*, *94*, 27.

[46] Tian, W. N., Braunstein, L. D., Pang, J., Stuhlmeier, K. M., Xi, Q. C., Tian, X. & Stanton, R. C. (1998). *J. Biol. Chem.*, *273*, 10609.

[47] Luzzatto, L. (2006). *Hematologica/Hemat. J.*, *91*, 1303.

[48] Scheibe, R., Geissler, A. & Fickenscher, K. (1989). *Arch. Biochem. Biophys.*, *274*, 290.

[49] Bezáková, L., Grančai, D., Obložinský, M., Vanko, M., Holková, I., Paulíková, I., Garaj, V. & Gáplovský, M. (2007). *Acta Facult. Pharmac. Univers. Comen.*, *54*, 48.

In: From Seed Germination to Young Plants
Editor: Carlos Alberto Busso

ISBN: 978-1-62618-653-8
© 2013 Nova Science Publishers, Inc.

Chapter 3

Light-Induced Exaggeration of the Hypocotyl Hook: Its Developmental Basis and Significance

Chizuko Shichijo[1]* and Tohru Hashimoto[2]

[1]Department of Biology, Graduate School of Science, Kobe University,
Rokkodai, Nada-ku, Kobe, Japan
[2]Uozaki Life Science Laboratory, Uozakiminami-5-chome,
Higashinada-ku, Kobe, Japan

Abstract

The hypocotyl hook forms in the dark and opens in the light. This was a widely accepted notion until recently, and many sophisticated studies have been made on the basis of this notion. However, the discovery of light-induced hook exaggeration (LIHE) in tomato has posed the problem of how to understand LIHE in the framework of the hypocotyl development during epigeal germination. This article attempts to provide answers to the problem. LIHE is caused by phytochrome A and other non-identified phytochromes (phy), involving very low- and low-fluence responses as well as red-light-high-irradiance responses. Besides phy signals, LIHE requires the endosperm remaining within the seed coat. This is evidenced by the observation that if the endosperm, together with the seed coat, is removed no LIHE takes place any more, but the hook opens immediately in response to light. Further, LIHE is likely to be under the control of the dark-formed original hook, occurring at the same site and in the same orientation as the original hook. Surveys showed that similar LIHE is scattered among epigeal dicotyledons, being found in 10 out of 22 species, including tomato. The LIHE-exhibiting seed species are characterized by an underdeveloped small embryo embedded in abundant endosperm in hard seed coats. On the contrary, the 12 species lacking LIHE showed a fully developed embryo with traces of or a soon-disappearing endosperm in the soft seed coat, with the exception of persimmon, which lacks LIHE despite the LIHE type of seed structure. Field simulation experiments with tomatoes implicate that LIHE, due to

---

* Corresponding author: E-mail address: shichijo@kobe-u.ac.jp.

light streaming through the soil, contributes to successful emergence, releasing the hard seed coat. Finally, the possible relation between light-exaggerated and original hooks is discussed as a perspective for future studies. This article presents many photos and several movies to evidence the above descriptions.

Introduction

An apical hook is the arch-shaped, transient structure of the hypocotyl top observed in the seed germination process of most, if not all, dicotyledonous plants. It forms at an early stage of germination in the soil, and opens after emergence above the surface soil in response to light. It contributes to safe emergence of a seedling from the soil by protecting the apical meristem from mechanical injury [1] or by facilitating its passage throughout the soil [2]. Accordingly, the hook must be tightly held until the seedlings emerge from the soil. At emergence, seedlings need to quickly open the hook and expand the cotyledons to prepare themselves to grow autotrophically under sunlight. Accordingly, it is natural that seedlings can (i) produce more ethylene in the dark and in response to mechanical stress and exaggerate the hook, (ii) open the hook and accelerate the expansion of the cotyledons by perceiving light. Extensive work has been made to support this scenario mainly using *Arabidopsis* [3-15], bean [16-21], lettuce [22], pea [23-26], and soybean [27]. During the course of the studies, the action spectrum for hook opening in pea [28] was determined, and the involvement of phytochrome (phy) in the response was shown [29]. It was also clarified that for maintaining or exaggerating the hook, ethylene functions with indoleacetic acid [5, 7, 10, 12-15, 20-21, 26, 30] and gibberellins [11, 15, 22]. Nevertheless, none of the studies found the hook-exaggeration induced by light before studies by Shichijo et al. (2010a, b) [31-32]. This was likely because previous studies used plant species lacking this particular photo-response.

In 2000, we used the red light-induced anthocyanin synthesis with the high-pigment tomato mutant (*hp-1*) to demonstrate with tomato the cryptic red-light signal discovered in *Sorghum* [33-34]. Unfortunately, the *hp-1* mutant (synthesizing anthocyanin in high concentrations) [35-40] was not as sensitive as *Sorghum*, and the original attempt was almost hopeless. Instead, the present first author (C.S.) noticed that the seedling top was enormously wound under red light (R) when searching for any new phenomena in the same mutant. First she thought that it might be a deformity due to some genetic deficiency in the mutant, but the same phenomenon also occurred in wild tomato cultivars. This is the origin of the light-induced hook exaggeration (LIHE). It occurs widely in many species, but not in others, as described below.

Now we are faced with the problem of how to understand the newly found LIHE within the framework of hook formation in epigeal dicotyledons. In an attempt to solve this problem, this article (i) reviews the information published to date on LIHE in tomato, (ii) presents the process of hypocotyl hook formation in the dark, (iii) shows the hypocotyl hook exaggeration and non-exaggeration in the light in relation to seed structure, and (iv) proposes the significance of LIHE in seedling development deduced from photographs and time-lapse movies. It demonstrates some phenomena discovered first by taking movies, and (v) finally discusses the possibly distinct natures of the original and the light-exaggerated hooks.

Materials and Methods

General Methods for Germination and Seedling Culture

Seeds were sterilized for 30 to 90 min with kitchen bleach diluted 25 times, imbibed in running tap water, and allowed to germinate on filter paper in Petri dishes. Uniformly germinated seeds were selected, transferred in general to manufactured porous gravel, Filtration Gravel® (commercial name; GEX, Osaka, Japan - subsequently referred as just Gravel). It was packed in transparent, colorless, wide-mouth bottles or plastic boxes, which were loosely covered with a cap or a lid to maintain humidity, unless otherwise stated. Temperature for seed germination varied with species, and will be given at the respective sites of appearance, but the temperature for growing seedlings was always 25.5 ± 0.5°C. To test the effects of light, transfer of the germinated seeds was made under a dim green safe light. Germinated seed transfer for the control was made under complete darkness.

Culture method of long radish and tomato seedlings

Long radish for Movies 1 and 3: Seeds, sterilized and imbibed as above, were germinated on filter paper in Petri dishes in the dark at 25.5 ± 0.5°C for one day; uniformly germinated seeds were transferred to the surface of or in wet Gravel in plastic boxes, as stated above in the general methods, and used for shoot production.

Tomato: Seeds were germinated in the same manner as radish for 3 days instead of one day. Uniformly germinated seeds were transferred to Gravel in bottles or plastic boxes or other substrata, and immediately or after 1 to 3 days cultured in the dark; seedlings were subjected to experiments.

For testing the effects of different sowing depths in long radish and tomato, germinated seeds were placed on a bed of Gravel packed flat in a plastic box. Thereafter, the bed was separated into two equal sections by erecting a piece of transparent plastic sheet, and finally, covered with Gravel to obtain the desired study depths.

Determination of hook angle: To quantify the magnitude of a hypocotyl hook, the angle formed by the extended lines of the lower and the upper arms of the hook was determined, defining as null when the hook would open completely to straighten up [31].

Light source and lighting: White light (WL) was supplied from "Day Light" fluorescent lamps. Red (R) and far red (FR) light were from respective colored fluorescent lamps through relevant filters [31].

Photographs and movies: Digital cameras (Nikon Coolpix P6000, Tokyo, Japan) were used, which were capable of automatically shooting repeatedly at a pre-set interval with automatic focusing and exposure. The movies presented here were always taken at 10-min intervals.

Results

Formation of the Hypocotyl Hook

When long radish, *Raphanus sativus* cv. *longipinnatus*, representing the epigeal germination of most dicotyledons, germinates, first the root protrudes from the seed coat and elongates downward to fix the seedling. Thereafter, the hypocotyl emerges[1], bending positively with respect to gravity to form the hook (Figure 1). This is similar to that reported for cress, cucumber, and sunflower [41]. The orientation of the hook (plane of a hook) with respect to gravity is not affected by the direction of seed placement (Movie 1). Soon, as the part of the hypocotyl lower than the hook elongates upright in the dark, the hook angle converges mostly to 120 to 140° with respect to the vertical, defining as null the fully opened hook (Figure 1a). Not only in the dark but also in the light, the hook forms once without fail before it opens in response to light (Figure 1b). The strict gravity dependency of hook formation is more clearly seen in persimmon (Figure 2). When a persimmon seed is placed upright with the micropyle end at the bottom, the hypocotyl elongates straight upward without forming the hook for a while, but soon begins to bend sensing a difference of gravity due to an accidental slanting of the hypocotyl (Movie 2 Note the seedling at the front center). These observations support the view [41] that the hypocotyl hook is formed in response to gravity.

In long radish, as the hypocotyl grows, the cotyledons also grow to some extent in the dark, and more markedly in the light (Figure 1). The repulsion by the cotyledons facilitates release of the seed coat. If in soil, the friction of the cotyledons with it, arising from hypocotyl elongation helps to accomplish that releasing. Thus, the seed coat is released from the seedling either on the surface of a substratum or in soil (Movies 1, 3). Snapdragon (*Antirrhinum majus*) is an interesting example that the repulsion of the growing cotyledons abruptly splits off the seed coat to release it (Movie 4). Seeds that have a hard coat, such as persimmon, need to germinate deep in the soil to successfully release the seed coat.

Light-Induced Exaggeration of the Hypocotyl Hook (LIHE)

When tomato seeds germinate in the dark, the hypocotyl emerging from the seed coat forms the hook in a manner similar to radish and persimmon. At first, both lower and upper arms of the hook elongate 5 to 7 mm, forming an inverted V-shape with an angle of 140 to 160° (Figure 3a, 2nd from the left). Subsequently, the lower arm predominantly elongates and the hook tends to open partially, but is hindered due to limited space if in the soil. In parallel, the endosperm (which was first solid) gradually digests to mucilage, and the cotyledons elongate sluggishly, preparing the seed coat with the residual endosperm for releasing [32] (Figure 4b, c). Hence, if seeds germinate at some depth (more than 8 mm) [31] (Figure 5), the seed coats are released within the soil as the lower arm of the hook elongates to push up the seedling apex above the soil surface. This is what happens in tomato in the dark.

---

[1] When germination is hypogeal such as in pea or broad bean, the hypocotyl grows first for a limited period, then root elongation starts [42].

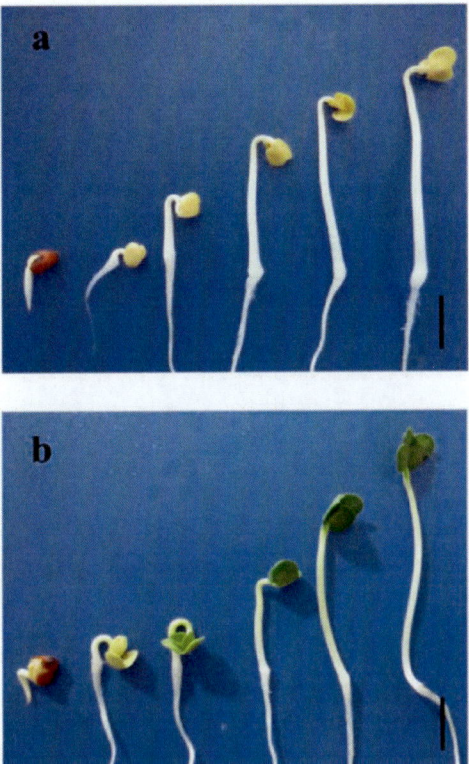

Figure 1. Long radish. Development of the hypocotyl hook in the dark (**a**) and light (**b**). From left to right: 1.0, 1.5, 2.0, 2.5, 3.0, and 3.5 days after sowing; lighting: continuous R (Rc): 1.7 µmol m^{-2} s^{-1}; scale bar: 10 mm. Seeds were sown successively at intervals on Gravel and immediately transferred to the dark (**a**) or light (**b**) until simultaneously photographed. Cf. Movie 1.

Figure 2. The hypocotyl hook of persimmon forms in the same orientation regardless of whether seeds are placed flat or vertical to the surface of a vermiculite bed. Seeds, collected from ripe fresh fruits in November to December and stored humid in the cold, were sown in mid-February and grown for 16 days under continuous white light (WL) until photographed. Lighting: WL, 10.5 µmol m^{-2} s^{-1}. Cf. Movie 2.

When transferred to continuous light conditions, dark-grown 3-d-old seedlings steadily exaggerate the hook during the subsequent 3 days, and then gradually recede (Figure 3b, c). Movie 5 shows the uphill process of LIHE observed with 5-d-old seedlings. Interestingly, when transferred to light, seedlings first tend to open the hook, but soon close it again, and exaggerate it (See also Movie 6).

Effective in hook exaggeration are WL, R, and even FR given in a pulse(s) or continuously. The magnitude of exaggeration depends on the quality and quantity of pulse light as well as the intensity of continuous light (Note the difference in hook angle by different light intensities in Figure 3b, c) [31].

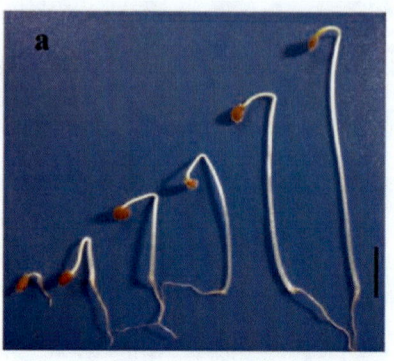

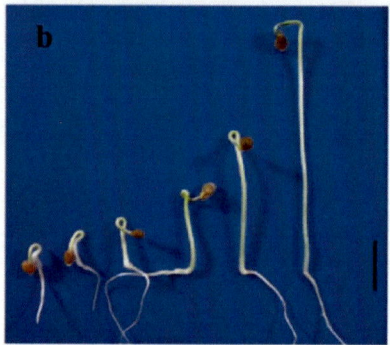

Figure 3. Tomato. Development of the hypocotyl hook in the dark (a) and light (b, c). From left to right: 1.0, 1.7, 2.0, 2.8, 3.0, and 3.8 days after the start of experiment;. Scale bar: 10 mm. For simultaneous photographing, seeds were successively sown at intervals and 3 days later, germinated seedlings (3-d-old) were successively transplanted on the surface of Gravel in boxes, which were kept under Rc, 1.7 μmol m^{-2} s^{-1} (b) and 37 μmol m^{-2} s^{-1} (c) until photographed. For the dark control (a), since even green safe light is effective, transplanting of seedlings was omitted, seeds were successively sown each time in a new bottle of Gravel, and kept in the dark at the same temperature as in b and c.

Figure 4. Tomato. As germination proceeds, the abundant solid endosperm embedding the embryo (**a**, 3.5-d-old) digests to a mucilage, and the cotyledons enlarge pushing out the hypocoty out from the seed coat cavity (**b**, 4.5-d-; **c**, 5.2-d-old). Thus, the seed coat encasing the endosperm becomes ready to be smoothly released. The same changes happen when the hook exaggerates in response to light (**d**, 5.0-d-old; placed under Rc for the last 1.0 d). Top picture for every figure: a whole seedling; bottom: anatomy of the seed part. The lowest bottom in **a**: the back-lighted picture of the middle clearly shows the underdeveloped embryo being embedded by the abundant endosperm. Lighting: Rc, 1.7 µmol m^{-2} s^{-1}; scale bars: black, 1 mm, and orange, 5 mm.

Figure 5. Tomato seedlings under darkness, germinated at a shallow depth (8 mm), fail to release the seed coat (right), whereas those at 15 mm depth are successful in doing it. Seeds were placed on the same level in a pot, and then covered with soil layers of different thickness; note the different levels of the soil surface. Photos: 7 d after sowing seeds (From Shichijo et al. 2010 [31]).

Involvement of Phytochromes in LIHE

Hook exaggeration in wild-type tomato induced by a single R pulse is partially reversed by an immediately subsequent FR pulse repeatedly to the same level, which corresponds to the exaggeration level caused by a single FR pulse alone [31]. Also, the same is true with the high-pigment (anthocyanin), phy-complete mutant, *hp-1*. With phy A-deficient mutant, *phyA hp-1*, however, an FR pulse causes no exaggeration and a complete R/FR reversion is repeatedly accomplished [31] (Figure 6). This indicates that *phy A* and other non-identified phy(s) are involved, the latter being responsible for the observed R/FR reversibility. Fluence-response curves, determined with a wild type and *hp-1* as well as *phyA hp-1* and phy B1 mutant, *phyB1 hp-1* using a single R pulse ranging from 10^{-4} to 10^{4} µmol m^{-2}, show the involvement of very low- and low-fluence responses. The fluence rate-dependency observed under Rc implies the operation of an R-high-irradiance response of phy [31]. Thus, although phy species, other than phy A, remain to be identified, it is clear that phy's mediate LIHE.

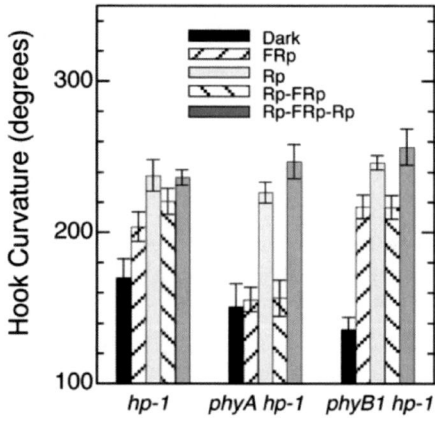

Figure 6. R/FR reversibility of LIHE, an important criterion for examining the phy action, is partial in phy-whole, high-pigment tomato mutant *hp-1* and the phy B1-deficient mutant *phyB1 hp-1*, but complete in the phy A-deficient mutant *phyA hp-1*. This indicates that phy A and other phy's are involved in LIHE (From Shichijo et al. 2010 [31]).

Variation of LIHE Responsiveness with Seedling Development

As tomato seedlings develop, the responsiveness to Rc, which is maximum shortly before the inverted V-shaped stage, decreases exponentially, but 4 days later (8-d-old, about 50-mm-tall seedlings), it still retains about a quarter of the maximum (Figure 7). A similar trend, including a sharp decline from the 4th to the 5th day, is also observed with a single R pulse (data not shown).

Essential for responsiveness is that the seed coat keeping the endosperm remains unreleased. Once it is released, the seedlings lose the responsiveness completely and, instead, come to open the hook in response to light (Figure 8; Movie 6). In seedlings just germinated (3.5-d-old), the endosperm is still abundant in the seed coat cavity (Figure 4a). It gradually decreases as the seedlings grow (Figure 4b, 4.5 d), but a fairly large portion of it remains as a mucilage even when the seedlings become more than 15 mm tall after 5.2 d in the dark

(Figure 4c). The close correlation of the responsiveness with the remaining amount of endosperm strongly suggests that the endosperm, rather than the seed coat, supports the responsiveness of the LIHE.

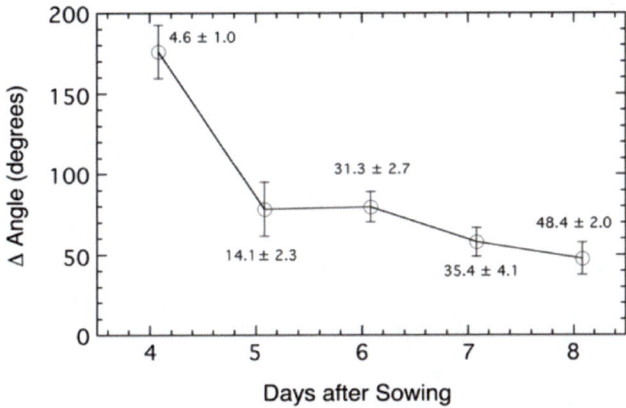

Figure 7. The responsiveness for LIHE in tomato is highest at germination and declines exponentially with seedling age. The ordinate indicates the increase in hook angle during the first 24 h after dark-grown seedlings are transferred under Rc, 35 μmol m^{-2} s^{-1}. Numbers near symbols are the mean ± SE of seedling height (mm; lower arm of the hooks) when transferred (Adapted from Shichijo et al. 2010 [31]).

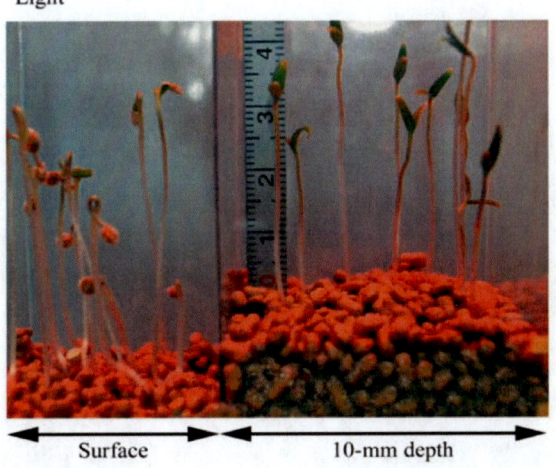

Figure 8. After releasing the seed coat (together with the endosperm), tomato seedlings exhibit no LIHE and, instead, open the hook in response to light (right side), while those holding the seed coat show LIHE (left side). Germinated seeds were planted on the surface (left side) or at 10-mm depth (right side) of Gravel, and grown in the dark for 3 d before being exposed to Rc + WL, 18 and 8 μmol m^{-2} s^{-1}, respectively, for 22 h until photographed. Cf. Movie 6.

Possible Function of the Dark-Formed Original Hook for LIHE

As described above, LIHE takes place literally by exaggerating the original hook formed in the dark. Firstly, the site of the hook, i.e. the center of the hook arch, does not move by the exaggeration, keeping the distance from the cotyledonary node, i.e. the length from the upper arm, almost unchanged; however, the lower arm elongates greatly (Table 1; Figures 3, 4). Secondly, although during the course of LIHE, the hook could be influenced by gravity in various directions, the orientation of the hook does not deviate, i.e. does not twist (Figures 3, 4d), and the exaggeration can easily go beyond 180° (Figure 9; Movie 5).

Naturally, the hook results from a differential growth between the outer and inner sides whether a difference in cell number or length is involved [9, 11]. As seedlings grow, the tissues of a hook on the outer as well as the inner sides are continuously replenished with new tissues from the apical meristem, and move down into the lower arm [32] (Figure 9). In such circumstances, it is noteworthy that LIHE exhibits the above-stated features. These findings suggest that LIHE takes place under an obstinate control of the dark-formed original hook, surmounting the possible influence of the existing gravity.

Table 1. Length of the upper and lower hypocotyls of tomato seedlings

	Light		Dark	
	Upper	**Lower**	**Upper**	**Lower**
5-d-old seedlings	5.40 ± 0.16	6.92 ± 0.47 (n = 60)	4.66 ± 0.15	7.89 ± 0.41 (n = 70)
6-d-old seedlings	5.64 ± 0.16	10.46 ± 0.68 (n = 40)	5.66 ± 0.15	13.51 ± 0.56 (n = 74)

Length: mean $\pm$ SE (mm); Light: Rc, 1.7 µmol m^{-2} s^{-1}; Dark-germinated 3-d-old seedlings were grown under Rc or in darkness for additional 2 and 3 days.

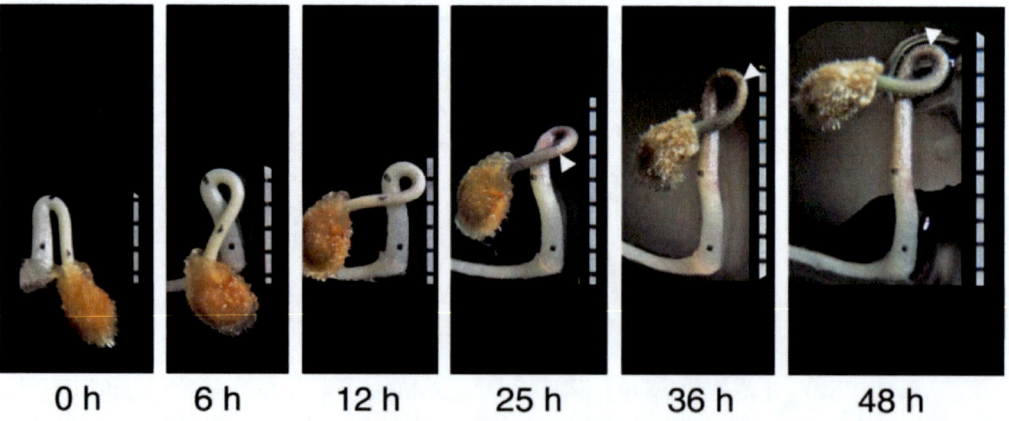

| 0 h | 6 h | 12 h | 25 h | 36 h | 48 h |

Figure 9. Downward flow of tissues through the exaggerated hook from the apical meristem. The hook stays at almost the same distance from the cotyledonary node, although the tissues constituting the hook are continuously replenished with new tissues from the apical meristem and enter the lower part of the hypocotyl. The time indications are periods from the transfer of seedlings to light, Rc, 1.7 µmol m^{-2} s^{-1}. Arrows indicate blurred markings. Photos by M. Takahashi-Asami (From Shichijo et al. 2010 [32]).

LIHE in Other Species

The discovery of LIHE in tomato stimulated us to search for seed species showing a similar hook exaggeration. The results included 22 species, mostly vegetable and ornamental, 10 species including tomato exhibiting LIHE, while radish and 11 other species did not (Table 2). The latter group of species behaves like radish. As to the presence or absence of LIHE, no taxonomical relationship was found.

Table 2. Epigeal dicotyledons surveyed up to date for light-induced hook exaggeration (LIHE)

A) Plants exhibiting LIHE

Family	Species	Common name
Apiaceae	*Cryptotaenia japonica*	Mitsuba, Japanese wild parsley
	Docus carota	Carrot
	Petroselinum crispum Parsley	
Basellaceae	*Basella alba*	Malabar-, Indian-, or Buffalo spinac
Malvaceae	*Abelmoschus esculentus*	Okra
Solanaceae	*Capsicum annuum* var. abbreviatum	Red Chili pepper
	Capsicum annuum var. grossum	Bell pepper, Green pepper
	Solanum lycopersicum	Tomato
	Solanum melongena	Eggplant
Tiliaceae	*Corchorus olitorius*	Jute

B) Plants lacking LIHE

Family	Species	Common name
Amaranthaceae	*Gomphrena globosa*	Globe amaranth, Bachelor button
Asteraceae	*Glebionis coronaria*	Crown daisy
	Tagetes sp.	Marigold
Brassicaceae	*Arabidopsis thaliana*	Arabidopsis
	Brassica campestris	Field mustard
	Brassica oleracea	Cabbage
	Brassica oleracea var. italica	Broccoli
	Raphanus sativus var. longipinnatus	Long radish
Fabaceae	*Trifolium* sp.	Clover
Ebenaceae	*Diospyros kaki* cv. Hiratanegaki	Persimmon
Portulacaceae	*Portulaca oleracea*	Common purslane
Scrophulariaceae	*Antirrhinum majus*	Snapdragon

As typical examples in the former group, bell pepper, eggplant, jute (*Corchorus oritorius*), mitsuba (*Cryptotaenia japonica*), okra and parsley are presented in Figure 10, where marked hook exaggerations are recognized under Rc (upper right for each species) in contrast to controls in the dark (upper left). Characteristic features of seed structure common to the group are the underdeveloped thin cotyledons embedded in the endosperm, abundant in the seed coat cavity (Figures 4, 10). Jute embryo (bottom right) looks big, but it is thin, being sandwiched by endosperm from the front and back. Although okra embryo is well developed, a fairly large amount of endosperm is visible on the inner side of the seed coat, if the embryo is removed (Figure 10 Okra, two photos at the bottom, right).

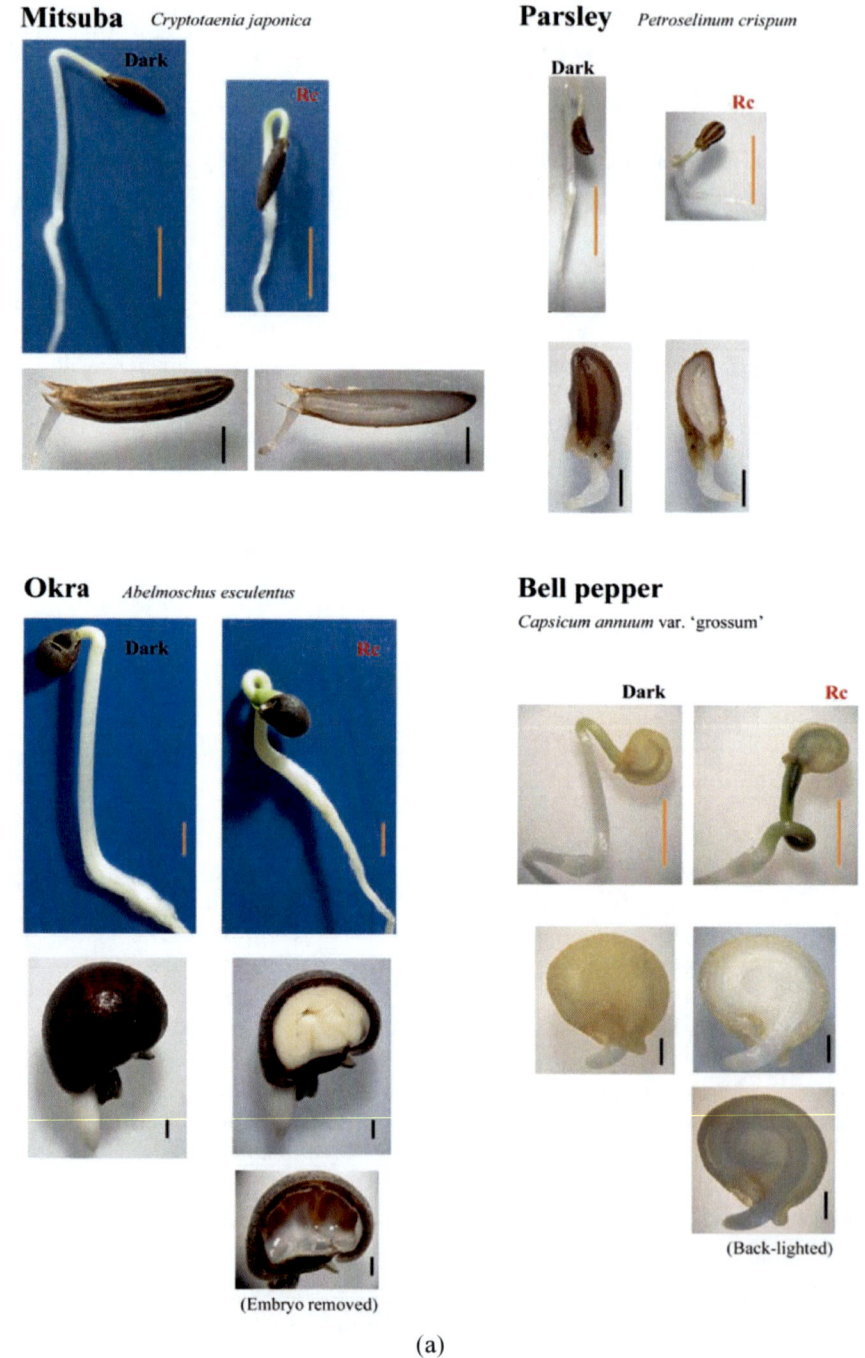

(a)

Figure 10. Continued on next page.

Eggplant *Solanum melongena*   **Jute** *Corchorus olitorius*

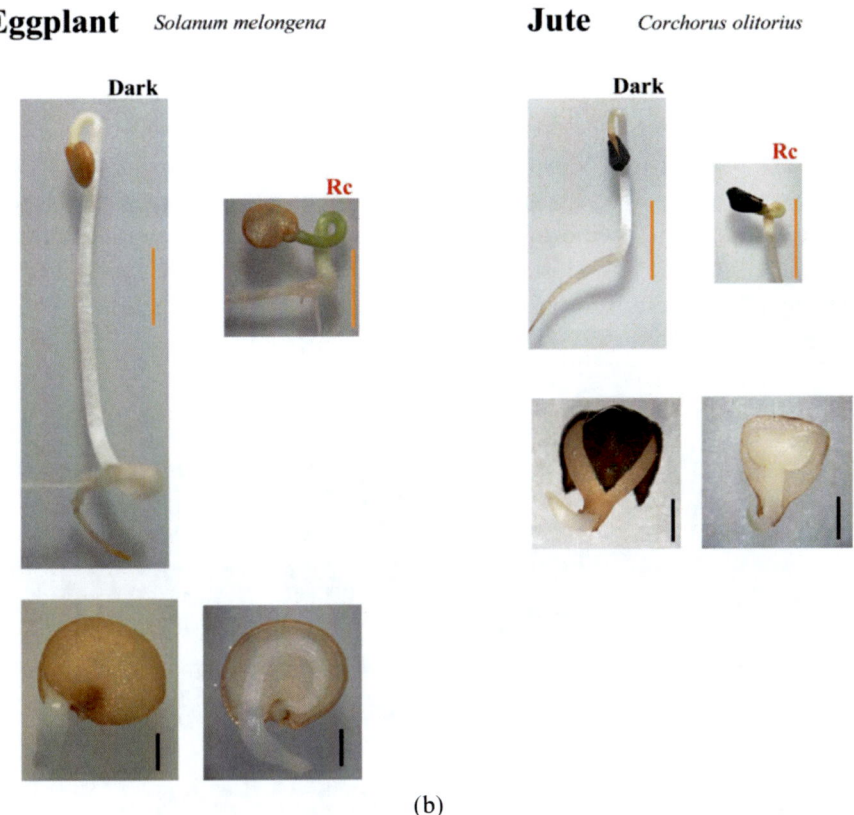

(b)

Figure 10. Representative species exhibiting LIHE. For each species: a whole seedling grown in darkness (top left) and in the light (top right), and a germinated whole seed to be transplanted for LIHE test (bottom left) and their anatomies (bottom right). Seeds were germinated in the dark at the indicated temperatures, and transplanted onto the substratum surface, and subjected to LIHE test for the indicated periods; bell pepper, eggplant, jute, and okra: 25°C; mitsuba: 20°C; parsley: 15°C; eggplant, 4.5 d; bell pepper, 3 d; mitsuba and okra, 2 d; and jute and parsley, 1 d. Substratum: Gravel in bottles for all species except for eggplant and bell pepper placed on wet filter paper. LIHE test: Rc, 38 μmol m^{-2} s^{-1} at 25.5 ± 0.5°C. Scale bars: black, 1 mm, and orange, 5 mm.

As the latter group, Figure 11 shows the cases of cabbage, crown daisy (*Glebionis coronaria*), globe amaranth (*Gomphrena globosa*), and marigold (*Tagetes* sp.), where the hooks straighten up under Rc instead being exaggerated. In the dark, the typical hooks are kept as already seen in radish (Figure 1). Their structural characteristics are that the embryos, particularly the cotyledons, are well-developed and occupy almost the whole seed coat cavity with a trace of endosperm within; the seed coat is easy to tear, and hence, readily released during the early germination process. Exceptions to the common features are persimmon, globe amaranth, and *Portulaca*. A persimmon seed has the very small embryo in the abundant endosperm encased by the thick, hard seed coat when sown, but it exhibits no LIHE (cf. Movie 2). It takes a few weeks for the seed to germinate and for the hook to be raised upward (cf. Figure 2 caption). It was first suspected that during the germination period the endosperm might be consumed up for the growth of the embyo and disappear until the hook forms, but dissection showed that a fairly large portion of the endosperm remained around the cotyledons within the seed coat. Thus, persimmon is truly exceptional to the rule. Globe amaranth (Figure 11) and *Portulaca oleracea* also have a large endosperm at the center of the seed, embraced by the cotyledons (photograph at bottom right), but it rapidly disappears before the stage of the upper two photographs.

Thus, seed structure characteristics for the LIHE group are: (i) the embryo is underdeveloped, and abundant endosperm fills the seed coat cavity. Even when seeds start to germinate and form the hypocotyl hook, the developing cotyledons and the endosperm share the seed coat cavity; (ii) the seed coat is hard to tear. The first characteristic agrees with the conclusion that the presence of endosperm is essential for LIHE as already demonstrated with tomato. The second characteristic has important connections to the possible significance of LIHE, and we will refer to it below.

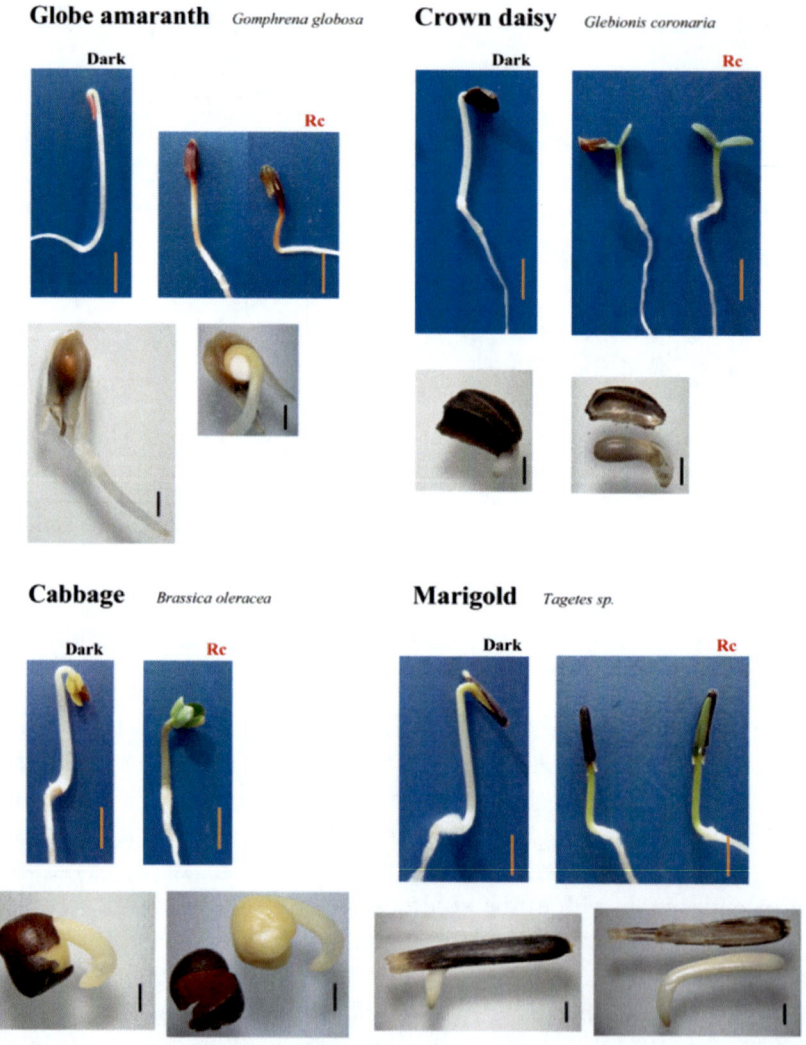

Figure 11. Representative species lacking LIHE. Other explanations are the same as in Figure 10, unless otherwise stated. Germination temperature and period; cabbage and globe amaranth: 25°C; crown daisy and marigold: 20°C; cabbage: 3 d; crown daisy, globe amaranth, and marigold: 2 d.

Significance of LIHE in the Field

What is the significance of LIHE in the field, although it seems to be a drawback to the seedling? The possible significance examined first is that when a seedling emerges through hard soil, LIHE may contribute for the seedling to break through it. To test this, imbibed tomato seeds were placed in a 10-mm depth of granite gravel and allowed to emerge under Rc and in darkness. Both lots showed good emergence of seedlings; however, it was one day earlier in the dark than in the light, rejecting the possibility [31]. The same experiment was conducted using the manufactured gravel "Filtration Gravel®" in place of granite gravel, which gave a similar result (Data not shown). The one-day delay of emergence in the light is ascribable to a suppressed increase in height due to LIHE and direct action of light to suppress growth (cf. Figures 3, 4). The time lag can also be noticed in Movie 7 which presents complicated seedling movements before emergence.

Another possible significance is that LIHE may act on removing the seed coat, which is hard to split, a characteristic of the LIHE species. Seeds were allowed to germinate on the surface as well as at various depths of soil or Gravel. The results were: If germinated at a depth greater than 8 mm, seedlings released the seed coat both in the light and in the dark. Therefore, no LIHE took place [31] (Figures 5, 8). By contrast, if germinated at an 8-mm depth, seedlings released the seed coat only in the light, but not in the dark [31] (Figure 12). At depths shallower than 5 mm or on the surface, the seed coat was released neither in the dark nor in the light, although LIHE took place in the light [31] (Figures 8, 12). Similar results were reproduced with eggplant, but the critical depth was 5 mm instead of 8 mm for tomato (Data not shown). Thus, LIHE contributed to the seed coat release when seeds germinated at a certain medium depth, probably by allowing time for the seedlings to release their seed coat, preventing a likely seed coat desiccation. During this time, the cotyledons elongated, and the endosperm further digested to facilitate the slipping-off of the hard-to-split seed coat. Moreover, the burial depth is important to provide the seed coat with the friction needed to be smoothly released (Figure 13; Movie 7).

Figure 12. Tomato seedlings release the seed coat only when planted at 8-mm depth in the light (Light, right), whereas they fail to do it when at 8-mm depth in the dark (Dark, right) and on the surface in the light and dark. This suggests that at 8-mm depth, LIHE contributes to releasing the seed coat. Germinated seeds were transplanted to the indicated depths, and kept in the dark for another day until the lighting experiment was initiated. Lighting: Rc + WL, 19.3 and 12.2 μmol m^{-2} s^{-1}, respectively. Cf. Movie 7.

Figure 13. Struggle of tomato seedlings in Gravel to release the seed coat by LIHE under WL. The time indications are the period from the transfer of seedlings under light. Following hook formation at 20 h, seedlings perform LIHE at 34 h to 46 h, followed by rapid elongation of the lower part of the hypocotyl, leaving the seed coat behind (arrows). Subsequently, vigorous opening motion of the exaggerated hook occurs. The continuous movements are viewed in Movie 7. For this Figure, germinated seeds were transplanted to an 8-mm depth, and grown under WL: 34 µmol m^{-2} s^{-1}. Scale bar: 10 mm.

Discussion

This article showed that LIHE, first discovered in dark-germinated tomato seedlings, is not a curious phenomenon limited to tomato, but it occurs widely in various dicotyledonous species distributed among families having no particular taxonomical relationships. The common features of seed species exhibiting LIHE are, without exception among 10 of the 22 study species (Table 2; Figures. 4, 10): (i) the abundant endosperm, (ii) underdeveloped embryo, and (iii) hard-to-split seed coat. The unused part at least, of the endosperm, is essential for LIHE to take place. The embryo and the seed coat benefit from LIHE for the successful release of the seed coat; these characteristics further secure the safe development of seedlings (cf. *Significance of LIHE in the Field*). The remaining 12 species lack the features 1 to 3 (persimmon needs confirmation, though), and exhibit no LIHE.

When the hook opens in response to light, phy(s) operates in the LIHE species, as shown with tomato [31], and in the non-LIHE plants [4, 24, 28]. For LIHE to occur, not only phy's but also the endosperm are essential (cf. *Variation of LIHE Responsiveness*), although other unknown factor(s) might also be involved. Seedlings of non-LIHE species, by contrast, carry no more endosperm when they form the original hook (Figure 11). Accordingly, we can assume that the endosperm availability at the time of hook formation distinguishes LIHE from non-LIHE species.

The time-lapse movies disclosed interesting movements during the development of seedlings. When transferred to light, dark-germinated tomato seedlings (LIHE species) first opened the hook for a short period (about 6.3 h) before starting to exaggerate it (Movie 5). On the contrary, dark-germinated radish seedlings (non-LIHE species), at early stages, tend to close the hook momentarily before starting to open it in response to light (Movies 1, 3). These curious opposite responses make us suspect that both LIHE and non-LIHE species possess both properties for hook opening and exaggeration as a response to light. Which property dominates would depend on various factors at the developmental seedlings stages, but the most prominent among them must be the endosperm.

LIHE can take place at the same time when the original hook forms (Figure 3, 7). If, however, LIHE is induced after the original hook is complete, it is recognized that the orientation of the exaggerated hook corresponds to that of the original hook (Figures 4, 9; Movie 5). Therefore, it is most likely that the LIHE is under the control of the latter. If the exaggerated hook, as discussed above under *Possible Function of the Dark-Formed Original Hook for LIHE*, is insensitive or poorly sensitive to gravity, it would be possible to assume that the exaggerated hook is not the original hook simply exaggerated, but a distinct entity.

Still another finding with movies is that the original hook forms not only in the dark but also in the light due to open a hook. Furthermore, it seems to form before the cotyledons open (Figure 1b; Movies 1, 2). The original hook seems to be a key step programmed to be completed before the seedling proceeds to another step of development. The hook might not only protect the apical meristem and facilitate the passage of the apex through soil, but also might play a role in determining the subsequent development of the seedlings. The above-assumed control of the orientation of an exaggerated hook is an example of the assumed roles.

Thus, the discovery of LIHE in tomato provided the turning point to pay renewed attention to the hypocotyl hook. The intriguing properties and involved mechanisms of the hypocotyl hook continue to be a research subject in our laboratories. The discussion here includes hopeful speculations as a guide for future studies.

Movie 1. Hypocotyl hook of long radish, its formation and opening in the light. The hook forms following gravity even in the light before it opens. Seed coats are easily released by the repulsion of the growing cotyledons. Shooting period: 2.1 d, starting with 1-d-old seedlings transplanted on the Gravel surface; lighting: Rc + WL, 8.3 and 5.4 μmol m^{-2} s^{-1}, respectively.

Movie 2. Hypocotyl hook formation from persimmon seeds placed at various orientations. Hooks are uniformly formed following gravity regardless of orientation. Note the seedling in the front center, which has arisen from the seed placed exactly upright with the micropyle end towards the bottom, grows upright until it forms the hook by sensing gravity. Seed sowing and seedling culture are the same as in Figure 2. Shooting period: 12.5 d, starting on 3 March.

Movie 3. In long radish, seed coats are readily released by the repulsion of growing cotyledons regardless of whether sown on the surface or at 10-mm depth of Gravel. Shooting period: 2.8 d, starting with 1-d-old, dark-germinated seedlings; lighting: Rc + WL, 19.3 and 12.2 μmol m^{-2} s^{-1}, respectively.

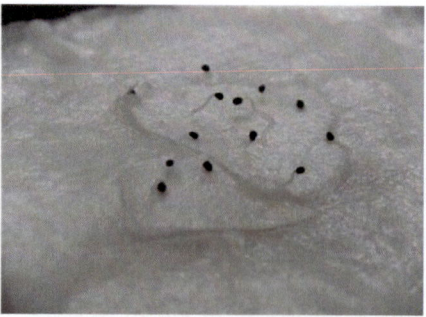

Movie 4. Snapdragon (*Antirrhinum majus*) abruptly releases the seed coat by repulsion of the cotyledons. No LIHE is observed. Seeds were germinated and grown on filter paper placed on water-soaked cotton in a Petri dish in the dark at 25.5 ± 0.5°C for 4 days, and then transferred to WL, 10.5 μmol m^{-2} s^{-1} to start the movie. Shooting period: 6.17 d.

Movie 5. Tomato. Uphill process of LIHE under continuous light. Hook angle continues to increase for about 2 d or more. Dark germinated seeds were placed on the substratum, kept further for 2.0 d in the dark, and then transferred to light when the movie started. Lighting: Rc + WL, 3.3 and 10 μmol m^{-2} s^{-1}, respectively. Substratum: plastic net placed on water-soaked cotton in a Petri dish. Shooting period: 1.8 d.

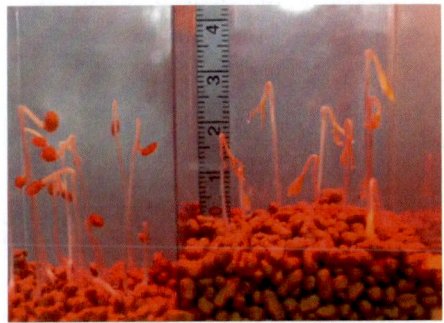

Movie 6. Tomato. Seedlings after releasing the seed coat (together with the endosperm) do not exaggerate, but open the hook (right side), while those retaining the seed coat exhibit LIHE (left side). Interestingly, the hook opening proceeds very fast at the beginning, and almost stops when the cotyledons expand, then re-starts. A few seedlings which released the seed coat (on the right side) exhibit some extent of LIHE, which might be due to residual endosperm on the cotyledon surface. Shooting period: 21.8 h. Other explanations are the same as in Figure 8.

Movie 7. Struggle of tomato seedlings in Gravel to release the seed coat by executing LIHE. Lighting: Rc + WL, 28 and 6 μmol m^{-2} s^{-1}, respectively. Shooting period: 4.0 d.

References

[1] Sasse, J. (2006). *Brassinosteroid and the apical hook - An ongoing story in plant architecture*. L. Taiz and E. Zeiger (Eds.) Plant Physiology online. A companion to Plant physiology, fifth edition.

[2] Taiz, L. & Zeiger, E. (2010). *Plant Physiology*, fifth edition. Sinauer Associates, Inc., Sunderland.

[3] Guzmán, P. & Ecker, J. R. (1990). *Plant Cell, 2*, 513.

[4] Liscum, E. & Hangarter, R. P. (1993). *Plant Physiol., 101*, 567.

[5] Lehman, A., Black, R. & Ecker, J. R. (1996). *Cell, 85*, 183.

[6] Raz, V. & Ecker, J. R. (1999). *Development, 126*, 3661.

[7] Harper, R. M., Stowe-Evans, E. L., Luesse, D. R., Muto, H., Tatematsu, K., Watahiki, M. K., Yamamoto, K. & Liscum, E. (2000). *Plant Cell, 12*, 757.

[8] Knee, E. M., Hangarter, R. P. & Knee, M. (2000). *Physiol. Plant., 108*, 208.

[9] Raz, V. & Koornneef, M. (2001). *Plant Physiol., 125*, 219.

[10] Li, H., Johnson, P., Stepanova, A., Alonso, J. M. & Ecker, J. R. (2004). *Develop. Cell, 7*, 193.

[11] Vriezen, W. H., Achard, P., Harberd, N. P. & Van Der Straeten, D. (2004). *Plant J., 37*, 505.

[12] Stepanova, A. N., Robertson-Hoyt, J., Yun, J., Benavente, L. M., Xie, D. Y., Doležal, K., Schlereth, A., Jürgens, G. & Alonso, J. M. (2008). *Cell, 133*, 177.

[13] Vandenbussche, F., Petrášek, J., Žádníková, P., Hoyerová, K., Pešek, B., Raz, V., Swarup, R., Bennett, M., Zažímalová, E., Benková, E. & Van Der Straeten, D. (2010). *Development, 137*, 597.

[14] Žádníková, P., Petrášek, J., Marhavý, P., Raz, V., Vandenbussche, F., Ding, Z., Schwarzerová, K., Morita, M. T., Tasaka, M., Hejátko, J., Van Der Straeten, D., Friml, J. & Benková, E. (2010). *Development, 137*, 607.

[15] Gallego-Bartolomé, J., Arana, M. V., Vandenbussche, F., Žádniková, P., Minguet, E. G., Guardiola, V., Van Der Straeten, D., Benkova, E., Alabadí, D. & Blázquez, M. A. (2011). *Plant J., 67*, 622.

[16] Kang, B. G., Yocum, C. S., Burg, S. P. & Ray, P. M. (1967). *Science, 156*, 958.

[17] Kang, B. G.& Ray, P. M. (1969). *Planta (Berl.) 87*, 193.

[18] Kang, B. G. & Ray, P. M. (1969). *Planta (Berl.) 87*, 206.

[19] Schierle, J. & Schwark, A. (1988). *J. Plant Physiol., 133*, 325.

[20] Schwark, A. & Schierle, J. (1992). *J. Plant Physiol., 140*, 562.

[21] Schwark, A. & Bopp, M. (1993). *J. Plant Physiol. 142*, 585.

[22] Poovaiah, B. W. & Leopold, A. C. (1974). *Plant Physiol., 54*, 289.

[23] Goeschl, J. D., Rappaport, L. & Pratt, H. K. (1966). *Plant Physiol., 41*, 877.

[24] Goeschl, J. D., Pratt, H. K. & Bonner, B. A. (1967). *Plant Physiol., 42*, 1077.

[25] Peck, S. C., Pawlowski, K. & Kende, H. (1998). *Plant Cell, 10*, 713.

[26] Du, Q. & Kende, H. (2001). *Plant Cell Physiol., 42*, 374.

[27] Samimy, C. (1978). *Plant Physiol., 61*, 772.

[28] Withrow, R. B., Klein, W. H. & Elstad, V. (1957). *Plant Physiol., 32*, 453.

[29] Shropshire Jr., W., in: Mitrakos, K. & Shropshire, W. Jr.. (Eds.), (1972). Phytochrome, Academic Press, London, 161-181.

[30] Friml, J., Wiśniewska, J., Benková, E., Mendgen, K. & Palme, K. (2002). *Nature, 415*, 806.

[31] Shichijo, C., Ohuchi, H., Iwata, N., Nagatoshi, Y., Takahashi, M., Nakatani, E., Inoue, K., Tsurumi, S., Tanaka, O. & Hashimoto, T. (2010a). *Planta, 231*, 665.

[32] Shichijo, C., Takahashi-Asami, M., Nagatoshi, Y. & Hashimoto, T. (2010b). *Plant Sig. Behav., 5*, 1266.

[33] Shichijo, C. & Hashimoto, T. (1997). *J. Photochem. Photobiol. B: Biol., 38*, 70.

[34] Shichijo, C., Onda, S., Kawano, R., Nishimura, Y. & Hashimoto, T. (1999). *Planta, 208*, 80.

[35] Peters, J. L., van Tuinen, A., Adamse, P., Kendrick, R. E. & Koornneef, M. (1989). *J. Plant Physiol., 134*, 661.

[36] Peters, J. L., Schreuder, M. E. L., Verduin, S. J. W. & Kendrick, R. E. (1992). *Photochem. Photobiol., 56*, 75.

[37] Peters, J. L., Széll, M. & Kendrick, R. E. (1998). *Plant Physiol., 117*, 797.

[38] Kerckhoffs, L. H. J., de Groot, N. A. M. A., Van Tuinen, A., Schreuder, M. E. L., Nagatani, A., Koornneef, M. & Kendrick, R. E. (1997). *J. Plant Physiol., 150*, 578.

[39] Kerckhoffs, L. H. J., Schreuder, M. E. L., Van Tuinen, A., Koornneef, M. & Kendrick, R. E. (1997). *Photochem. Photobiol., 65*, 374.

[40] Lieberman, M., Segev, O., Gilboa, N., Lalazar, A. & Levin, I. (2004). *Theor. Appl. Genet., 108*, 1574.

[41] MacDonald, I. R., Gordon, D. C., Hart, J. W. & Maher, E. P. (1983). *Planta, 158*, 76.

[42] Obroucheva, N. V. (1999). *Seed Germination: a guide to the early stages*, Backhuys Publishers, Leiden.

In: From Seed Germination to Young Plants
Editor: Carlos Alberto Busso

ISBN: 978-1-62618-653-8
© 2013 Nova Science Publishers, Inc.

Chapter 4

ROS Generation in Willow Seeds: A Delicate Equilibrium between Seed Ageing and Normal Seedling Growth

Humberto Causin[*1], *Gonzalo Roqueiro*[2]
and Sara Maldonado[1]

[1]Departamento de Biodiversidad y Biología Experimental,
Facultad de Ciencias Exactas y Naturales, Universidad de Buenos Aires,
C1428EGA C.A.B.A, Argentina
[2]Instituto de Recursos Biológicos, INTA-Castelar, Hurlingham, Argentina.
Present address: Estación Experimental Agropecuaria San Juan, Instituto de Tecnología
Agropecuaria, Pocito, San Juan, Argentina

Abstract

Willow (*Salix* spp.) seeds are desiccation-tolerant, like orthodox seeds. However, they lose viability in a few weeks when stored at room temperature. They also differ with most orthodox seeds in that the chloroplasts of the embryo tissues conserve intact their chlorophyll and endomembrane system. The loss of viability is preceded by a decrease in normal germination (NG) percentage (i.e., the percentage of embryos having intact cotyledons and primary roots after germination), a phenomenon that correlates to an increase in the production of reactive oxygen species (ROS) and free radicals. Light and oxygen play an important role in free-radicals generation producing photooxydative damage in embryo tissues. Subcellular and biochemical analyses of experimentally aged seeds indicate that the thylakoid membranes of the chloroplasts are probably the first target of the ROS. During imbibition, an increase in the generation of superoxide anions (O_2^-) in cotyledonary tissues is among the first symptoms associated to willow seed ageing. In agreement with this view, a short term treatment of aged seeds with different ROS scavengers may significantly increase their NG percent age. However, the continuous presence of O_2^- (but not H_2O_2) scavengers in the germination media inhibits

[*] Corresponding author: E-mail address: ssvhfc@gmail.com.

root growth, suggesting that an adequate generation of superoxide anions should be maintained for normal growth of this organ. In fact, normal seedlings show an intense production of O_2^- in the root apical zone. The activity of NADPH-oxidase appears to be involved in this process, although the presence of endogenous O_2^- sources cannot be discarded. The exogenous supply of either abscisic acid (ABA) or inhibitors of the gibberellic acid (GA) synthesis suppressed root growth without inhibiting O_2^- production. This indicates that, even though O_2^- generation in the root apex would not depend on changes in ABA or GA availability, an adequate endogenous ABA/GA ratio is necessary for the ROS stimulation of root growth. O_2^- production is also very active at the sites of root hair budding. Nevertheless, root hair density and growth proved to be affected by changes in H_2O_2 rather than in O_2^- levels. Possible mechanisms of action of the different ROS in seed ageing as well as root growth and development of willow seedlings are discussed.

Introduction

During seed maturation, the embryo arrests its own growth and enters a quiescent state until environmental conditions become favorable for germination. Dehydration of tissues and organelle de-differentiation are typical processes associated to the seed maturation phase in most economically-important crop species. However, a wide range of postharvest responses of seeds may be found, especially when considering species from tropics and sub-tropics [1]. According to their storage behavior, seeds that are tolerant to relatively high desiccation levels (e.g., 7 to 12 % tissue water content) and survive in the dehydrated state for long periods, depending on storage conditions, are said to show orthodox behavior. At the other extreme, seeds that are damaged by dehydration, usually chilling-sensitive, and that cannot be stored effectively for useful periods are denominated recalcitrant. Seeds that do not tolerate high desiccation levels and/or are chilling-sensitive even in the dehydrated state, are categorized as showing an intermediate behavior [2-4].

Willow seeds were provisionally classified as recalcitrant because their viability is lost within a few days when stored at room temperature [5, 6]. More recently, Hong et al. (1996) [4] classified seeds from 28 species of *Salix* as orthodox, taking into account that seeds of the 28 species studied had been maintained without loss in viability for some years at subzero temperatures and low moisture contents. In fact, willow seeds possess the ability to tolerate chilling and desiccation to very low water contents, as do orthodox seeds. However, they differ from them in two aspects: (i) they lose viability in a few weeks when stored at room temperatures [7], and (ii) their chloroplasts in the embryo tissues do not dedifferentiate during maturation drying, retaining chlorophyll and maintaining their endomembrane system intact [8].

In fresh green tissues, chlorophyll excited by light can activate molecular oxygen to form singlet oxygen ($^1O_2$) and superoxide anions (O_2^-). Both molecules are highly unstable and can give rise to other reactive oxygen species (ROS) like the hydroxyl radical (^-OH) and hydrogen peroxide (H_2O_2), among others [9-13]. The mitochondrial respiratory chain is also one major source of ROS, although the activity of different types of oxidases and peroxidases can significantly contribute to their generation. The uncontrolled accumulation of ROS is highly toxic because they can react with the majority of biomolecules, thus resulting in oxidative damage and, eventually, cell death. To prevent this, cells are equipped with various antioxidants, including the enzymes superoxide dismutase (SOD), catalase (CAT), ascorbate peroxidase (APX), guaiacol peroxidase (POX), glutathione peroxidase (GPX), and the powerful antioxidants ascorbic acid, reduced glutathione and α-tocopherol, among others [14-17]. Free-radical mediated lipid peroxidation, protein inactivation and/or degradation, damage to nucleic acid integrity, non-enzymatic protein glycosilation and lipid autoxidation have been identified as the major causes of seed ageing in different plant species [18-21]. Because the generation of ROS (as well as the activation of the antioxidant defenses) is strongly dependent on the metabolic activity of the cells, the lack of photosynthetic tissue together with low tissue water content and respiratory rate are among the main traits associated to the high longevity of orthodox seeds.

Few studies on oxygen activation have been carried out on dry green tissues, such as those of Salicaceae seed embryos. Even so, it is possible to assume that electron transference does not occur, since photosystem I would not be operative in this state. Hence, the paths for energy dissipation of activated chlorophyll in willow seeds exposed to light probably include the activation of molecular oxygen by energy transference [22, 23]. Because the antioxidant machinery is not effective prior to imbibition, the generation of ROS by this mechanism would lead to oxidative damage, thus decreasing seed viability and normal germination percent age. To test this hypothesis a series of experiments were undertaken using *Salix nigra* seeds as a model system.

ROS Generation during Aging and Early Imbibition of Willow Seeds

In *Salix nigra* seeds, the loss of viability during seed aging is preceded by a decrease in the normal germination (NG) rate (i.e. the percentage of embryos with intact cotyledons and primary roots after germination), and an increase in the mean germination time (MGT) (Table 1). When non-aged seeds are exposed to increasing light intensity in air for 72 h prior to imbibition, NG and MGT reveal a decrease and increase, respectively (Table 1). A high correlation is found between both traits ($R^2 = 0.97$, $\alpha \leq 0.01$), as well as between each of them and stable free radical (SFR) production, as determined by electronic spin resonance (ESR) spectroscopy (NG *vs.* FR $r^2 = 0.94$ and MGT *vs.* FR $r^2 = 0.98$, $\alpha \leq 0.01$; Table 1), supporting the hypothesis that seed aging may be associated with an increment in ROS generation.

Now, control seeds also show an important amount of SFRs prior to imbibition, representing about 50% of the amount found in the seed lots whose NG decreased to nearly 0 % (Table 1). However, soon after imbibition and during early germination, the activity of several antioxidant enzymes becomes consistently higher in non aged controls as compared to photo-oxidized seeds [24]. Hence, both the lower initial SFRs level together with a higher antioxidant activity may explain why the detected SFRs do not affect NG in control seeds. The SFRs found in the control seeds prior to imbibition could have been generated by the superoxide radical formed as a consequence of the deregulation of the respiratory electron transport chain in the later stages of development [12, 15, 25]. As shown in Table 1, NG percentage decreases when seeds are exposed to air in darkness. This response seems to reflect the damage produced by the reaction of oxygen with pre-existing radicals, especially considering that the seeds were treated for three days at 25°C, a time period that although brief, is conducive to aging [7]. The key role of O_2 in willow seed aging is further supported by the fact that NG percentage in seeds exposed to N_2 in darkness does not significantly differ from the control. The increment of the EPR signal in seeds treated with light and N_2 suggests that light-activated chlorophyll might react directly with the substrate to form free radicals in the absence of atmospheric oxygen [26]. Nevertheless, the formation of SFR is clearly enhanced in the presence of air due to photooxidation.

Despite the marked effects exerted by light and O_2 pre-treatments on NG percentage, no significant differences in total germination (TG) percentage compared to the control were observed in the above-described experimental conditions. In orthodox seeds, NG usually precedes TG; however, as aging increases, these two parameters (NG and TG) decrease at nearly the same rate, so that the declines occur almost simultaneously [27-29]. In photo-oxidized *S. nigra* seeds, NG and TG percentages do not decline simultaneously, but rather TG percentage begins to decline once the former reaches zero [30] This might be explained taking into account how the photooxidation process takes place in willow seeds. Since they have a thin and transparent seed coat [8], the attack of the SFRs induced by light in the presence of oxygen is particularly strong in the superficial tissues of the embryo. In fact, the damage is clearly visible in the chloroplasts on the abaxial side of cotyledons, especially in their thylakoid membranes, while only slight damages are detected in chloroplasts from the central and adaxial mesophyll [30]. Similarly, no damage is observed in the central tissues of the axis nor in the shoot apical meristem, which is surrounded by cotyledons (Figure 1). The photooxidative damage produced in the seed manifests itself during germination, producing smaller seedlings with under-developed roots and cotyledons with brown-red areas in their abaxial surfaces and borders (Figure 1), all aspects considered in the evaluation of NG. Because oxidative damage spreads from the outermost to the innermost tissues, TG percentage would only begin to decrease once the shoot and root apical meristems are reached. In typical orthodox seeds, which are not susceptible to light, the aging process occurs at the same time in all of the tissues causing the simultaneous decrease of TG and NG percentages and, consequently, the overlap of falling curves [27].

In light-aged seeds, a strong decrease in the polyunsaturated fatty acids of the glycolipid fraction (a main component of the thylakoid membranes) is found (Figure 2), which suggests that the thylakoid membranes are among the first targets of the oxidative process. The FR and ROS produced in the chloroplast would then extend the oxidative damage to the plasma membrane, as indicated by the decrease in the polyunsaturated fatty acids of the phospholipids fraction. This is consistent with the increase in plasmalemma permeability, as measured by both ESR spin probe and conductivity [30]. In addition, other cellular components are also damaged by the ROS attack [24, 30].

The decrease in NG percentage in photo-oxidized willow seed is powered by a marked increase in the oxidative burst that occurs at early imbibition [24]. The oxidative burst is one of the first biochemical processes that occur during seed imbibition due to the mobilization of pre-existing free radicals [31]. In orthodox seeds, this phenomenon has been related to a beneficial rather than a deleterious process [32], and in some species it has been correlated to seed vigor [33]. The negative effect exerted by the oxidative burst in photo-oxidized willow seeds may be related to the atypical characteristics of these seeds: i) a high basal ROS level in dry seeds, due to the high sensitivity to light of willow seedlings; ii) the presence of a high endomembrane content, which results in more targets for ROS attack; and iii) a decreased antioxidant enzymatic activity due to protein oxidation [24].

Table 1. Normal germination (NG), total germination (TG), mean germination time (MGT), and stable free radical (SFR) content in non-aged *Salix nigra* seeds (control), or seeds aged during 3 d at 45% RH, 25°C at the indicated storage conditions. Seed germination was conducted during 7 d, under dim red light (16 h photoperiod) at 21°C

Treatment	NG	TG	MGT	SFR
	(%)	(%)	(days)	(amplitude of EPR signal
Control	97±3	100	1.00 ± 0.00	1908 ± 20
Dark + N$_2$ atmosphere	90 ±4	100	1.00 ± 0.01	1960 ± 16
Dark + air	76±6	100	1.04 ± 0.02	1962 ± 117
Light (1.75 µmol m^{-2} s^{-1}) + N$_2$ atmosphere	72±2	100	1.15 ± 0.01	2200 ± 97
Light (1.75 µmol m^{-2} s^{-1}) + air	18±1	100	1.51 ± 0.04	2916 ± 165
Light (16.1 µmol m^{-2} s^{-1}) + air	4±3	100	1.70 ± 0.10	3580 ± 126

NG: Germination is considered normal if the seedlings stand erect and develop cotyledons, a hypocotyl and a root. Seedlings not meeting these criteria are classified as abnormal.

TG: is the sum of NG and abnormal germination.

MGT: is calculated by Σ(*ni,ti*) divided by Σ*ni, ni* being the number of germinated seeds at time *ti* at the beginning of seed imbibitions.

SFR: is determined by ESR direct measurements. Here the signal intensity is found using the total height of the free radical peak in the first derivative spectrum signal intensity. Since all the ESR spectra had equivalent line shapes and line-widths, the ESR signal intensities were considered proportional to the radical concentrations. Then, radical concentrations were estimated from the total heights of the respective peaks and expressed as relative measures, determining the relationship within each spectrum.

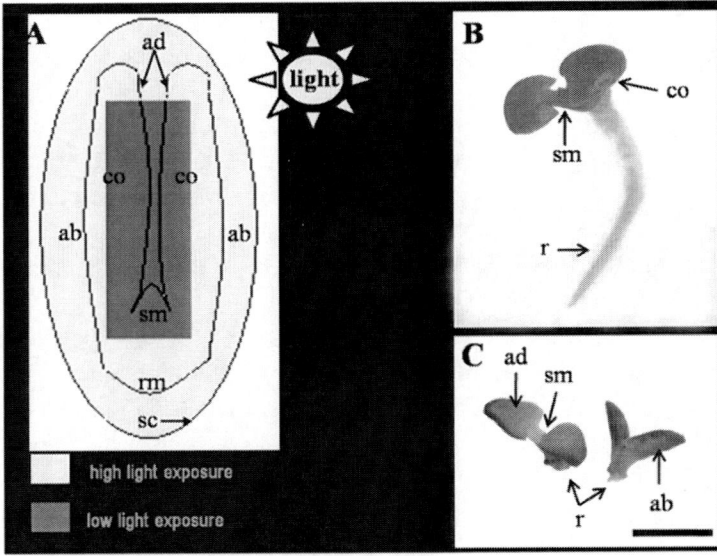

Figure 1. (A) Damages to *Salix nigra* seeds caused by light exposure. (B, C) Seven-day-old *Salix nigra* seedlings: (B) seedling from control seeds (non-aged); (C) seedling from a seed aged for 3 d, under high light intensity at 25°C. Abbreviations: ab, abaxial face of cotyledons; ad, adaxial face of cotyledons; r, root; sm, shoot apical meristem; rm, root apical meristem; sc, seed coat; co, cotyledon. Scale bar = 2 mm.

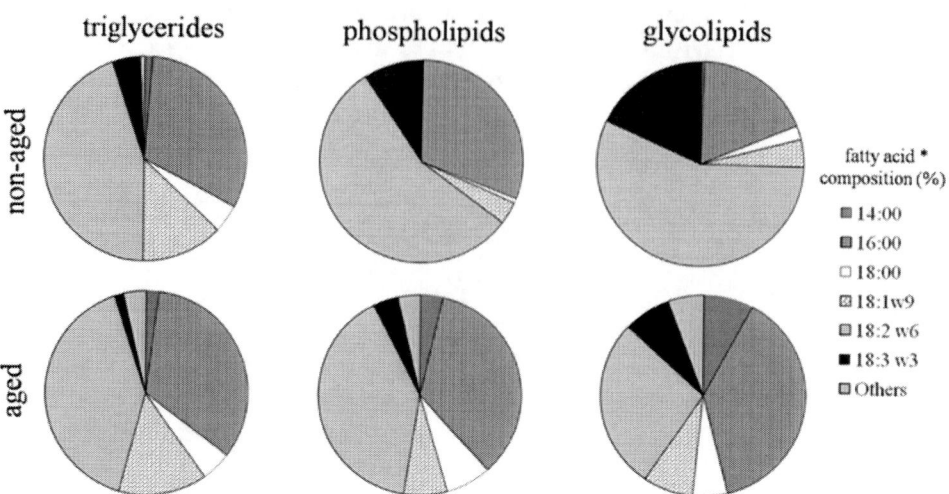

*Fatty acid notation: C14:00 myristic; C16:00 palmitic; C18:00 stearic; C18:1 w9 oleic; C18:2w6 linoleic; 18:3w3 linolenic.

Figure 2. Effect of light intensity on fatty acid composition of triglycerides (reserve lipids), phospholipids (component of all membranes), and glycolipids (component of thylakoids) in *Salix nigra* seeds aged for 3 d, under high light intensity at 25°C.

An increase in O_2^- levels at early imbibition would be one of the main responsible of the decrement of NG percentage observed in aged seeds. In fact, when aged seeds were exposed to different FRs scavengers during early imbibition, NG percent was significantly increased in the presence of O_2^- scavengers or NOX (a O_2^- generating enzyme) activity inhibitors, while only minor effects were exerted by the presence of either H_2O_2 or ˙OH scavengers (Figure 3)

The high O_2^- production would be a consequence of chloroplast alteration during seed ageing in the presence of light. In fact, when seeds are aged at low light intensity a high O_2^- production can be detected at the cotyledon edges, while in seeds aged at high light intensity O_2^- generation spreads all over the abaxial cotyledon surface [24]. It is worth mentioning that ROS production by altered chloroplast can be induced by different abiotic stresses [25, 34, 35].

Despite the positive effect exerted by the presence of O_2^- scavengers during early imbibition, a more prolonged treatment markedly inhibited root growth (and hence NG) (Figure 3), suggesting that O_2^- might be required for normal growth of this organ.

The Role of ROS in the Regulation of Root Growth and Development in Willow Seedlings

A seen in Figure 4a, an active production of superoxide anions (as revealed by nitro blue tetrazolium staining) can be detected in the embryo root apex of non-aged seeds within the first hour after imbibition. In some cases, O_2^- production is first detected in the cells at the base of the hypocotyl, which originate the hypocotyl hairs [36]. The microscopic analysis of longitudinal sections of root apexes stained with NBT at different days from sowing showed that, in the hypocotyls hairs, O_2^- generation is intense at the moment of root hair "budding", but it rapidly decreases as the hair elongates. On the contrary, O_2^- generation is very marked in cells from (or close to) the root apical meristem, part of the ground meristem beyond the apical region, and in cells from the protodermal layer in both the apical and elongation zones. Very little or no NBT staining is detected in the root cap and the region of the procambium. (Figure 4b, c).

An active production of superoxide anions in the apex of growing roots had been already reported for some plant model systems like maize [37], cucumber [38] and *Arabidopsis* [39]. However, in all these cases the highest amounts of O_2^- were observed in the central growing zone (i.e., 2-5 mm behind the root tip, depending on the species), with very little detection in the meristematic region.

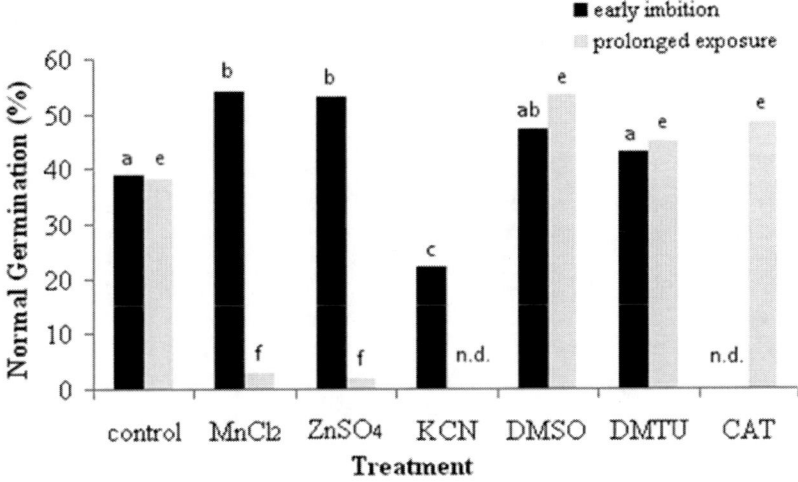

Figure 3. Effect of scavengers of O_2^- (MnCl$_2$), HO (DMSO), H_2O_2 (DMTU and CAT) or inhibitors of NAD(P)H oxidase (ZnSO$_4$) and peroxidase (KCN) activities, on NG of *Salix nigra* seeds aged for 3 d under high light intensity at 25°C. The aged seeds were imbibed during 45 minutes in water alone (control) or solutions containing the different reagents, and then germinated during 72 h in either agar-agar alone (early imbibition treatments), or in the presence of the respective reagent (prolonged exposure). Data were analyzed separately for each group. Within each time from imbibitions treatment, the same letter indicate lack of significant differences at $p \leq 0.05$ (Tukey HSD test). n.d., not determined.

As previously mentioned, the pharmacological downregulation of O_2^- generation in the root apex markedly inhibited root growth of willow seedlings (e.g., Figure 5a). Despite there is ample evidence indicating that O_2^- generation is an essential process in the control of root growth, the mechanism involved still remains controversial. Several experimental approaches support the notion that polysaccharide splitting due to the action of hydroxyl radicals produced in the cell wall from a peroxidase-catalyzed reaction between O_2^- and H_2O_2, is directly responsible for cell wall loosening and cell expansion in the growing zone of roots [37, 39-41]. Superoxide anions would be generated in the apoplast by plasma membrane NADPH oxidase/s (NOX), while H_2O_2 might be produced either in the apoplast as well as in different endogenous cellular compartments [39]. NOX-dependent ROS formation required for root hair growth in *Arabidopsis thaliana* was shown to depend on the activity of the ROP (R̲HO o̲f p̲lants) GTPase system [42]. Besides affecting cell wall extensibility, ROS production may also stimulate the influx of Ca^{2+} via plasma membrane hyperpolarization-activated calcium channels (HACC), so that ROS regulated calcium gradients can contribute to the growth of root hairs, and eventually the root as a whole [43-45]. Interestingly, Ca^{2+} itself may amplify NOX-dependent ROS production since NOX contains an EF-hand motif indicative of calcium stimulation [45, 46]. Studies with the NOX inhibitor Zn^{2+} as well as SDS PAGE followed by western blot analysis of protein extracts from different regions of the root indicate that NOX activity is also involved in the production of superoxide ions in willow seedlings [47]. Moreover, pharmacological treatments with either the calcium channel inhibitor La^{3+} or the calcium scavenger EGTA also indicate that Ca^{2+} is required for the stimulation of O_2^- generation in the root apex. Nevertheless, because the localization of the blue formazan in many of the NBT stained cells is not restricted to the apoplast as expected if plasma membrane NOX were the only source of superoxide ions (Figure 1b), the contribution of intracellular O_2^- generation sources cannot be discarded.

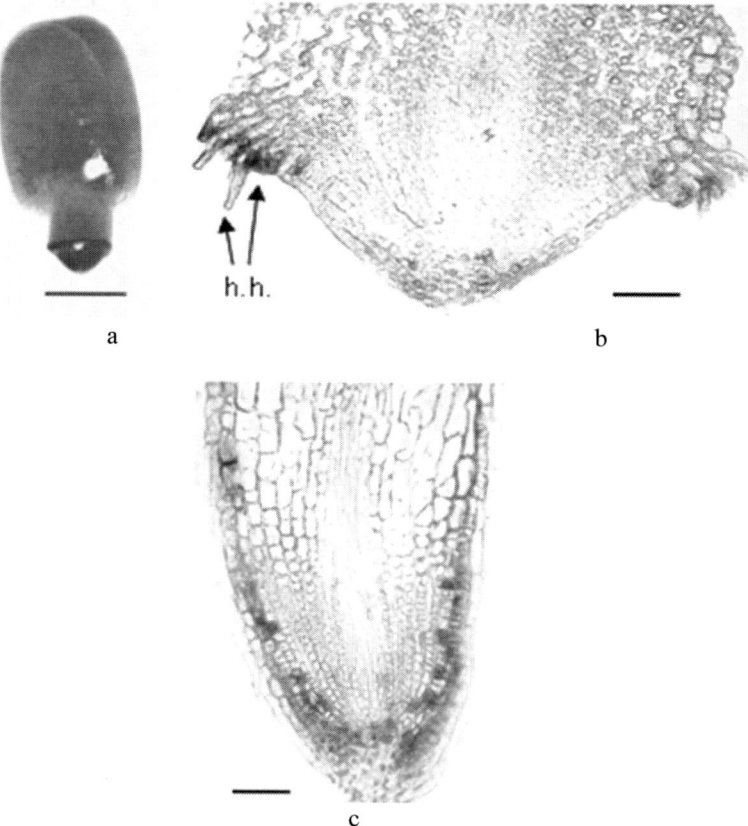

Figure 4. NBT staining (dark zones) of *Salix* seedlings at 1 h after seed imbibition (a), and light microscopy of root apexes of *Salix* seedlings stained with NBT at 24 h (b), and 80 h (c) after sowing. Bars, a) 0.5 mm; b-c) 50 μm; h.h., hypocotyl hairs. Adapted from Causin et al. , 2012 [47].

In germinating radish seeds, the release of reactive oxygen intermediates by the embryo was shown to be inhibited by abscicic acid (ABA) and stimulated by gibberellins (GA) [40]. A similar effect by ABA and GA was reported for the release of ROS during the induction of programmed cell death in cereal aleurone cells [48]. These results support the view that ROS generation is a developmentally controlled process, and that these phytohormones play a key role in the regulation of ROS release during seed germination. When willow seeds were imbibed in water during 4 to 24 h and then germinated in the presence of different concentrations of ABA or inhibitors of GA synthesis, radicle growth (and hence normal germination) was completely inhibited despite O_2^- generation was not suppressed (Figure 5b-d).

This suggests that O_2^- formation in the root apex would be under a different control mechanism than ROS release by the embryo at the onset of germination, although an adequate balance between ABA and GA is necessary for the O_2^- stimulation of root growth in willow seedlings.

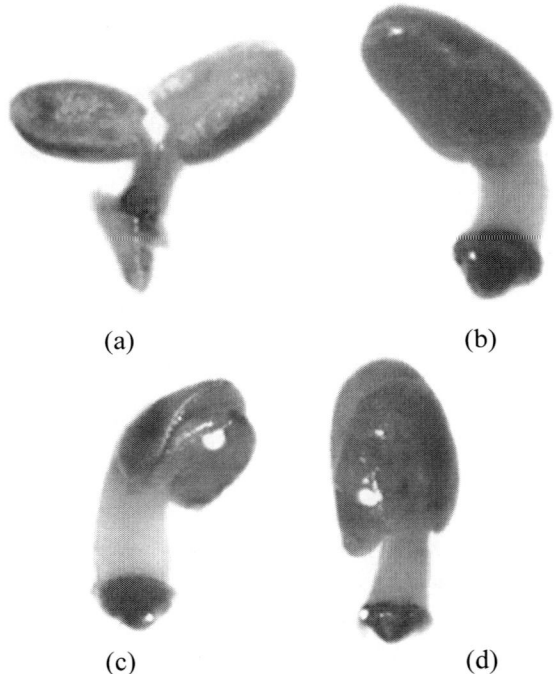

(a)   (b)

(c)   (d)

Figure 5. NBT staining of 3 day old *Salix nigra* seedlings germinated in the presence of a O_2^- scavenger (a), ABA (b) or GA inhibitors (c,d). Seeds were surface sterilized and, after rinsing, sown on 0.8 % agar-agar containing either 8 mM MnCl$_2$ (a), 40 µM ABA (b), 250 µM Pachlobutrazol (c), or 0.005% trinexapac ethyl (MODDUS 250 EC®) (d). Germination was conducted in a growth chamber at 22°C and16 h photoperiod.

While several experimental approaches support the notion that the apoplastic supply of superoxide anions and its dismutation product, H_2O_2, will contribute to elongation growth through the generation of ˙OH by cell wall peroxidases [37, 39], our results on willow seedlings suggest that other mechanisms should be taken into account. In fact, considering that: (i) the presence of ˙OH quenchers had no significant effect, while the external supply of H_2O_2 scavengers had a relatively slight effect as compared to O_2^- scavengers on root growth,

(ii) in the hypocotyl hair cells, NBT staining is intense at the moment of root hair budding, but it gradually disappears when the hair begins to elongate, and (iii) NBT staining is not limited to elongating cells but it is also intense in cells from the meristematic zone [47]; then O_2^- generation should be related to other processes besides cell elongation.

Because superoxide anions are potentially harmful, root apical cells should be provided with very active O_2^- detoxifying mechanisms in order to prevent oxidative damage. Western blot analysis of extracts from the apical and middle (i.e. completely differentiated) region of 3 day-old seedling roots using a "cytosolic" Cu/Zn SOD antibody, revealed a much higher amount of this protein, as well as more antibody reactive bands in the apex than in the middle region [47]. The fact that "cytosolic" Cu/Zn SOD was more abundant in the apex (i.e. where most of the O_2^--generating activity was located) is consistent with the hypothesis that both apoplastic and intracellular O_2-generating sources may contribute to O_2^- formation, given that isoforms of this metal-protein are present not only in the cytoplasm, but also in the nucleus and the apoplast [49, 50]. Nevertheless, the presence of elevated amounts of Cu/Zn SOD protein was not correlated to a marked production of H_2O_2 as expected if O_2^- detoxification merely relied on SOD dismutation activity [47]. Interestingly, an analysis of the presence of carbonyl groups in the proteins of the same extracts revealed that the proportion of carbonylated proteins was about two folds higher in samples from the apical than the middle zone of the root [47]. Hence, even though H_2O_2 accumulation may in part have been prevented by the action of root peroxidases, the high content of carbonylated proteins detected in the apex may be indicative of an alternative role for the elevated O_2^- production in this zone. In fact, it has been recently suggested that peptides derived form proteolytic breakdown of oxidized proteins can act as specific secondary ROS messengers in different signaling cascades [51]. Under this assumption, it is possible that superoxide production in actively-growing and/or differentiating cells of the root apex contribute to the generation of peptide signals involved in the regulation of the organ growth.

While the presence of O_2^- scavengers markedly altered root growth, the pharmacological modulation of H_2O_2 pools exerted dramatic effects on the formation of root hairs. In fact, root hair density increased in the presence of H_2O_2 scavengers, while it was suppressed when we attempted to increase root H_2O_2 concentration by germinating the seeds in the presence of either exogenous H_2O_2 or inhibitors of peroxidase activity [47] (Figure 6 a-c).

Hystochemical analysis with 3,3′-diaminobenzidine (DAB) revealed that willow roots have a high peroxidase activity, particularly in the region with root hairs along the axis, as well as in the root hair tips. However, they did not show an active H_2O_2 generation in any particular region [47], as opposed to reports for other plant model systems [37, 39, 52, 53]. Studies on some of these species also indicate a marked influence of H_2O_2 on root hair formation. However, the presence of exogenous H_2O_2 usually stimulated rather than inhibited root hair development. This response is consistent with the fact that high H_2O_2 production rate and peroxidase activity were detected in the differentiation zone of the apex and in growing root hairs, and is in line with the notion that the simultaneous presence of cell wall peroxidases, O_2^- and H_2O_2 is involved in growth arrest and root hair formation through the production of $^{\cdot}OH$ radicals. In willow seedlings, the presence of $^{\cdot}OH$ quenchers did not alter the effect of exogenous H_2O_2 as it would have been expected if its action merely depended on the production of $^{\cdot}OH$ radicals [47]. Hence, given the stimulatory effect of H_2O_2 scavengers and the negative effect of H_2O_2 and peroxidase inhibitors on root hairs formation, it is likely that, in this species, a key role for root peroxidase activity is the maintenance of H_2O_2 within a low concentration range for their appropriate development.

The action of H_2O_2 may involve different mechanisms, including calcium and/or calmodulin signaling, the activation of G-proteins, the activation of phospholipids signaling, and mitogen-activated-protein kinases (MAPKs) cascades [16, 54, 55]. As for ˙OH quenchers, neither the presence of the calcium scavenger EGTA nor of excess Ca^{2+} in the growth medium reverted the inhibitory effect of H_2O_2 on root hair growth [47]. However a partial recovery was observed when H_2O_2 was supplied together with the MAPKK inhibitor PD 98059 (Figure 6d), suggesting that its action would involve, at least in part, MAPK cascades.

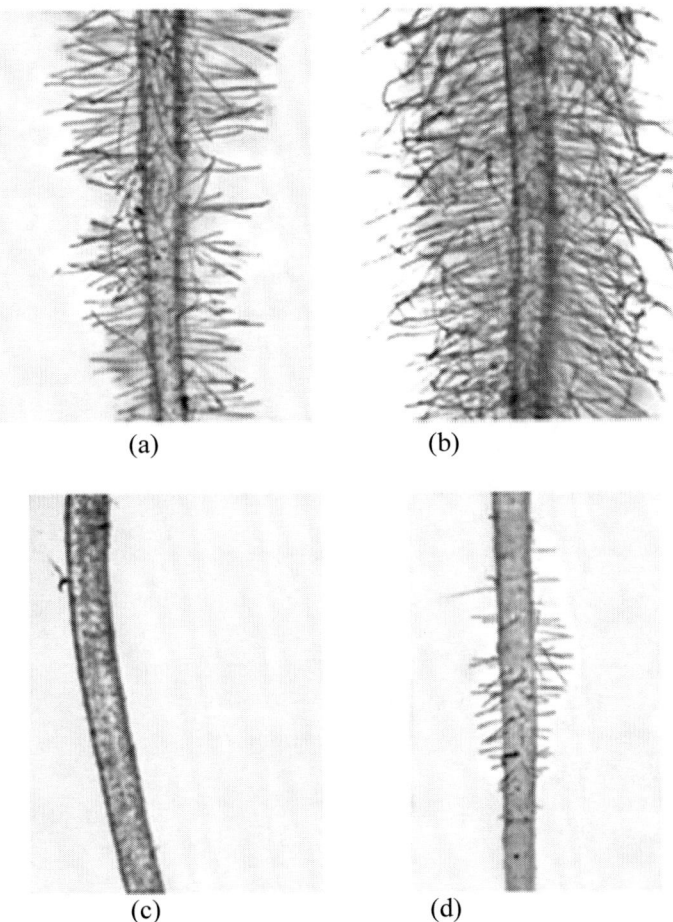

(a)   (b)

(c)   (d)

Figure 6. Roots of 4 d-old salix seedlings grown in agar-agar alone(a), or after 72 h exposure to 55 EU mL^{-1}Catalase (b), 1 mM H_2O_2 (c), or 1 mM H_2O_2 + 15 µM PD 98059 (d). Prior to imaging, roots were stained with 0.05% methylene blue to facilitate the visualization of root hairs. Treatments were applied at 24 h from sowing. EU= enzymatic units.

Conclusion

Because of their atypical characteristics, willows seeds are highly sensitive to photooxidation when stored in the presence of light and oxygen, at ambient temperature.

Under these conditions, the production of FR is enhanced, causing deleterious effects on endomembrane systems and the integrity of key macromolecules. While the increment in FRs becomes one of the major causes of the rapid loss of normal germination, the maintenance of an adequate generation of ROS in specific regions of the root is required for the normal development of this organ in germinating seeds. This apparent paradox is in agreement with the hypothesis about the existence of an "oxidative window" for germination [56]: This window defines a critical level of FRs not to overcome, which would otherwise affect normal germination and eventually prevent it, and a ROS threshold level below which root growth (and/or normal development) cannot occur. Within this oxidative window, specific ROS may play different roles, including defense against pathogens, modification of cell wall properties and cell signaling.

References

[1]   Pammenter, M. W. & Berjak, P. (1999). *Seed Sci. Res.*, *9*, 13-37.
[2]   Roberts, E. H. (1973). *Seed Sci. Technol.*, *1*, 499- 514.
[3]   Ellis, R. H., Hong, T. D. & Roberts, E. H. (1990). *Seed Sci. Technol.*, *18*,131-137.
[4]   Hong, T. D., Linington, S. & Ellis, R. H. (1996). *Seed storage behaviour: a compendium* (electronic version). International Plant Genetic Resource Institute, Rome.
[5]   King, M. W. & Roberts, E. H. (1979). *The storage of recalcitrant seeds.* Rome: International Board for Plant Genetic Resources.
[6]   Pence, V. C. (1995). in: Y. P. S. Bajaj (Ed.), Springer-Verlag, Berlin.
[7]   Maroder, H. L., Prego, I. A., Facciuto, G. R. & Maldonado, S. B. (2000). *Ann. Bot.*, *86*, 1017-1021.
[8]   Maroder, H., Prego, I. & Maldonado, S. (2003). *Trees*, *17*, 193-199.
[9]   Asada, K. (1994). In: C. H. Foyer & P. M. Mullineaux (Ed.), CRC Press, London: 77-104.
[10]   Dalton, D. A. (1995). In: S. Ahmad (Ed.). Chapman & Hall, New York. 298–355.
[11]   Foyer, C.H. (1996). *Biochem. Soc. Trans.*, *24*, 427-433.
[12]   Hideg, E., Barta, C., Kalai, T., Vass, M., Hideg, K. & Asada, K. (2002). *Plant Cell Physiol.*, *43*, 1154–1164.
[13]   Chen, S. G., Ying, C. Y., Qiang, S., Zhou, F. Y. & Dai, X. B. (2010). *Biochim. Biophys. Acta*, *1797*, 391-405.
[14]   Asada, K. (1999). *Annu. Rev. Plant Physiol. Plant Mol. Biol.*, *50*, 601-639.
[15]   Asada, K. (2006). *Plant Physiol.*, *141*, 391-396.
[16]   Mittler, R., Vanderauwera, S., Gollery, M. & Van Breusegem, F. (2004). *Trends Plant Sci.*, *9*, 490-498.
[17]   Apel, K. & Hirt, H. (2004). *Annu. Rev. Plant Biol.*, *55*, 373-399.
[18]   Priestley, D. A. (1986). *Seed ageing. Implications for Seed Storage and Persistence in the Soil.* Cornell University Press, Ithaca.
[19]   Wilson, D. O. & McDonald, M. B. (1986). *Seed Sci. Technol.*, *14*, 269-300.
[20]   McDonald, M. B. (1999). *Seed Sci. Technol.*, *27*, 177-237.
[21]   Narayana Murthy, U. M., Kumar, P. P. & Sun, W. Q. (2003). *J. Exp. Bot.*, *54*, 1057-1067.

[22] Golbeck, J. H. (1992). *Annu. Rev. Plant Physiol. Plant Mol. Biol.*, *43*, 293-324.

[23] Chitnis, P. R. (1996). *Plant Physiol.*, *111*, 661-669.

[24] Roqueiro, G., Maldonado, S., Ríos, M. C. & Maroder, H. (2012). *J. Exp. Bot.* Available: 10.1093/jxb/ers030.

[25] Chen, S., Yin, C., Strasser, Govindjee, R. J., Yang, C. & Qiang, S. (2012). *Plant Physiol. Biochem.*, *52*, 38-51.

[26] Benson, E. E. (1990). *Free radicals damage in stored plant germoplasm.* Rome: International Board of Plant Genetic Resources.

[27] Ellis, R. H. & Roberts, E. H. (1981). *Seed Sci. Technol.*, *9*, 373-409.

[28] Sivritepe, H. O. & Dourado, A. M. (1995). *Ann. Bot.*, *75*, 165-171.

[29] Sun, W. Q. & Leopold, A. C. (1995). *Physiol. Plant.*, *94*, 94-104.

[30] Roqueiro, G., Facorro, G. B., Huarte, M. G., Rubín de Celis, E., García, F., Maldonado, S. & Maroder, H. (2010). *Ann. Bot.*, *105*, 1027-1034.

[31] Boveris, A., Puntarulo, S. A., Roy, A. H. & Sánchez, R. A. (1984). *Plant Physiol.*, *76*, 447-451.

[32] Kranner, I., Roach, T., Beckett, R. P., Whitaker, C. & Minibayeva, F. V. (2010). *J. Plant Physiol.*, *167*, 805-811.

[33] Chen, W., Xing, D., Wang, J. & He, Y. (2002). *Luminescence*, *18*, 19-24.

[34] Wi, S. G., Chung, B. Y., Kim, J. S., Kim, J. H., Baek, M. H., Lee, J. W. & Kim, Y. S. (2007). *Micron*, *38*, 553–564.

[35] Xu, S., Li, J. L., Zhang, X. Q., Wei, H. & Cui, L. J. (2006). *Environ. Exp. Bot.*, *56*, 274-285.

[36] Pólya, L. (1961). *Nature*, *4759*, 159-160.

[37] Liszkay, A., van der Zalm, E. & Schopfer, P. (2004). *Plant Physiol.*, *136*, 3114-3123.

[38] Renew, S., Heyno, E., Schopfer, P. & Liszkay, A. (2005). *Plant J.*, *44*, 342-347.

[39] Dunand, C., Crèvecoeur, M. & Penel, C. (2007). *New Phytol.*, *174*, 332-341.

[40] Schopfer, P., Plachy, C. & Frahry, G. (2001). *Plant Physiol.*, *125*, 1591-1602.

[41] Passardi, F., Penel, C. & Dunand, C. (2004). *Trends Plant Sci.*, *9*, 534-540.

[42] Jones, M. A., Raymond, M. J., Yang, Z. & Smirnoff, N. (2007). *J. Exp. Bot.*, *58*, 1261-1270.

[43] Demidchik, V., Shabala, S. N., Coutts, K. B., Tester, M. A. & Davies, J. M. (2003). *J. Cell Sci.*, *116*, 81-88.

[44] Foreman, J., Demidchik, V., Bothwell, J. H. F., Mylona, P., Miedema, H., Torres, M. A., Linstead, P., Costa, S., Brownlee, C., Jones, J. D. G., Davies, J. M. & Dolan, L. (2003). *Nature*, *422*, 442-446.

[45] Takeda, S., Gapper, C., Kaya, H., Bell, E., Kuchitsu, K. & Dolan, L. (2008). *Science*, *319*, 1241-1244.

[46] Torres, M. A., Onouchi, H., Hamada, S., Machida, C., Hammond-Kossack, K. E. & Jones, J. D. G. (1998). *Plant J.*, *14*, 365-370.

[47] Causin, H. F., Roqueiro, G., Petrillo., E., Láinez, V., Pena, L. B., Marchetti, C. F., Gallego, S. M. & Maldonado, S. I. (2012). *Plant Sci.*, *183*, 197-205.

[48] Fath, A., Bethke, P., Beligni, V. & Jones, R. (2002). *J. Exp. Bot.*, *53*, 1273-1282.

[49] Giannopolitis, C. N. & Ries, S. K. (1977). *Plant Physiol.*, *59*, 309-314.

[50] Raychaudhuri, S. S. & Deng, X. W. (2000). *Bot. Rev.*, *66*, 89-98.

[51] Møller, I. M. & Sweetlove, L. J. (2010). *Trends Plant Sci.*, *784*, 1-5.

[52] Frahry, G. & Schopfer, P. (1998). *Physiol. Plant.*, *103*, 395-404.

[53] Córdoba-Pedregosa, M. C., Córdoba, F., Villalba, J. M. & González-Reyes, J. A. (2003). *Plant Physiol.*, *131*, 697-706.

[54] Pitzchke, A. & Hirt, H. (2006). *Plant Physiol.*, *141*, 351-356.

[55] Mittler, R., Vanderauwera, S., Suzuki, N., Miller, G., Tognetti, V. B., Vandepoele, K., Gollery, M., Shulaev, V. & Van Breusegem, F. (2011). *Trends Plant Sci.*, *16*, 300-309.

[56] Bailly, C., El-Maarouf-Bouteau, H. & Corbineau, F. (2008). *CR Biol.*, *331*, 806-81.

In: From Seed Germination to Young Plants
Editor: Carlos Alberto Busso

ISBN: 978-1-62618-653-8
© 2013 Nova Science Publishers, Inc.

Chapter 5

The Role of Root Border Cells in the Formation of a Root-Microenvironment System in Wheat Seedlings

Anatoly I. Bozhkov[*], *Yuliia A. Kuznetsova,*
Nataliia G. Menzyanova and Marina K. Kovaleva
Institute of Biology, Kharkov Karazin's National University, Kharkov, Ukraine

Abstract

Carbohydrates composed up to 60% of the microenvironment components on 1-day-old wheat seedling roots. The root microenvironment included also proteins of different molecular weights, peptides and amino acids, neutral lipids and phospholipids. Root exudate composition changed when seedling cultivation conditions also changed. Together with macromolecular components, there was a rather heterogeneous population of border cells in the microenvironment of wheat seedling roots. There was protease activity in the microenvironment (i.e., this system is dynamic). In model experiments, it was shown the ability of the root excretion system self-regulation. There was an inverse correlation between the root growth rate and the rate of formation of its microenvironment (i.e., the slower the root growth the more exudates it excreted to form its microenvironment). The quantitative and qualitative composition of the root microenvironment depended on the root growth rate. These relationships were shown on different models of experimental regulation of root growth rate (i.e., adding of different sodium fluoride-NaF-concentrations in the growth medium, natural ageing of seeds, etc.). The role of border cells in the formation of the root microenvironment was studied. The root growth inhibition caused by adding NaF from 1 to 10 mM did not relate to the induction of free radical oxidation processes; content of thiobarbituric acid-active products and carbonylated proteins did not change. Border cell numbers increased two-fold. The border cell number increase in the root environment was not accompanied by an increased content of carbohydrates and proteins in the root exudates. We also studied the biological activity of root exudates from 1-day-old wheat seedlings. These root

---

[*] E-mail address: julyashka81@yandex.ru.

exudates showed antibacterial features against *Staphylococcus aureus* and *Streptococcus pyogens*, displaying antioxidant activities. The possible mechanisms of autoregulation in the "root-microenvironment" system are discussed.

Introduction

Plants might excrete various exudates (e.g., amino acids, sugars, proteins, organic acids and phenol compounds) during their vital activity period [1-7]. The qualitative and quantitative composition of these exudates varies strongly and depends on plant species, growth stage, culture medium and growth conditions [8-10]. On the *Triticum aestivum* hydroponic culture model, the exúdate composition changed from the 1st to the 3rd day of growth, i.e. this process is highly dynamic [9]. The dynamic nature of the microenvironment formation let us suppose that the functional root-microenvironment system is formed on the initial stages of root growth. The exudates of the microenvironment might provide the signaling between plants, bacterium, fungi [11-17]. This is exhibited in the stimulation of reproduction on one bacterium species and the inhibition on another [11, 13]; on the inhibitory effects on one plant species and the stimulatory on another ones, or in formation of phytocenosis [18]. The exudates can protect the plant from different toxic compounds particularly heavy metal salts [19-23].

Although allelopathy was described by Molisch as long ago as 1913, it has not been studied enough [24]. At the same time, knowledge that contribute to determine the quantitative and qualitative root microenvironment composition is of interest in two aspects. Firstly, this knowledge is necessary for contributing to the understanding of the (i) mechanisms that regulate of plant growth and development, (ii) control of the formation of phytocenosis, and (iii) biological information on the plant-fungi and plant-bacteria inter-actions. Secondly, the various known root technologies should focus producing biologically active compounds, and use them in pharmacy, cosmetology and medicine [25]. The (i) processes of microenvironment formation, (ii) rate of exudate composition changes, (iii) reasons of these changes and (iv) regularity of formation of root-microenvironment systems are still poorly understood. In connection with this, we studied the formation of the "root-microenvironment" system on the early stages of seedling growth in hydroponic culture in the present work.

We hypothesize the root microenvironment forms immediately after germination, and that a dynamic change settles subsequently in that system.

Materials and Methods

Plant Material

Experiments were conducted with seedlings of winter wheat (*Triticum aestivum* L. cv. Donetskaya-48). Seeds harvested in 2003 and 2004 were washed in water for 2 h, incubated in 0.005% $KMnO_4$ for 10 min, and soaked in distilled water in a flat glass container at 25°C for 21–22 h. The germinating seeds were transferred to petri dishes (50 seeds per dish), and 4 ml of distilled water were added to each dish. Twenty four hours later, we measured root length, collected the root exudates, determined their composition, and counted borders cells (BC). The pea seedlings were cultivated in the same way as wheat seedlings.

The culture medium containing the root exudates was transferred from the petri dishes to test tubes, and the seedlings in the dishes were additionally washed with sterile distilled water, and combined with the culture medium. The mixture thus obtained was used for determining the composition of root exudates.

Modes of Root Exudate Collection

There were two modes of root exudate collection in the experiments:

Running mode: root exudates were removed from the petri dishes every 24 h, after one, two and three days of cultivation. Fresh water was added to the washed seedlings which were cultivated during 24 h periods.

Cumulative mode: root exudates were accumulated in dishes during one, two or three days of growth. Thus, at this mode of collection there were three experimental groups: 1st group – after one day of cultivation; 2nd group – after two days of cultivation; 3rd group – after three days of cultivation.

Growth of Seedlings at Different Stocking Densities

To estimate the influence of stocking density on root exudate composition, germinating seeds were transferred to petri dishes at 25, 50, 100, 150, 200, 250, 300 and 350 seeds per dish. Exudates were collected on the 1st, 2nd, and 3rd day of growth by the running mode.

Adding of Sodium Fluoride to the Seedling Cultivation Medium

To estimate the influence of sodium fluoride on seedling growth rate, root exudates and root border cell number in 5 ml of 1, 5, 10, or 20 mM solution of chemically pure NaF were added to the dishes containing 50 seeds. After one day, root exudates were collected; and root length, exudate composition and border cells number were determined.

Determinations

Total protein content: The content of total protein in root exudates was assayed using the method of Lowry et al. [26].

High-molecular proteins: High molecular weight (HMW) proteins were precipitated with 76% ethanol for 12 h at 4°C. The precipitate was dried and dissolved in 1 N NaOH, and proteins were assayed in the root exudate aliquots following Lowry et al. [26].

Carbohydrates: Carbohydrate content was determined according to Molisch [27]. With this purpose, 0.5 ml of phenolic reagent was added to the aliquots of root exudates. Samples were actively agitated, and concentrated H_2SO_4 (2.5 ml) was added. In 20 min, extinction was determined at 488 nm. The level of total carbohydrates was calculated using a standard curve constructed for glucose, and it was stated in mg of glucose per 100 seedlings.

Total lipids: To separate the lipid aliquots from the root exudates, they were sequentially extracted by mixtures of organic spirits: chloroform - methanol (1:2, v/v) and chloroform – methanol – water (1:2:0.8, v/v) [28]. Aliquots of lipid extracts were stripped to dryness. Lipid extracts were soluted in 0.1 ml chloroform and were separated by thin-layer chromatography on plates Silufol (Czechia) – the neutral lipids in a mixture hexane:ether (4:1), and phospholipids in mixture chloroform – methanol – water (65:25:4). The fractions were eluted by a mixture chloroform:methanol (1:1). After that, the samples were stripped to dryness and then mineralized in H_2SO_4 at 120°C for 20 minutes [29]. The extinction was measured at 400 nm. The content of lipids was determined from a standard curve, and calculated as mg per 100 seedlings.

Amino acids content: The amino acid content in root exudates was determined by the ninhydrin method [30]. To the root exudate aliquots, the same volume of 10% TCA was added. The sample was thereafter incubated for 12 h at +4°C for protein precipitation. After that, the samples were centrifuged at 3000 g for 20 min. The amino acid content was determined in the supernatant. Water was added to 0.6 ml aliquots of supernatant; besides 1 ml of 1% ninhydrin and 1 ml of 10% pyridine were added, and the samples incubated in a boiling water bath for 30-40 min. Then the samples were cooled, and the extinction was measured at 570 nm. The amino acid content was determined from a standard curve and expressed in mcg of glycine per 100 seedlings.

F in roots: The content of F was determined in the roots of 1-day-old seedlings developing in the presence of NaF [31]. With this purpose, the seedlings were first repeatedly washed off from the culture medium with distilled water. Roots were detached from seedlings and dried at 25°C to constant weight. Half milliliter of strong nitric acid was added to the sample of dried roots (50 mg), and the sample was incubated at 4°C for 12 h. Samples were thereafter digested in strong nitric acid at 120°C for 20 min. After digestion, acidity of the samples was adjusted to pH 3.0 with 1.5 M sodium citrate. The content of F in the samples was determined using a fluoride-selective electrode, and expressed in moles of F – per mg dry weight of roots.

BC in situ: We determined the number of BC surrounding the root in situ. To this end, 20 roots of 1-day-old seedlings taken from each treatment were fixed in 2% glutaraldehyde for 1 h, and stained with 0.6% Trypan Blue for 15 min. Root apices were placed into a drop of water on a slide, covered with a cover glass, and the number of BC in the apex zone was counted under a light microscope (LOMO, Russia). The results were expressed as the BC number per root.

Isolation of border cells: In the case of BC isolation, 0.5 ml distilled water was added to a compartment of a plastic culture plate fixed on a magnetic stirrer. The apical root zone 1.5–2.0 cm in length (without detaching the root from the caryopsis) was immersed into actively stirred water for 1-2 min remove the apex gel sheath with BCs. In one compartment, from 20 to 50 roots were treated. The washing medium was transferred from the compartment to the test tubes and centrifuged at 3000 g for 15 min sediment BCs. The pellet containing BCs was fixed with 2% glutaraldehyde and stained with 0.06% Trypan Blue. The number of BCs removed was determined in a Goryaev hemocytometer and expressed in cells per root.

Microviscosity on the root gel sheath: Viscosity of the solution of gel sheath components (supernatant after BC sedimentation by centrifugation) was determined by a capillary method.

To this end, 0.5 ml of solution containing the components of the gel sheath was taken in a capillary, and the time for its flowing out from the capillary was determined. The viscosity of the solution was calculated according to the formula:

$$\eta = \frac{(t - t_0)}{t_0} \eta_0 + \eta_0 ,$$

where η_o is a viscosity of distilled water, t_o is time taken by distilled water to travel along the capillary; t is time taken by the solution containing the components of the gel sheath to travel along the capillary. The microviscosity was estimated per 50 roots.

Influence of root exudates on unicellular microalgae: The influence of wheat seedling root exudates on growth rate and biomass accumulation of the microalga Dunaliella viridis was evaluated. The algologically pure periodic culture of D. viridis was cultivated on Artari medium in glass cone flasks, reinoculating every three weaks. The culture volume was 20 ml. The Artari medium was prepared using 1%, 3%, 5%, 10% and 15% solutions of root exudates. The initial cell concentration was 1.3 mln of cells per ml. Cells were cultivated for 15 days, and then they were counted under a light microscope (LOMO, Russia) every two days in a Goryaev hemocytometer. Moreover, at counting in the Goryaev hemocytometer, the number of motionless cells per 100 cells in the field of vision was determined.

Antioxidant activity of root exudates: The antioxidant activity of root exudates was estimated by the (i) total antioxidative and (ii) antiradical activities.

The antioxidative activity was estimated by the ability of root exudates to inhibit the build-up of TBA-active products of lipid peroxidation (LP) in a system of vitelline lipoproteins [32]. With this purpose, 200 mcl of vitellus homogenized in a phosphate mixture 1:1; 300 mcl of root exudates (water to the control assay); 700 mcl of phosphate mixture (40 mM KH_2PO_4, 105 mM KCl, pH=7.45), and 100 mcl of ferric sulphate (II) were added. The samples were incubated for 15 min at 37°C. Subsequently, 0.5 ml of 20% TCA was added, and samples were centrifuged to 3000g for 10 min. The supernatant was added to 1 ml of 0.5% TBA-0.3% DDS (sodium dodecylsulfate), and it was incubated in a boiled water bath for 15 min. The extinction was measured at 570 nm. The toThe effectiveness of scavenging OH· radicals (antiradical activity) by root exudates was determined by their ability to inhibit the deoxyribose degradation by OH-radicals, generated in a reaction mixture: $Fe2+$, EDTA, H_2O_2 and ascorbic acid [33]. Twenty mcl of $FeCl_3$ (8.1 mg in 10 ml of water) were added to a 500 mcl reaction mixture [24 мM KH_2PO_4, pH=7.4; 20 mcl 1.2% deoxyribose; 20 mcl EDTA (1.61 mg in 10 ml of water); 20 mcl ascorbic acid (5.28 mg in 10 ml, pH=7.4), 12 mcl 3% H_2O_2, and 200 mcl of root exudates]. Incubation of this mixture was at 37°C for 1 hour. The reaction was stopped adding 0.5 ml 10% TCA, and the assay was centrifuged to 3000g during 10 minutes. The supernatant was added to 0.5 ml 0.5% TBA, and it was incubated 15 minutes at 100°C. The extinction was measured at 530 nm. The antiradical activity was expressed as percent to the control.

Antibacterial activity of root exudates: The antibacterial activity of root exudates for 1-2- and three-day-old seedlings collected differently modes was estimated. The method of paper disks was used for this estimation [34]. The Staphylococcus aureus and Streptococcus pyogenes were used as test-organisms. The test-organisms were plated and dishes were parched at 37°C for 15-20 min. The disks of bibulous paper were autoclaved for 30 min at 1.5

atm. The sterile disks were moistened in root exudates and were laid on the surface of nutrient agar having the test-organisms. Sterile disks moistened by sterile water were used as control. The dishes with the test-organisms were incubated at 37°C for 24 h. After 24 h, the diameter of bacteriostasis zones around disks was determined.

Free proline: The free proline was determined by method of Bates et al. [35]. Ten seedling roots were cut and homogenized in 10 ml tris-HCl buffer (0.05 M, pH=7.4) at 4°C. The homogenate was centrifuged at 10 000 g during 20 min at 4°C. One ml of supernatant reacted with 1ml acid-ninhydrin reactive (1.25 g of ninhydrin in 30 ml of glacial acetic acid and 20 ml of 6 M phosphoric acid) and 1 ml glacial acetic acid, and it was incubated for 1 hour at 100°C. The reaction was terminated in an ice bath (4°C). The reaction mixture was extracted with 2 ml of toluene, mixed vigorously and centrifuged during 10 min at 3000 g. The upper phase was collected and its absorbance read at 520 nm using toluene as a blank. The proline concentration was determined from a standard curve, and calculated on a fresh weight.

TBA-active products: One ml of root homogenate was added to 1 ml of 10% trichloroacetic acid (TCA) and centrifuged at 3000 g for 10 min [36]. One ml of 0.8 % TBA was added to 1 ml of supernatant and the reaction mixture was incubated at 100°C for 10 min. Two ml of butanol were added and the mixture was vigorously mixed and then centrifuged to 3000 g for 10 min. The upper mixture phase was collected and its absorbance read at 530 nm. The blank sample was 2 ml of water + 1 ml of 0.8 % TBA.

Carbonylated proteins: The method of estimation of the oxidative modification of proteins is based on the interactive reaction between the oxygenized amino acids of proteins and the 2,4-dinitrophenylhydrazine (2,4-DNPH), with formation of 2,4-dinitro-phenylhydrazones [37]. The content of carbonylated proteins was determined in pellets after homogenization of roots for proline determination. Half a ml 10 mM 2,4-DNPH solution in 2 N HCl (test sample), (in control sample 0.5 ml of 2N HCl) was added to the pellets. Samples were incubated at room temperature for 1 hour, mixing every 10-15 min. After that in control and test samples, 0.5 ml of 20% TCA was added, and protein was precipitated by centrifugation at 3000 g for 15 min. Pellets were washed three times using 1 ml of a mixture ethanol:ethyl-acetate (1:1, v/v) to remove unbound 2,4-DNPH. After washing, pellets were dissolved in 2 ml of 8 M carbonyl diamide (pH 2.3 buffer). The absorption spectrum of solutions was analyzed using a double-beam spectrophotometer "Specord UV VIS" (Germany) with interval 363–373 nm. The quantity of carbonylated proteins was expressed in nM per mg protein using the molar extinction coefficient $22 \cdot 10^3$ M^{-1} cm^{-1}.

Copper in water: Copper content in water from different sources used for wheat seedling cultivation was determined by atomic-absorption spectroscopy.

Statistical Analysis

All experiments were conducted at least 3 times, and every experimental series included at least 3 replications for each study variable. Means and standard errors were determined. Differences between sampling means were tested using the non parametric Mann-Whitney test. The coefficient of rank correlation was determined by Spirman.

Results and Discussion

Macromolecular Composition of the Microenvironment of 1-day-old Seedlings

Twenty hours after germination, root length of wheat "Doneckaya-48" was from 0.7 to 1 cm and formed a rhizosphere (Figure 1). At this time, 60-70% of all compounds in the rhizosphere were carbohydrates (Table 1). Carbohydrates dyed well by trypane blue giving evidence of polysaccharides. The polysaccharides possible formed the specific structure of the microenvironment, and provided the functional activity that border cells lost the mechanical contact with the root apex.

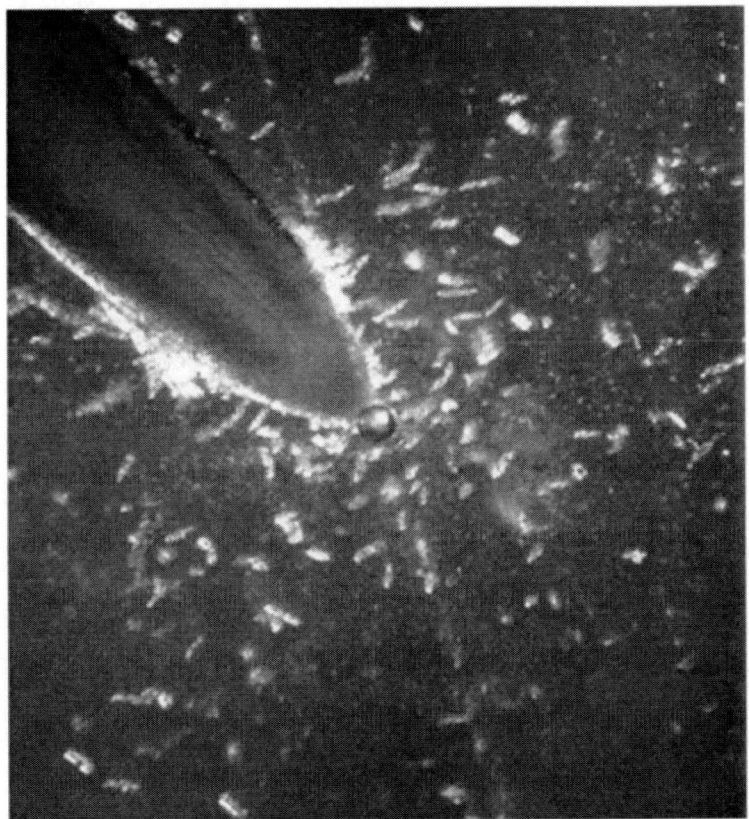

Figure 1. The rizhosphere of 1-day-old wheat seedlings.

Table 1. Macromolecular compounds in root exudates of 1-day-old wheat seedlings (HMP – high-molecular proteins, AA – amino acids)

	Total proteins	Carbohydrates	Lipids	HMP	AA
Mcg per root	1.57 ± 0.38	10.63 ± 0.68	0.50 ± 0.07	0.14 ± 0.05	4.82 ± 0.08
% of exudates	8.9	60.2	2.8	0.8	27.3

Table 2. The composition of root exudates of 1- to 3-day-old wheat seedlings

Day of cultivation	Proteins		Lipids		Carbohydrates	
	Mg per 100 seedlings	%	Mg per 100 seedlings	%	Mg per 100 seedlings	%
	2.43±0.23	31.7	0.50±0.05	6.6	4.73±0.09	61.7
2	4.77±0.14	42.0	0.41±0.05	3.6	6.17±1.40	54.4
3	9.99±0.14	61.0	0.41±0.05	2.5	5.94±0.50	36.5

The great number of amino acids was found in root exudates – to 20-25% of all the compounds. The proteins accounted for 8-10% and lipids – for 2-5% (Table 1).

One-day-old seedling exudates had from low- (12 kDa) to high- (12 to 45 kDa) molecular weight proteins. Proteins from 12 to 45 kDa molecular weight accounted for no more than 9% of total proteins.

The protein variability in the microenvironment can be explained by a number of reasons. Root cells excreted a specific group of low-molecular proteins, executing a regulatory role. Such high variability of proteins resulted from proteolysis in the rhizosphere. The protease activity was rather high in the root exudates of 1-day-old seedlings and it accounted for the 400 mcg of glycine per 100 seedlings per hour. These results suggest that there are highly fluid processes of change in the root rhizosphere composition of 1-day-old seedlings.

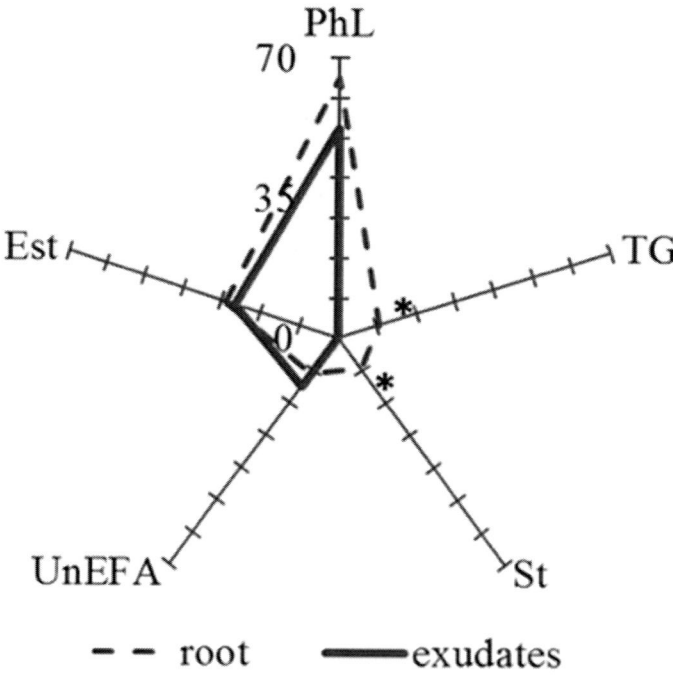

Figure 2. The lipid composition of roots and exudates of 3-day-old wheat seedlings. Est-steroids ethers; PhL- phospholipids; TG- triacylglycerides; St- steroids; UnEFA- unetherified fat acids. *Contents of St and TG significantly differed between exudates and roots.

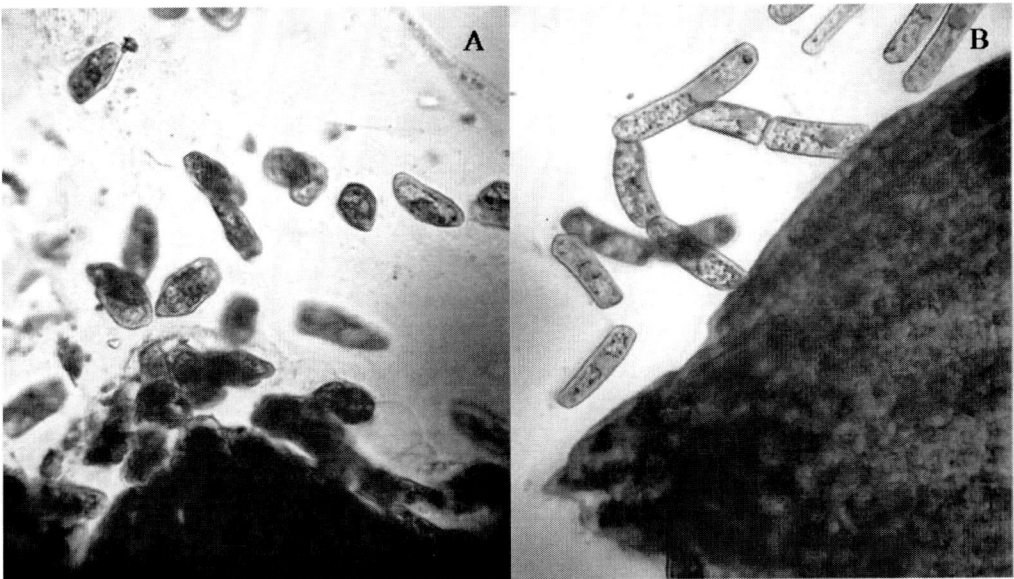

Figure 3. The morphology of root border cells in wheat: A- spherical near the crown of the root apex; B- elongated, rod-shaped near the lateral surfaces of the root apex.

The lipid components in the rhizosphere of 1-day-old seedlings can be attributed to root cell degradation. In this regard, we studied the lipid composition of root exudates in 1-day-old wheat seedlings (Figure 2). With this purpose, we collected the root exudates of 1-day-old seedlings and determined its lipid composition.

The rhizosphere lipid composition differed from that in the roots. While there were no triglycerides and plant steroids in the exudates, they accounted for 8.7 and 5.9 % of all lipids, respectively, in the roots. At the same time, contents of phospholipids, unetherified fat acids and steroid ethers were similar in both roots and exudates (Figure 2). This suggests that lipids in the rhizosphere are not the product of root cells degradation.

Dynamics of the Microenvironment Composition from the 1st to the 3rd Day of Root Growth

The next series of experiments was designed to investigate the content of carbohydrates, proteins and lipids in seedling root exudates from the 1st to the 3rd day of root growth. From the 1st to the 3rd study day, the root exudate composition differed significantly (Table 2).

The carbohydrate content of root increased more than 1.2 fold from the 1st to the 3rd day of seedling growth (Table 2). During this period, lipids decreased by 18% in the root exudates. Protein in the root exudates increased by more than 3-fold between the 1st and the 3rd day of growth (Table 2). The root microenvironment included not only the various low- and high-molecular weight compounds but also the specialized border cells.

Morphological Characteristics of Root Border Cells

There were from 30 to 40 cells per root on gel sheaths of roots of 1-day-old wheat seedlings in situ. Cells were spherical or elongated, rod-shaped. A few spherical cells were near the crown of the root apex, and elongated, rod-shaped cells were near the lateral surfaces of the root apex (Figure 3). Elongated, rod-shaped cells were often aggregated (Figure 4).

Border cells on gel sheaths of roots were washed away by top water. While border cells maintained alive, aggregations kept as such. In this case, there were near 80-90 cells per root (i.e., twice more than in situ). These results indicate that bounds were stronger between border cells than between border cells and roots.

Roots begin to produce border cells after their mechanical removal and after an hour they appeared again in gel sheaths.

It was reported that border cells number is greater in the rhizosphere of pea than in that of wheat seedlings [38]. Thereafter, the new experimental approach was designed to estimate border cell number on pea seedling roots. Indeed, in the rhizosphere of 1-day-old pea seedlings there were at least ten times more border cells than in that of wheat. The shape of the pea border cells differed from those in wheat (Figure 5). Therefore, there are species particularities in quantitative and qualitative border cells characteristics.

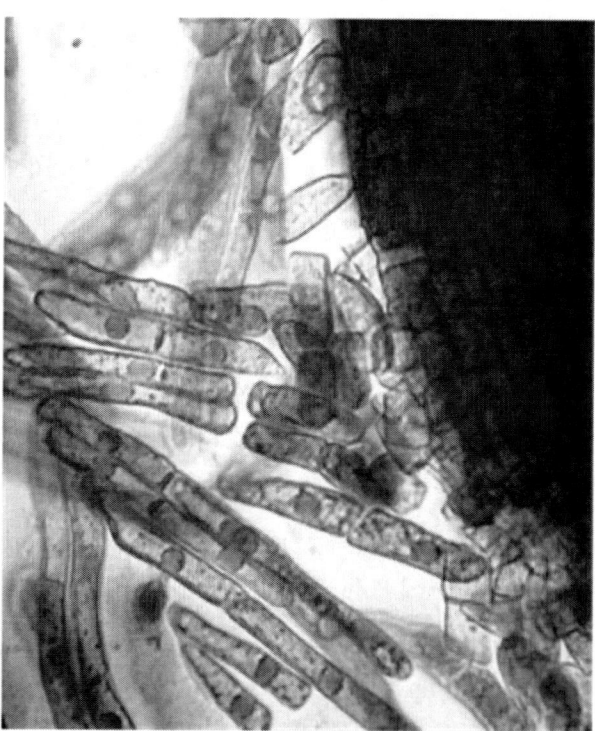

Figure 4. Root border cells were often aggregated in wheat.

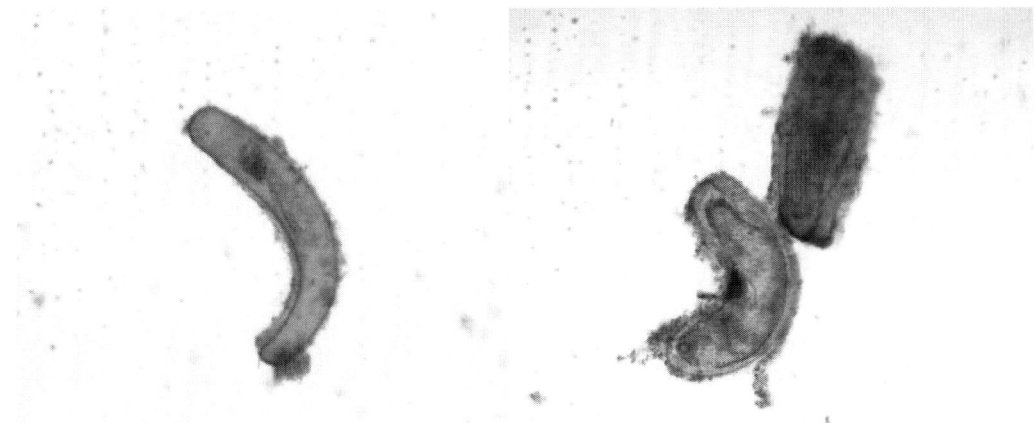

Figure 5. The root border cells on pea.

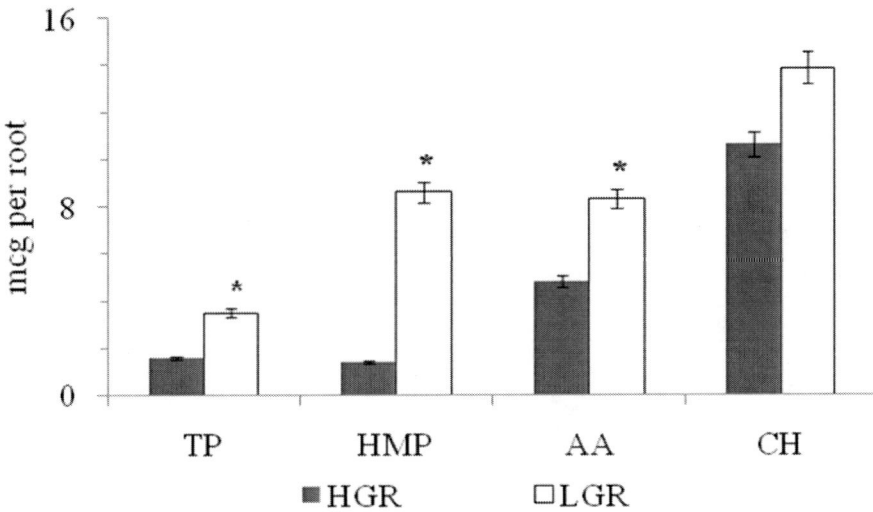

Figure 6. Total proteins (TP), high-molecular weight proteins (HMP), amino acids (AA) and carbohydrates (CH) excreted by HGR and LGR roots of 1-day-old wheat seedlings (values are per 100 seedlings in the case of HMP). * indicates significant differences in the study variables between LGR and HGR roots. Vertical bars on histograms represent ± 1 SE of the means.

One might assume that the microenvironment is formed from the beginning of root growth, and is possible that its macromolecular and cell composition might be completed by the 3rd day of growth germination. Formation of the microenvironment can be divided on three stages: (i) initiation, (ii) a steady-state level (a constant qualitative and quantitative composition of macromolecules and border cells), and (iii) maintenance of a qualitative and quantitative composition of the microenvironment during its vital activity.

It is clear that the microenvironment formation is a function of the growing root. The formed microenvironment can influence the excretory system and possibly the metabolic processes of the root.

Interrelationship between the Root Growth Rate and the Microenvironment Formation

One could assume that there might be a direct relationship between root growth rate and the rate of excretion of microenvironment components. Studies on this aspect are important to understand the physiology of a growing root, and to get root exudates.

Seeds are known to lose their germinating capacity after long-term storage and to grow slowly. Therefore, further experimental series were conducted with seeds having different growth rates caused by different storage histories.

Roots with a length of 0.9±0.1 cm were related to roots with low growth rates (LGR). On the other hand, roots with a length of 1.8±0.2 cm were related to roots having high growth rates (HGR). As a result, LGR roots excreted twice more protein and showed a six-fold higher molecular weight protein content than roots with high growth rates (Figure 6). Exudates of LGR roots contained 1.7 times more amino acids than HGR roots. Also, LGR roots excreted 24% more carbohydrates HGR roots.

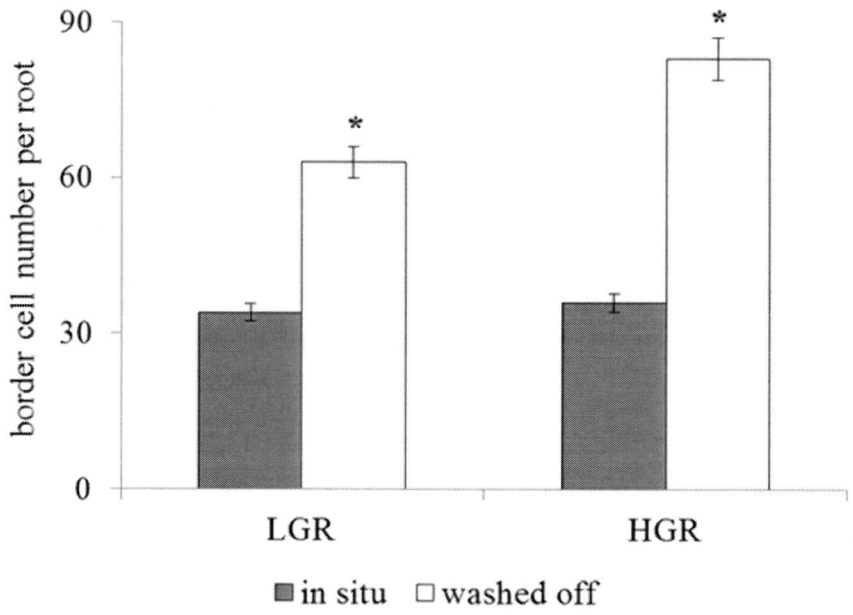

Figure 7. Root border cell number (per 1 root) counted *in situ* and washed off the roots in 1-day-old wheat seedlings with LGR and HGR. *the border cell number counted *in situ* was significantly lower than that on washed off roots. Vertical bars on histograms represent ± 1 SE of the means.

Therefore, there was an inverse relationship between root growth rate and its excretory activity. The slower the root growth the greater exudate excretion. In particular, when root growth rate decreased twice, there was a two-fold increase in protein excretion. There were no differences in carbohydrate excretion between LGR and HGR roots.

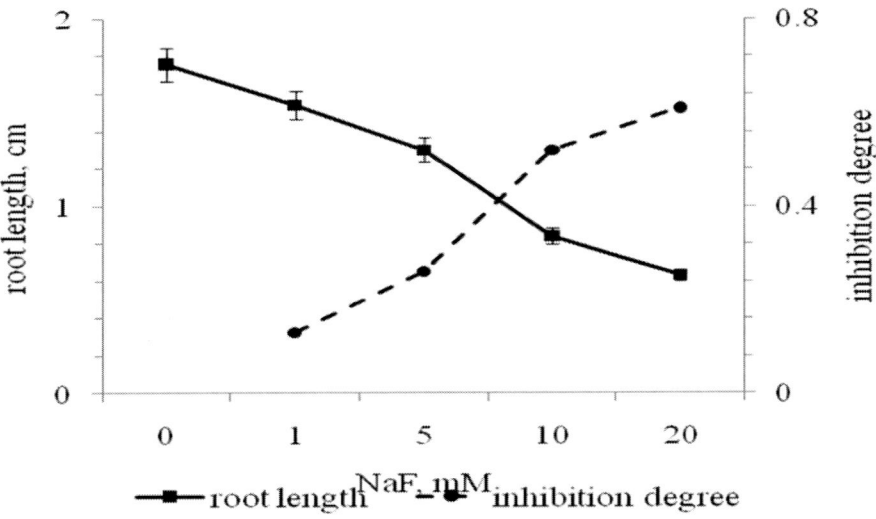

Figure 8. Root growth inhibition on 1-day-old wheat seedlings under increasing NaF concentrations.

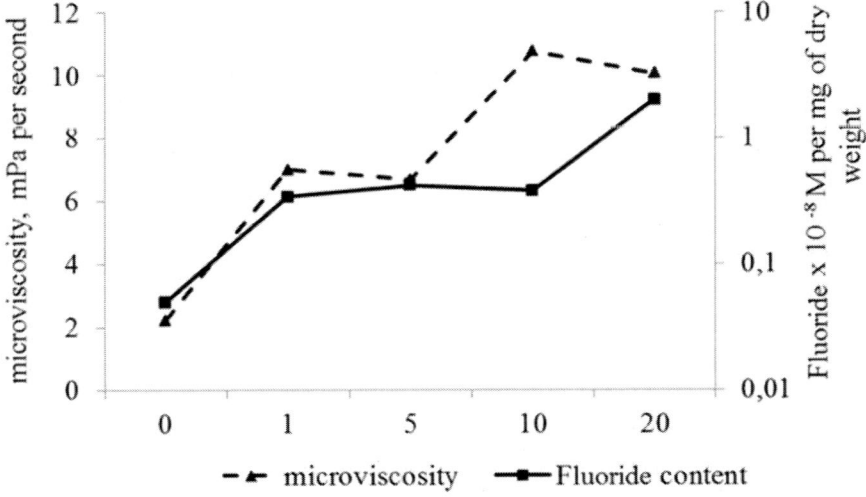

Figure 9. Fluoride ion content on 1-day-old roots of wheat seedlings, and viscosity of gel sheaths of 1-day-old wheat seedlings after 1-day incubation in the medium containing sodium fluoride.

The border cell number was found similar in the rhizosphere of both LGR and HGR roots (Figure 7). However, the border cells number of washed off LGR and HGR roots was significantly greater than that on roots measured in situ (Figure 7).

Therefore, there was no correlation between exudate quantity and border cell number in the root microenvironment of LGR and HGR 1-day-old seedlings. At the stage of microenvironment formation, the excretory activity was determined by that of the root cells rather than that of the border cells.

The root growth rate is regulated by different factors, and in particular by phytotoxicants. The inorganic fluorine bounds Ca^{2+} in soil, preventing the entry of Ca in roots; calcium deficit inhibits root growth and plant development [39, 40]. At the same time, fluorine may inhibit the membrane ATP-ase activity – the respiratory enzyme [41]. We found an almost linear root growth inhibition with increasing sodium fluorine concentrations in the medium (Figure 8).

One might assume that the wheat root growth inhibition in water culture could be caused either by the fluorine penetration in the root cells or by additional inhibition because of regulatory influences on the microenvironment components. To answer this question, the content of fluorine was determined in roots of one-day-old wheat seedlings growing in a NaF medium.

The fluorine content was 0.05×10^{-8} M.mg^{-1} of dry weight of roots in one-day-old seedlings (Figure 9). Content of fluorine increased 6.8 times on roots of one-day-old seedlings that grew in a medium containing 1mM NaF. When the medium contained 5, 10 or 20 mM NaF, content of fluoride in roots increased 8.4, 8.4 or 40 times compared to the controls (Figure 9).

The viscosity of gel sheaths of one-day-old roots cultivated with NaF at different concentrations increased 3.14 times from the control to 1 mM NaF (Figure 9). It also increased 1.5 times from 1 and 5 mM to 10 mM. The viscosity was the same at 10 than at 20 mM of NaF (Figure 9).

Fluoride content in one-day-old roots increased 6.8 times at 1 mM NaF compared with the control. Increases of NaF concentrations in the medium to 5 or 20 mM were accompanied by an 8.4 or 40-fold increase, respectively, in the fluoride content in roots compared to the control. However, the fluoride content in the roots was the same at 5 than at 10 mM of NaF in medium.

Therefore, fluorine in the root microenvironment induces both root growth inhibition and changes of physico-chemical microenvironment characteristics. It could help to explain the nonlinearity in fluorine accumulation in roots. The microenvironment holds and binds the toxic compounds.

Influence of the microenvironment on the root excretory activity and root growth rate If proceeding from the conception of autoregulatory activity of the root microenvironment, one may assume that its formation process would go through the following stages: 1. The formation of an excretory root activity will be characterized by an exponential increase in the quantity of exudates; 2. Attainment of a stationary phase in the extraction activity; 3. An unbalanced stage characterized by the interaction of a ''root-microenvironment'' system, aimed at the conservation of the stationary level.

The cumulative dry weight of exudates obtained during 1, 2 and 3 days of growth in water culture was determined to test our assumption (Figure 10).

During the 1st day of growth, 100 seedlings excreted 3.2 mg of exudates; root excretion during the 2nd and 3rd day of growth was 11.5 mg and 20.7 mg, respectively (Figure 10A). Viewing these data as specific exudates excretory rates, one may assume that there was an exponential growth from the 1st to the 2nd day (Figure 10B). By the 3rd day, the stationary stage in "root-microenvironment" system was coming in (Figure 10B).

At the stationary level, 100 seedlings excreted 20.7 mg of dry exudates in the microenvironment. Attainment of this stationary level of exudate content in the microenvironment can be explained by autoregulation (i.e., inhibition of excretion after attainment of a certain critical exudate quantity). In this case, the regular removal of exudates from the microenvironment would determine an increased excretory activity of roots.

The next experimental series implied the exudate removal every 24 hours during the 1st, 2nd and 3rd day of cultivation (i.e., a running cultivation was used) (Figure 11).

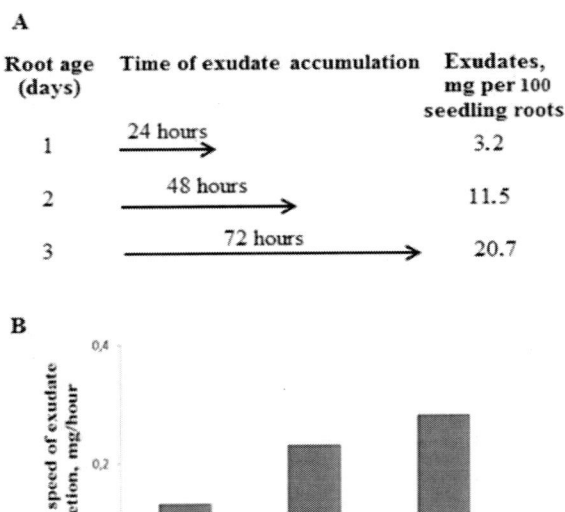

Figure 10. (A) Scheme of cumulative exudate collection from wheat seedlings; (B) specific cumulative excretory activity at the 1st, 2nd and 3rd day of cultivation.

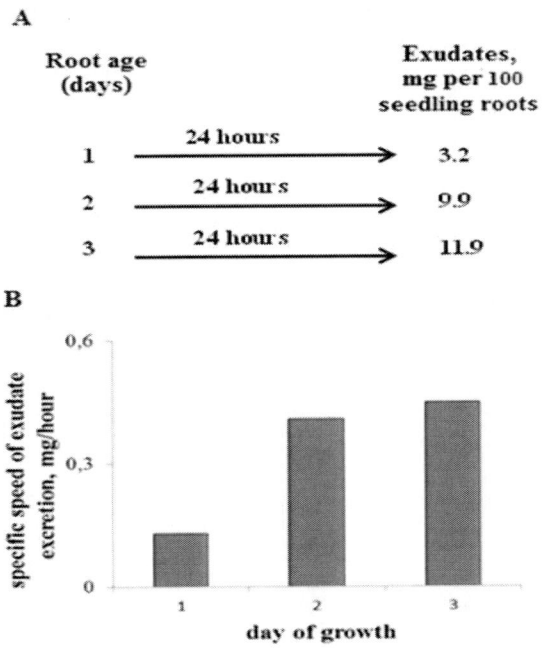

Figure 11. (A) Scheme of exudate collection from cultivation of wheat seedlings at the running mode; (B) specific exudates excretory activity at the 1st, 2nd and 3rd day of running cultivation.

After 3 days of cumulative cultivation, 100 seedlings excreted 20 mg of exudates (Figure 10A). However, 25 mg of exudates excretion (i.e., a 25% increase) were obtained after using running cultivation during a 3-day period (Figure 11A). From the 1st to the 2nd day, the specific exudate excretory activity increased 1.7 times in the case of cumulative cultivation (Figure 10B). During the same time period, however, that variable increased 3 times in the case of running cultivation (Figure 11B). By the 3rd day, however, the specific exudates excretory activity attained the stationary level in the case of both the cumulative and the running cultivations. However, at the stationary level, the specific exudates excretory activity was twice as much in the case of running than in that of cumulative cultivation (Figures 10B, 11B).

Therefore, in the case of running cultivation at a daily removal of exudates the excretory activity increased, but not significantly; by the 3rd day, the stationary stage was coming in. It is evident that in this case the metabolic activity of the root was the limiting step in the "root-microenvironment" system.

It is important to clarify if the system of exudate collection influenced their qualitative composition. Therefore, in the next experimental series the content of total protein and carbohydrates in exudates was determined under different modes of collection.

The protein content of exudates of 2-day-old seedlings increased twofold compared to that content on 1-day-old seedlings, both at the cumulative and running cultivation systems (Figure 12). When comparing the 1st with the 2nd day, the carbohydrate contents increased 30 and 43% at the cumulative and running cultivation systems, respectively. However the exudates collected on the 3rd day of growth, compared to those collected on the 2nd day, differed significantly on protein but not carbohydrate content at the cumulative and running cultivation systems (Figure 12). The protein content in exudates was near 4 mg per 100 seedlings lower at the running than at the cumulative collection mode; in contrast, the content of carbohydrates was 1.5 mg per 100 seedlings more at the running than at the cumulative mode of cultivation (Figure 12).

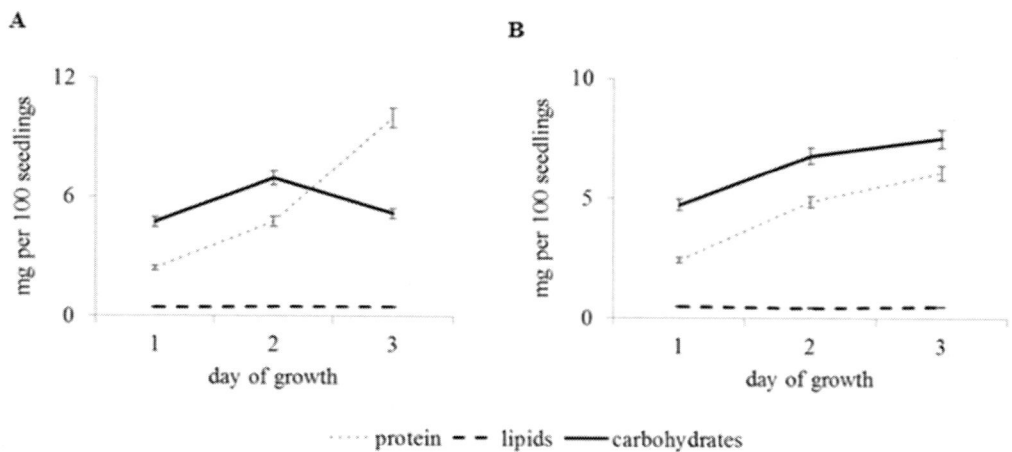

Figure 12. Protein, lipid and carbohydrate contents in root exudates of wheat seedlings at the accumulative (A) or running (B) mode of cultivation on the 1st, 2nd and 3rd day of growth. Note the change of scale in the Y axis.

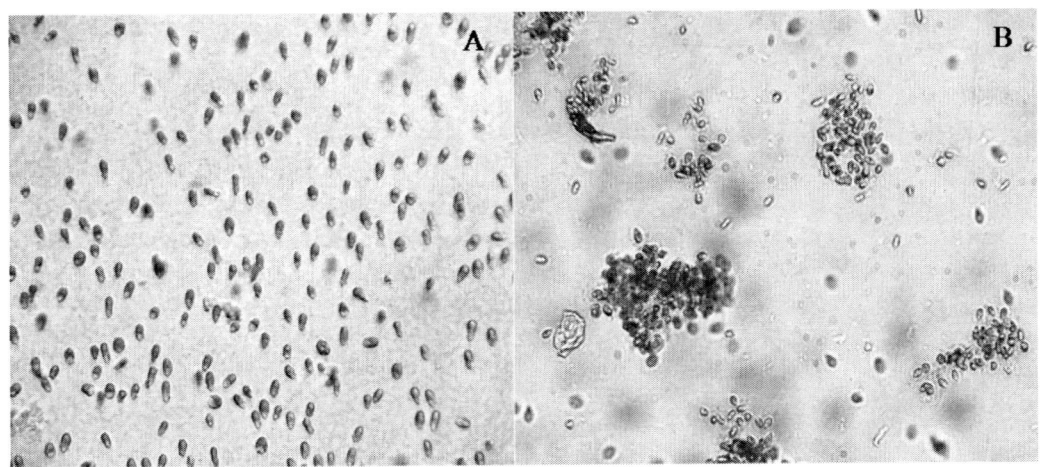

Figure 13. *Dunaliella viridis* culture. (A) control; (B) aggregates after adding root exudates.

Our results suggested that: (i) the removal of root exudates during root growth increases its excretory activity; (ii) the increase of specific exudates excretion rates is exhibited mostly in 2-day-old seedlings, while a stationary level is achieved when seedlings are 3-day-old. However, specific exudate excretion rates at the 3rd day of growth were almost two-fold higher at the cumulative cultivation system; (iii) the mode of exudates collection influenced the exudate composition: daily exudate removal contained more carbohydrates and less protein compared to the cumulative collection mode. Thus, changing the root microenvironment may influence the exudate composition, and possibly the root growth rate. The natural model-experimental change of stocking density – was used to explore that idea.

The protein content in exudates of 1-day-old seedlings was 1.57 mg per root at stocking density 50 seedlings per Petri dish (Table 3a). Increases of stocking density to 100-150 seedlings per dish were accompanied by a twofold increase in protein content. The protein content in exudates of 1-day-old seedlings was the same at 50 than 200-350 seedlings per Petri dish. This might have been the result of an inhibition effect or the degradation of excreted proteins. In root exudates of 2-day-old seedlings, the total protein content increased as the number of seedlings per Petri dish also increased (Table 3b). The greatest protein content in exudates of 3-day-old seedlings was shown at 25 seedlings per Petri dish, and it decreased linearly to 100 seedlings per dish (Table 3c). Total protein content was similar at 50 than at 350 seedlings per Petri dish (Table 3c).

Therefore, two parameters influenced the protein content in root exudates: the time of cultivation, from the 1st to the 3rd day of growth, and the seedling stocking density. These parameters are interrelated and this relation is not linear. The obtained results suggested that there is: (i) a saturation phase of protein content in the root microenvironment, and (ii) an autoregulation in the "root-microenvironment" system.

The influence of stocking density on carbohydrate content in root exudates was determined in the next experimental series. The carbohydrate content did not depend on stocking density in exudates of 1-day-old seedlings (Table 3a). However, the carbohydrate content in exudates of 2- and 3-day-old seedlings significantly decreased as the density increased from 50 to 100-150 seedlings per Petri dish (Table 3b, c). Further increases of stocking density did not influence the carbohydrate content of root exudates. Therefore, the seedlings growing in a shared space are forming a common system of regulation of root exudates excretion.

Table 3. (a) Content of total protein and carbohydrates (mg per seedling x 10-2) in root exudates of 1-day-old wheat seedlings at different stocking densities; (b) Content of total protein and carbohydrates (mg per seedling x 10-2) in root exudates of 2-day-old wheat seedlings at different stocking densities; (c) Content of total protein and carbohydrates (mg per seedling x 10-2) in root exudates of 3-day-old wheat seedlings at different stocking densities

3	25	50	100	150	200	250	300	350
(a) Total protein	1.70	1.57	3.30	3.50	1.32	1.09	1.35	1.54
Carbohydrates	2.79	4.04	2.90	1.87	2.21	1.35	1.19	1.02
(b) Total protein	0,00	2,01	2,24	2,18	3,59	4,39	2,97	4,34
Carbohydrates	16,37	5,91	4,83	3,27	3,24	3,45	2,55	2,59
Seedlings per Petri dish	25	50	100	150	200	250	300	350
(c) Total protein	25.71	14.77	7.77	9.95	8.50	15.85	10.10	13.64
Carbohydrates	12.22	6.27	4.83	3.60	4.31	4.32	5.03	3.41

Estimation of Biological Activity in Wheat Exudates

It has been reported that lipid component of exudates has antibacterial features [9]. We estimated the antibacterial activity of root exudates of 1-, 2- and 3-day-old seedlings. Growth of Staphylococcus aureus and Streptococcus pyogenes was inhibited by the root exudates of wheat seedlings (Table. 4). The antibacterial effect depended on seedling age: it was the greatest in 1-day-old seedlings, and decreased not significantly by the 3rd day.

The antibacterial effects of the root exudates remained after a heat treatment (100°C, 15 minutes). These effects are probably caused by low-molecular coctostabile components: terpenoids, alkaloids, flavonoids, tannins that show antibacterial activity [42-46].

From the great range of components in the root exudates, it may be expected polyfunctionality of biological activity in the root exudates. The culture of the unicellular microalga Dunaliella viridis can be used to screen for biological active compounds. Adding exudates to a 1% concentration in D. viridis culture was accompanied by nonmotility of 5% of the cells. After 10 minutes the number of motionless cells increased to 10 % and after 30 minutes to 25%. This motionless cell percentage maintained after 1 hour, and after 24 hours they became mobile. However, at this time about 5% of the motionless cells died; this was verified after staining using trypan blue.

Increasing the root exudates concentration from 1 to 3.5 or 10% did not modify the alga cell motility. At the same time, addition of root exudates to the test-culture of D. viridis was accompanied by cell aggregation (Figure 13). This aggregation appeared after 10 minutes, remained unmodified during some hours, and disintegrated after 24 hoursin the case of 1, 3 or 5% concentration of root exudates. However, it remained in the case of addition of root exudates to 10 %. Therefore, the addition of root exudates of 1-day-old wheat seedlings in culture medium of D. viridis to a 1% concentration was accompanied by a reversible cell aggregation, and approximately a 5% cell death. These morphological changes were accompanied by the growth inhibition of microalga culture under continuous cultivation.

Table 4. The bacteriostasis zones of the microorganisms (mm) *Staphylococcus aureus* and *Streptococcus pyogenes* as an estimation of the antibacterial activity of root exudates of 1-,2- and 3-day-old wheat seedlings using the method of paper disks

Microorganism	Experimental variants	The bacteriostasis zone (mm) at the root exudates of wheat seedlings		
		1st day of growth	2nd day of growth	3rd day of growth
Staphylococcus aureus	control	confluent growth		
	exp	23±2	20±3	19±2
Streptococcus pyogenes	control	confluent growth		
	exp	35±2	30±4	28±2

Plant compounds may not only have antibacterial and algaecide activity but also show antioxidant activity (AOA) [47-50]. The general AOA of the root exudates water solution (0.6 ml per 1 ml of reaction medium) inhibited the free-radical oxidation of vitelline lipoproteins by 21-24% (Table 5). The antioxidant α-tocopherol in this system inhibited the free-radical oxidation by 49% (i.e., it was two times more effective than the root exudates). Root exudates of 1-, 2 or 3-day-old seedlings have the same general AOA (Table 5).

The ability of root exudates to catch OH-radicals was determined in the model system. It was found that the activity of catching was 41 % for 1-day-old seedlings when 600 mcl of root exudates (2 mcg of dry weight) were added to 1 ml of reaction medium. That activity increased to 54% to the 3rd day of growth (Table 5). It has been reported that ethanol is an active catcher of OH-radicals, and its effectiveness in the model system was 47 %. Therefore, the root exudates of 1- to 3-day-old seedlings showed evident antioxidant properties.

To conclude, it is clear that wheat root exudates exhibit a polyfunctional biological activity. Such wide range of biological activity could be explained because of the content of different compounds in the root exudates. More importantly, the physico-chemical characteristics and the exudate composition can essentially change both during root growth and storage. The wide range of biological activity of exudates shows that they can influence the root microenvironment and finally regulate the physiological processes in the root. Furthermore, root exudates can be a promising source of biologically active substances that could be used in various practices.

Table 5. The total antioxidant activity and effectiveness of scavenging of OH·radicals of root exudate water solutions from 1-, 2- and 3-day-old wheat seedlings

Compounds with antioxidant features		AOA, %	effectiveness of scavenging of OH radicals, %
Root exudates, day of growth	1	22.5±1.5	41.6±2.3
	2	21.1±2.3	47.8±2.5
	3	24.1±1.7	54.0±1.5
α –tocopherol [7 mcg per ml]		49.4±5,6	—
ethanol [5 mg per ml]		—	47.5±2.3

Influence of Water Quality on Root Growth Rate and Excretory Activity of One-day-old Seedlings

Our results suggest that the system "root-microenvironment" is able to respond to changes of cultivation conditions. This should be taken into account in research focusing on (i) the root excretory activity of exudates, and (ii) the physiological mechanisms of growth regulation. At the same time, one may expect that the "root-microenvironment" system can be used in the capacity of biological test for water quality estimation and determining the presence of toxic compounds in a water solution. Subsequently, the influence of water from different sources was determined on some characteristics of the "root-microenvironment" system.

The experiment was conducted with water from sources № 1 and № 2. These two are used as sources of drinking water. Distilled water was used as control. Root length of one-day-old seedlings growing on distillated water was 1.18 ± 0.06 cm. This root length was reduced by 0.5 cm in the source water № 1, and by 0.94 cm in the source water № 2 (Figure 14). Such significant inhibition of root growth during the 1st day was due to the presence of toxic compounds and heavy metal ions in the source waters. The study water sources showed the greatest difference in Cu^{2+} content compared to the control. Copper content in distilled water was 0.77×10^{-7} M, but it was 1.8 or 7 times greater in the water sources № 1 or № 2, respectively (Figure 14).

Copper ions can induce changes in the cell metabolism of plants and animals [51-54]. Roots growing on water from the sources № 1 or № 2 have an increased content of TBA-active products of 2.5 or 1.5 times, respectively, compared with the control (Figure 15). However, the content of carbonylated proteins did not change compared to control (Figure 15).

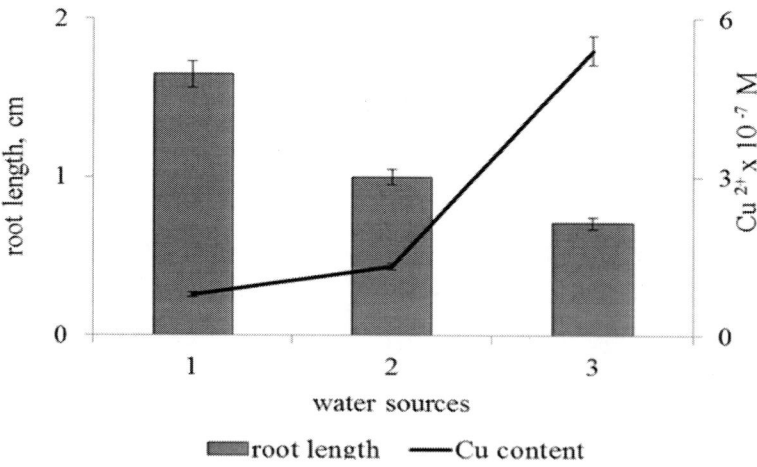

Figure 14. Root length of 1-day-old wheat seedlings at cultivation on water from different sources, and Cu content in such waters. 1– distillated water, 2– water from the 1st source, 3– water from the 2nd source.

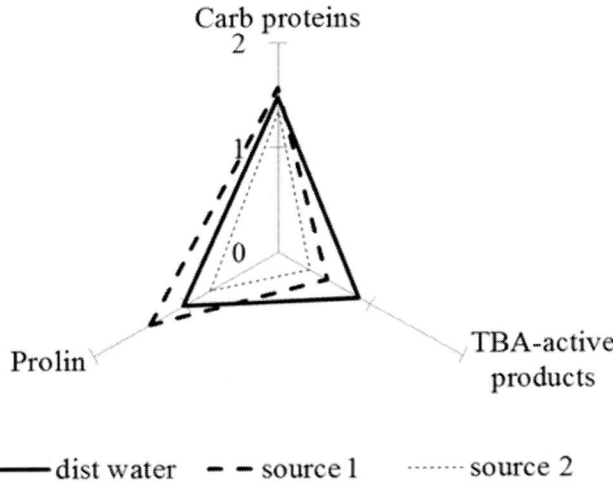

Figure 15. Content of free proline, TBA-active products or carbonylated proteins in 1-day-old roots at cultivation on distillated water (control) and o water from different sources.

The lack of a relationship between the content of copper ions in water and the TBA-active product content in roots of one-day-old seedlings can be explained by the induction of free proline generation, which can act as an antioxidant. Root growth inhibition increased by 1.4 or 1.9 times after cultivation on water from the sources № 1 or № 2, respectively, compared to distillated water (Figure 15). Therefore, wheat growth on water from different sources showed small differences in chemical composition. However, there was a marked root growth inhibition and change in the prooxidant-antioxidant system.

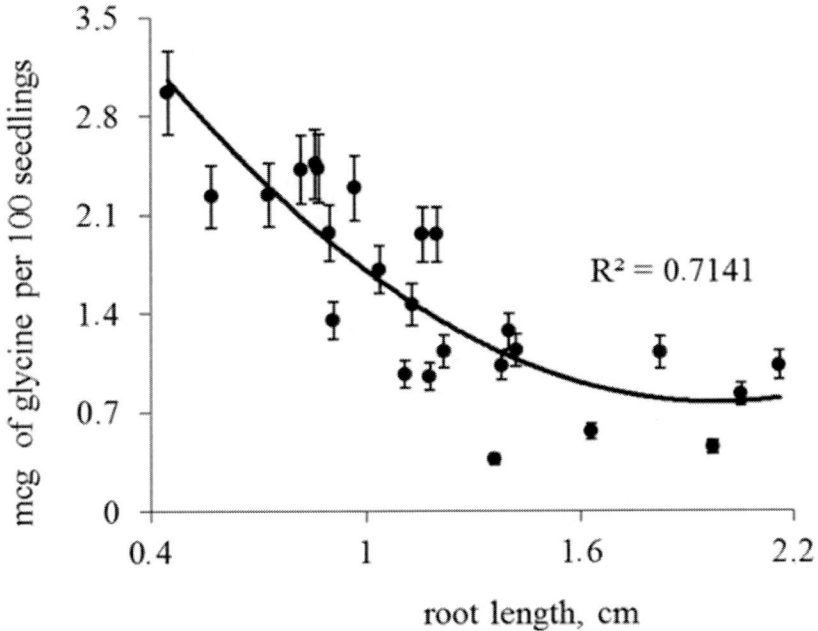

Figure 16. Relationship between root length and amino acid content in root exudates of wheat seedlings.

We were then interested in estimating the relationship between root growth rate and its excretory activity on this model. The free amino acid content in root exudates at cultivation on water source № 1 changed not significantly (by 35%). At the same time, their content at cultivation on water source № 2 increased twofold compared to the control (Table 6).

Table 6. Content of free amino acids (mg of glycine per 100 seedlings) in root exudates collected at cultivation of seedlings on waters from the sources № 1 and 2 and distilled water

Amino acids content	Sources of water for cultivation		
	Distilled water	Source № 1	Source № 2
	1.29±0.23	1.75±0.20	2.52±0.19

The coefficient of determination between the amino acid content in root exudates and root length was 0.71 (Figure 16). That is, the greater the root growth rate the lower the amino acid excretion rate.

Therefore, the "root-microenvironment" system is very sensitive to changes of composition of the water medium. Our results suggest that the "root-microenvironment" system could be used as a test for water quality estimation, and that there are not significant changes of the water medium on its functional state.

References

[1] H. P. Bais, V. M. Loyola-Vargas and H. E. Flores, *In vitro Cell Dev. Biol. Plant. 37,* 730 (2001).

[2] D. V. Badri and J. M. Vivanco, *Plant Cell Environ. 32,* 666 (2009).

[3] P. R. Ryan, E. Delhaize and D. L. Jones, *Ann. Rev. Plant Physiol. Plant Mol. Biol. 52,* 527 (2001).

[4] A. Sonawane, U. Klöppner, S. Hovel, U. Völker and K. H. Röhm, *Microbiology 149,* 2909 (2003).

[5] F. Wen, H. D. VanEtten and G. Tsaprailis, *Plant Physiol. 143,* 773 (2007).

[6] H. Wu, T. Haig and J. Pratley, *J. Agric. Food Chem. 49,* 3742 (2001).

[7] B. J. Lugtenberg, L. V. Kravchenko and M. Simons, *Environ. Microbiol. 5,* 439 (1999).

[8] S. G. Mashkovska, N. G. Dydyk and V. L. Brechko, *Fiziologiia Rastenii 34,* 437 (2002).

[9] S. I. Bozhkov, N. G. Menzyanova and V. P. Leontovich, *Fiziologiia Rastenii 43,* 920 (1996).

[10] J. A. Lucas García, C. Barbas and A. Probanza, *Phytochem. Anal. 15,* 305 (2001).

[11] A. K. Liliensiek, D. Thakuria and N. Clipson, *Environ. Microbiol. 63,* 509 (2012).

[12] S. Shi, A. E. Richardson, M. O'Callaghan, K. M. DeAngelis, E. E. Jones, A. Stewart, M. K. Firestone and L. M. Condron,. *FEMS Microbiol. Ecol. 77,* 600 (2011).

[13] E. M. Knee, F. Ch. Gong and M. Gao, *Mol. Plant Microbiol. Interact. 14,* 775 (2001).

[14] J. A. W. Morgan, G. D. Bending and P. J. White, *J. Exp. Bot. 56,* 1729 (2005).

[15] Ch. Kong, W. Liang, X. Xu, F. Hu, P. Wang and Y. Jiang, *J. Agric. Food Chem. 52,* 2861 (2004).

[16] C. D. Broeckling, A. K. Broz and J. Bergelson, *Appl. Environ. Microbiol. 74,* 738 (2008).

[17] B. L. Huang, Y. D. Xu and Y. Wu, *Ying Yong Sheng Tai Xue Bao. 18*, 559 (2007).

[18] E. M. Estabrook and J. I. Yoder, *Plant Physiol. 116*, 1 (1998).

[19] X. Chang, C. Duan and H. Wang, *Ying Yong Sheng Tai Xue Bao. 11*, 315 (2000).

[20] S. Doncheva, Z. Stoyanova and K. Georgieva, *J. Plant Nutr. Soil Sci.* **169, 247** (2006).

[21] P. S. Kidd, M. Llugany and C. Poschenrieder, *J. Exp. Bot. 52*, 1339 (2001).

[22] S. C. Miyasaka, J. G. Buta and R. K. Howell, *Plant Physiol. 96*, 737 (1991).

[23] D. M. Pellet, L. A. Papernik and L. V. Kochian, *Plant Physiol. 112*, 591 (1996).

[24] O. A. Inderjit, L. A. Weston and S. O. Duke, *J. Plant Interact. 1*, 69 (2007).

[25] D. T. Tsao, *Adv. Biochem. Eng. Biotechnol. 78*, 1 (2003).

[26] O. B. Lowry, N. J. Rosebrough, A. L. Farr and B. J. Randall, *Biol. Chem. 193*, 265 (1957).

[27] F. Gerhardt, *The Methods of General Bacteriology,* Moscow (1983).

[28] G. A. Gribanov and S. A. Sergeev, *Vopr. Med. Khim. 21(6)*, 652 (1975).

[29] L. D. Bergelson, E. V. Dyatlovitskaya and Y. G. Molotkovskiy, *The Preparative Biochemistry of Lipids,* Nauka, Moscow (1981).

[30] A. I. Yermakov, *The Methods of Biochemical Plant Research,* Agropromizdat, Leningrad (1987).

[31] J. Ruan, L. Ma and Y. Shi, *Ann. Bot. 93*, 97 (2004).

[32] G. I. Klebanov, I. V. Babenkova and Y. O. Teselkin, *Laboratornoe delo 5*, 59 (1988).

[33] B. Halliwell, J. M. C. Gutteridge and O. I. Aruoma, *Anal. Biochem. 165*, 215 (1987).

[34] N. S. Egorov, *The Fundamentals of Antibiotics Study.* High school, Moscow (2004).

[35] L. S. Bates, R. P Waldern and I. D. Teare, *Plant Soil. 309*, 205 (1973).

[36] H. Ohkawa, N. Ohishi and K. Yagi, *Analyt. Biochem. 5*, 351 (1979).

[37] E. E. Dubinina, S. O. Burmistrov, D. A. Hodov and I. S. Porotov, *Vopr. Med. Khim. 41*, 24 (1995).

[38] M. B. Stephenson and M. C. Hawes, *Plant Physiol. 106,* 739 (1994).

[39] K. M. Arnesen, *Plant Soil 191,* 13 (1997).

[40] D. P. Stevens, M. J. McLaughlin and A. M. Alston, *Plant. Soil 200*, 119 (1998).

[41] J. Lunardi, A. Dupuis, J. Garin, L. Michel, M. Chabre and P. V. Vignais, *PNAS 85*, 8958 (1988).

[42] M. M. Cowan, *Clin. Microbiol. Rev. 12,* 564 (1999).

[43] S. Sakanaka, M. Kim and M. Taniguchi, *Agric. Biol. Chem. 53*, 2307 (1989).

[44] K. Vijaya, S. Ananthan and R. Nalini, *J. Ethnopharmacol. 49,* 115 (1995).

[45] M. A. Fernandez, M. D. Garcia and M. T. Saenz, *J. Ethnopharmacol. 53*, 11 (1996).

[46] J. R. S. Hoult and M. Paya, *Gen. Pharmacol. 27*, 713 (1996).

[47] M. M. Posmyk, C. Bailly and K. Szafranska, *J. Plant. Physiol. 162*, 403 (2005).

[48] D. Wozniak, E. Lamer-Zarawska and A. Matkowski, *Food/Nahrung. 48*, 9 (2004).

[49] A. Braca, G. Fico and I. Morelli, *J. Ethnopharmacol. 86*, 63 (2003).

[50] J. P. Lee, B. S. Min and R. B. An, *Phytochem. 64*, 759 (2003).

[51] A. Bozhkov, V. Padalko, V. Dlubovskaya and N. Menzianova, *Indian J. Exp. Biol. 48,* 679 (2010).

[52] C. Manzl, J. Enrich, H. Ebner, R. Dallinger and G. Krumschnabel, *Toxicology 196,* 57 (2004).

[53] G. Krumschnabel, C. Manzl, C. Berger and B. Hofer, *Toxicol. Appl. Pharmacol. 209,* 62 (2005).

[54] A. Schützendübel and A. Polle, *J. Exp. Bot. 53,* 1351(2002).

Applied Research

Salt-Marsh Ecosystems

In: From Seed Germination to Young Plants
Editor: Carlos Alberto Busso

ISBN: 978-1-62618-653-8
© 2013 Nova Science Publishers, Inc.

Chapter 6

Substrate Properties Affect the Recovery of *Spartina alterniflora* from Drought and Herbivory

Stephen M. Smith[*]

National Park Service, Cape Cod National Seashore, Wellfleet, MA, US

Abstract

Coastal salt marshes develop in sheltered intertidal zones where sediment deposition and accumulation of peat from halophytic vegetation comprise the structural framework of this ecosystem. Within salt marsh systems, plants grow on substrates that vary in peat content depending on marsh age, plant productivity, and a number of physical and biological processes. On Cape Cod (Massachusetts, USA), spatial heterogeneity in substrate peat content is further amplified by vegetation dieback. Seed germination and the establishment of seedlings may occur in dieback areas that have varying proportions of peat and their subsequent development into mature plants varies with this property. Furthermore, re-vegetation by seedlings can be limited by top-down and bottom up stresses. In this study, we assessed the capacity for *Spartina alterniflora* seedlings (smooth cordgrass; the dominant taxon in Cape Cod salt marshes) to withstand simulated drought and herbivory in dense peat vs. soft sand-mud. The results indicate that the latter is much more conducive to survival. This information is valuable from the standpoint of predicting the potential for passive (i.e., natural) or active (i.e., managed) re-vegetation of dieback areas.

Introduction

The physical properties of soils have long been known to play a significant role in the growth and morphology of plants [1,2] especially during the seedling stage of development [3]. In coastal salt marshes, the accumulation of dense, hard peat resulting from hundreds or thousands of years of root production can increasingly inhibit plant growth [4,5,6]. This occurs despite the fact that peat has higher redox potential, higher nitrogen content, and

[*] stephen_m_smith@nps.gov.

higher water permeability (all of which enhance salt marsh plant growth) than substrates lacking peat [7]. Similar findings have been reported in freshwater systems where plants may be stunted by high concentrations of peat [8,9]. Moreover, it is well-known that soil hardness and compaction in itself can dramatically influence the growth of plants in many different types of environments [10,11,12].

On Cape Cod, large areas of salt marsh vegetation have been converted to bare substrate due to overgrazing by an exploding population of native, herbivorous crabs (*Sesarma reticulatum*) [13]. Where vegetation losses are relatively recent, hard, dense peat remains. Where denudation occurred in the more distant past and erosion/decomposition have occurred the remaining substrate is mainly soft, inorganic substrate with particle sizes ranging from sand to mud. Some of these bare areas have been partially re-colonized by *S. alterniflora* seedlings (Figure 1). However, the growth forms of seedlings are remarkably different depending on the substrate type. Larger, more vigorous plants are almost always found in mud-sand substrates while short, stunted ones dominate peaty areas [7].

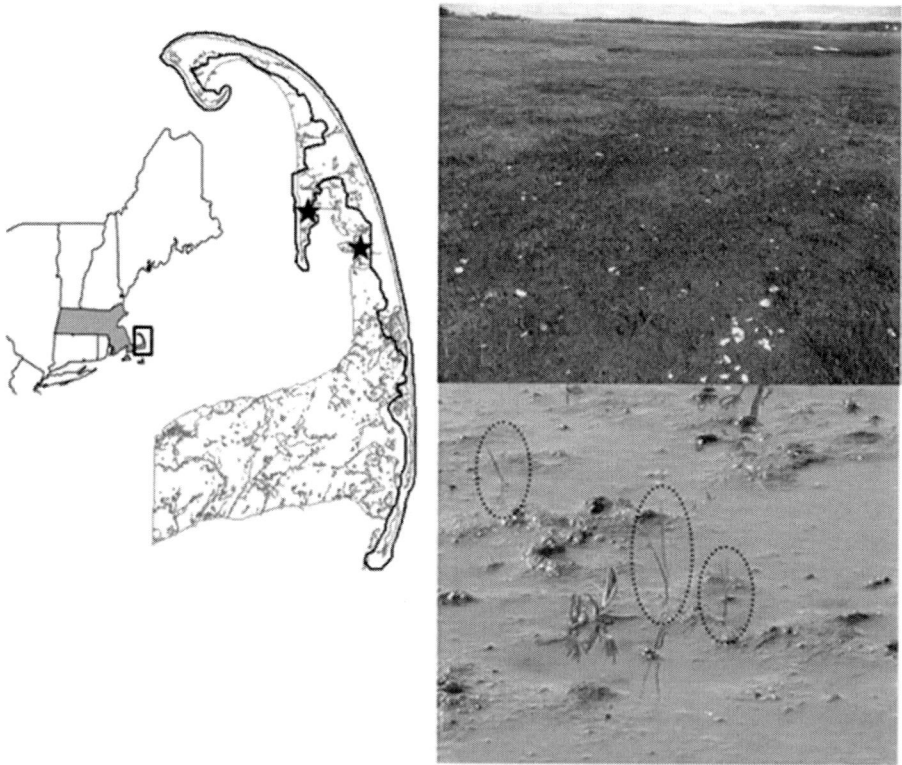

Figure 1. The Commonwealth of Massachusetts (shaded area in map on left) and outer Cape Cod (middle map; polygon is the boundary of CCNS; stars represent salt marshes where seedlings were collected). Photographs show the disappearance of *S. alterniflora* on hard peat substrate after a summer of herbivory and drought in 2011 (top right) and newly established *S. alterniflora* seedlings in a dieback area comprised of sand-mud in June 2011 (bottom right).

Such differences in plant morphology beg the question of whether substrate type also influences seedling resiliency to other stresses. Crab herbivory is highly variable in space and time but is an important stressor on newly established seedlings in that photosynthetic tissues are continually being removed. Anecdotal evidence from the recent past suggests that drought and herbivory together can result in widespread plant mortality on Cape Cod (Figure 1). While, the basis for the spatially heterogeneous patterns of decline or loss has not been elucidated, patterns of loss may be related to substrate type in that survival appears much higher in soft inorganic sediments than it does in harder peat-dominated areas.

In this study, we sought to determine the extent to which seedlings of *S. alterniflora* (the dominant species in Cape Cod salt marshes) were able to recover from experimentally-induced drought followed by simulated herbivory in peat vs. sand-mud. Because seedling establishment is a major way in which *S. alterniflora* can re-colonize bare areas [14,15], this kind of data enhances our ability to predict re-vegetation potential. Salt marshes comprise roughly 10% of the Cape Cod National Seashore (CCNS) and are an extremely important natural and socio-economic resource to coastal communities given that they are the most productive ecosystems on earth, provide habitat for a wide variety of marine species (many of which are recreational and commercially harvested), and protect coastlines from erosion and storm damage [16; and references therein].

Methods

S. alterniflora seedlings growing in peat vs. sand-mud were collected from two different marsh sites using 10-cm diameter, 30-cm-long butyrate tubes and transported back to the Cape Cod National Seashore greenhouse facility (locations are shown in Figure 1). Three collections of 10 replicate samples for each substrate type were done on June 20, June 30, and July 13. Additional cores (n=10 replicates for each substrate type) were collected to monitor substrate water loss rates. Initial values for soil hardness (i.e., before simulated drought and herbivory treatments) in all tubes was measured using a soil penetrometer (Lang Penetrometer, Inc, Gulf Shores, Alabama), which measures hardness as pounds (lbs) of force required to fully insert a 3-mm diameter steel probe into the substrate.

The tubes containing the individual seedlings and their surrounding substrate were left open at the bottom to allow for drainage and watered daily with fresh tap water for 1 week to allow for acclimation to the greenhouse environment. *S. alterniflora*, while a halophyte, grows best in fresh to very low salinity water [17,18,19,20]. The number of live leaves on each seedling was measured initially and plant biomass in grams of dry weight was calculated from a regression equation developed during the time of this study (biomass=0.048*number of live leaves$^{0.92}$; R^2 = 0.89). After 1 week of acclimation, watering was discontinued for a period of 30 days to simulate drought. Then all the aboveground foliage was clipped to within 1-cm of the substrate surface to simulate crab herbivory in the field where plants are typically grazed down to the shoot base [13]. At this point daily watering was resumed. After 30 days, plant recovery (i.e., re-growth) was assessed by measuring the plants for a final time. Any mortality (defined as the lack of any re-growth) was also recorded.

To determine relative differences in soil moisture loss in peat vs. sand-mud, the additional cores were thoroughly wetted with fresh tap water and the initial weights determined to the nearest 1/10 of a gram. Over the course of 1 month, at approximately 3-4 day intervals, all cores were reweighed. Water loss was then normalized to volume of soil, which differed slightly among replicate tubes.

Substrate hardness, total substrate moisture loss, and plant bioassay data all were log-transformed to meet the assumptions of normality and homoscedasticity and compared using T-tests (α=0.05).

Results

Average soil hardness values for peat vs. sand substrate (30 replicates each) were 11.41 ($\pm$ 2.82) and 3.87 ($\pm$ 2.71) lbs of force. In other words, peat substrate was approximately threefold harder than sand. Water loss rates exhibited little differences between peat and sand-mud. While there was some divergence over time, with peat losing water at a slightly higher rate, this difference was minor and not statistically significant (T=1.73, p=0.08). Peat and sand-mud lost a total of 125.5 ($\pm$8.9) g and 107.8 ($\pm$9.1) of water, respectively, over the course of 1 month. Translated to water loss per unit volume of soil, these values were 0.070 ($\pm$0.006) for peat and 0.066 for sand ($\pm$0.006).

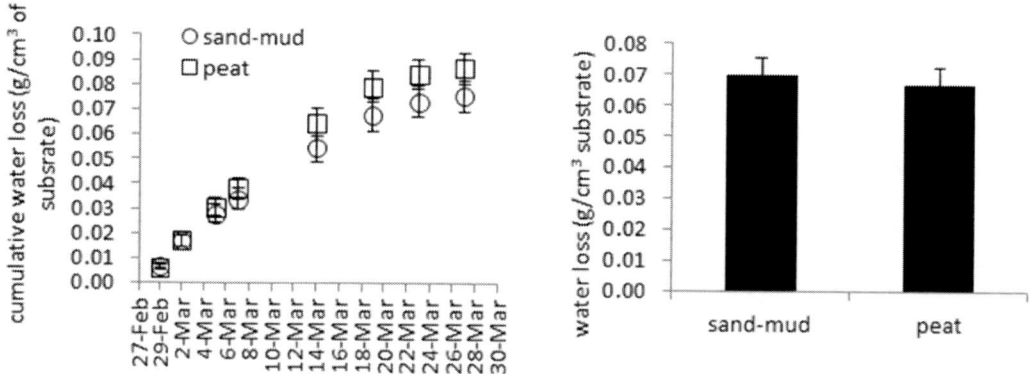

Figure 2. Progressive soil moisture loss over time (left) and cumulative totals (right) in peat vs. sand-mud substrate (symbols are the mean of n=10. histograms are the mean of n=10, vertical bars represent +/- 1 S.E.).

Table 1. Changes in seedling biomass and survival rate in peat vs. sand-mud substrate following recovery from simulated drought and herbivory (T=critical T value for T-tests; p = significance value; values for biomass and seedling survival are the mean of n=10; values in parenthesis adjacent to biomass are +/- 1 S.E.)

| | Biomass | | | | Survival | |
	Peat	Sand-mud	T	p	Peat	Sand-mud
Exp-1	-46% (61%)	129% (32%)	2.56	0.02	56%	78%
Exp-2	-79% (10%)	-33% (30%)	1.95	0.05	25%	50%
Exp-3	-82% (10%)	59% (20%)	1.75	<0.01	27%	100%
Grand means	-69%	52%			36%	65%
Standard error	-12%	47%				

Plant recovery from simulated drought and herbivory stress was significantly higher in sand-mud compared to peat in all three assays (Table 1). Although there was substantial variability among replicates and among assays, very large treatment differences overwhelmed this variability. Seedlings in sand-mud increased in biomass from their initial values in assays 1 and 3, but decreased in assay 2. In contrast, seedlings in peat all declined from the beginning of the experiments by an average of 69%. Survival essentially mirrored that of biomass although the magnitude of difference between sand-mud vs. peat substrate was less. In this regard, overall survival was 36% and 65% in peat vs. sand-mud, respectively.

Discussion

The specific mechanism by which peat inhibits salt marsh plant growth has remained unclear (and apparently unstudied) since the 1988 paper by Bertness [7]. The results of these experiments indicate that peat substantially diminishes the ability of *S. alterniflora* seedlings to withstand drought and herbivory. In essence, seedlings that manage to become established in dense peat areas are likely to suffer higher mortality under subsequent conditions of drought and herbivory. That abiotic factors can impact salt marsh plant responses to additional top-down and bottom-up stress factors has been reported previously. A relevant example comes from the work of Alberti et al. [21], who concluded that soil porewater properties (salinity, ammonium concentration, and anoxia) impacted *Spartina densiflora* (dense-flowered cordgrass) responses to nutrient inputs and crab herbivory in Argentina.

Simulated drought conditions in this experiment were somewhat exaggerated compared to a natural setting where most *S. alterniflora* vegetation in the low marsh zone will experience high tides on a regular basis. However, during neap tidal cycles coupled with no rainfall, soil moisture loss could be substantial during periods of exposure. In terms of the mechanism by which peat vs. sand-mud substrate influences plant growth and resiliency to stress, we propose that it is the physical impedance of root growth that is most important. This concept is supported by a number of silviculture and agriculture studies on soil compaction [22,23,24]. In our experiment, seedlings were treated with fresh water; in fact, cores were essentially flushed of residual salt with freshwater treatments before the experimental treatments were even applied. Thus, salinity was not a factor contributing to seedling decline in this study; neither were differences in drying rate of peat vs. sand-mud. Notwithstanding, under natural conditions drought can result in increased soil salinities [25]. Moreover, the magnitude and duration of drought is predicted to increase with global warming [26].

Although *Sesarma* crabs initially create areas of dieoff, the remaining peat undergoes a transition to soft sand as the former decomposes and is eroded away. In some cases, rapid peat loss can occur from ice scour [27,28]. The softer inorganic substrate (mainly sand-mud on Cape Cod) left behind is more conducive to vigorous seedling growth, which ultimately translates to a higher potential for re-vegetation of these sites. In contrast, where dense peat remains for longer periods of time it continues to provide a highly suitable habitat for *Sesarma* crabs that require firmer substrate to build and maintain their elaborate system of burrows [29]. Thus, the probability that seedlings will successfully colonize and survive in these peaty areas is diminished by higher localized grazing pressure from higher crab densities.

The combination of drought and herbivory is purported to be the cause for extensive salt marsh dieback in the southeastern United States [25,30]. This study suggests that substrate type either directly or indirectly influences seedling resiliency to top-down (herbivory) and bottom-up (drought) stresses. This information improves our ability to predict spatial patterns of vegetation recovery from dieback. It is also useful for selecting appropriate sites for active planting/seeding to accelerate re-vegetation and reduce further erosion of CCNS salt marshes.

References

[1]   D. Alameda and R. Villar, *Soil Till. Res.* 2, 325 (1987).

[2]   J. A. Murphy, S. L. Murphy, and H. Samaranayake, *Plant Environment Interactions 2nd Edition*, Taylor and Francis, New York (1999)

[3]   Pan, E. and N. Bassuk, *J. Environ. Hort.* 3, 158 (1985)

[4]   M. G. Barbour and W. D. Billings, *North American Terrestrial Vegetation*. Cambridge University

[5]   M. D. Bertness and A. M. Ellison, *Ecol. Mon.* 57, 129 (1987)

[6]   J. L. Bubier, S. Frokling, P. M. Crill and E. Linder, *J. Geophys. Res.* 104, 27683 (1999)

[7]   M. D. Bertness, *Ecology* 69, 703 (1988)

[8]   J. W. Barko and R. M. Smart. *J. Ecol.* 71, 161. (1983)

[9]   J. W. Barko and R.M. Smart. *Effects of Sediment Composition on Growth of Submersed Aquatic Vegetation, Technical Report*, U.S. Army Engineer Waterways Experiment Station, Vicksburg, MS. (1986)

[10]   A. Komiyama, T. Santiean, M. Higoa, P. Patanaponpaiboon, J. Kongsangchai, K. Ogino, *For. Ecol. Manage.* 81, 243 (1996)

[11]   J. Kuht and E. Reintam, *Agron. Res.* 2, 187 (2004)

[12]   Y. Xitian, D. Huiying , and Y. Yamaderaz, *Soil Sci. Water Conserv.* 3, 60 (2005)

[13]   C. Holdredge, M. D. Bertness, and A. H. Altieri, *Cons. Biol.* 23, 672 (2009)

[14]   D. R. Ayres, S. Klohr, D. L. Smith, D. R. Strong, and K. Zaremba, *Biol. Inv.* 6, 221 (2004) Press, United Kingdom (2000)

[15]   W. S. Metcalfe, A. M. Ellison, and M. D. Bertness, *Ann. Bot.* 58, 249 (1986)

[16]   S. Fagherazzi, R. Torres, C. Hopkinson, and D. van Proosdij, *EOS* 86, 57 (2005)

[17]   D. A. Adams, *Ecology* 44, 445 (1963).

[18]   F. Charles. *Ecology* 52, 908 (1971)

[19]   C. M. Crain, B. R. Silliman, S. L. Bertness, and M. D. Bertness, *Ecology* 85, 2539 (2004)

[20]   R. T. Parrondo, J. G. Gosselink, and C. S. Hopkinson, *Bot. Gaz.* 139, 102 (1978)

[21]   J. Alberti, J.M. Escapa, O. Iribarne, B. Silliman, and M. Bertness, *Ecology* 89, 155 (2008).

[22]   C. E. Bulmer and D. G. Simpson, *Can. J. Soil Sci.* 85, 667 (2005

[23]   M. O. A. Fawusi, *Sci. Hort.* 9, 329 (1978)

[24]   D. L. Mitchell, D. Hocking, and W. C. Kay, *Can. J. For. Res.* 2, 479 (1972)

[25] B. R. Silliman, J. van de Koppel, M. D. Bertness, L. Stanton, and I. Mendelsohn, *Science* 310, 1803 (2005)

[26] P. C. Frumhoff, J. J. McCarthy, J. M. Melillo, S. C. Moser, and D. J. Wuebbles, Synthesis report of the Northeast Climate Impacts Assessment (NECIA), Union of Concerned Scientists, Cambridge, MA (2007)

[27] B. A. Argow, Z. Hughes, and D. M. FitzGerald, *Cont. Shelf Res.* 15, 1294 (2011)

[28] M. N. Hardwick-Witman, *Est. Coast Shelf Sci.* 22, 379 (1986)

[29] M. D. Bertness, C. Holdredge, A. H. Altieri, *Ecology* 90, 2108 (2009)

[30] M. Alber, E. M. Swenson, S. C. Adamowicz, and I. A. Mendelssohn, *Est. Coast. Shelf Sci.* 80, 1 (2008).

Arid and Semiarid, Temperate Ecosystems

In: From Seed Germination to Young Plants
Editor: Carlos Alberto Busso

ISBN: 978-1-62618-653-8
© 2013 Nova Science Publishers, Inc.

Chapter 7

Effects of Microsite Conditions on Seedling Emergence at Grazed and Ungrazed Sites in Mountain Pampean Grasslands, Argentina

Alejandro Loydi[1] and Guadalupe Peter[2]*

[1] Departamento de Biología, Bioquímica y Farmacia,
Universidad Nacional del Sur. Centro de Recursos Naturales Renovables de la Zona Semiárida (CERZOS), CONICET. Bahía Blanca, Argentina
[2] CONICET. Centro Universitario Regional Zona Atlántica,
Universidad Nacional del Comahue, Sede Atlántica,
Universidad Nacional de Río Negro, Viedma, Argentina

Abstract

Seed germination and seedling establishment are the most critical periods among the developmental morphology stages in most plant species. Plant species diversity could be seen from the balance between the soil seed bank composition and the safe site availability. In Pampean grasslands of Argentina, plant species diversity is threatened by the advance of agriculture. Most of the original grasslands are lost as a result. In this context, it is increasingly important to know the mechanisms involved in the maintenance of plant diversity. Seedling emergence would be partly responsible for determining plant species diversity, and it may be determined by microsites conditions (competition level, ground cover, disturbances). We analyzed (i) the effect of above- and below-ground competition and litter application on seedling emergence in ungrazed areas, and (ii) the effect of vegetation cover, litter or dung, bare ground, and soil removal on seedling emergence in grazed areas. Within an exclosure to domestic livestock and a grazed site, different microsites were created, and natural seedling emergence was followed during two years. Above- and below-ground competition reduced grass emergence in ungrazed conditions, but did not affect forb emergence. Litter had no effect on emergence,

* E-mail address: aloydi@criba.edu.ar.

although it reduced grass survival. At the grazed sites, soil disturbance had a positive effect on emergence of seedling of all vegetation types, while litter application showed a positive effect on grass emergence. However, survival in this area was very low (<10%). Competition reduced seedling emergence, showing to be a major control factor for the regeneration of the dominant species in the ungrazed sites. Litter presence reduced survival of grasses, but not of forbs. At the grazed sites, presence of litter improved microsites conditions, increasing grass emergence; at the same time, soil removal increased emergence of seedlings of all vegetation types, probably by favoring root development. In natural grasslands, availability of safe sites has an important role in the establishment of new individuals and plant species regenerations.

Introduction

Establishment of new individuals is the most critical phase of plant life [1], and it depends on availability of seeds and a positive environment for seed germination and seedling survival [2]. Grubb (1977) [3] named these requirements for germination, emergence and survival as the regeneration niches, representing safe sites for seedling establishment [4]. Safe sites gather a series of abiotic (e.g. temperature, humidity, light) and biotic (e.g., herbivory, plant cover, density, litter presence, competition) conditions that allow establishment of plant species seedlings [5].

Litter effects on seedling establishment depend on litter accumulation [6]. In grassland ecosystems, these effects could vary from negative to slightly positive [7], but under harsh conditions (e.g., drought) litter presence could improve abiotic conditions (i.e., lower temperature fluctuations, higher water availability) thus promoting establishment [8-11]. Another important factor on seedling establishment is competition with neighboring vegetation. Generally, competition reduces the establishment of seedlings [12-14] but under stressful environmental conditions, the surrounding vegetation may exert facilitative effects [15-18].

Disturbances at different temporal and spatial scales are also important determinants of safe site availability. Grazing is one of the most common disturbances on grassland ecosystems, and its effects on seedling establishment are variable. Grazing can decrease the competitive effects of neighboring vegetation [19,20] and prevent excessive litter accumulation [21], promoting seedling establishment in some cases [22]. Dung presence has a similar effect [23,24]. Likewise, herbivores could remove superficial ground, creating new safe sites for seed germination [20,23,25]. On the other hand, grazing could reduce seedling recruitment by predation of seeds or seedlings [19,26,27], or by direct destruction of available safe sites [5]. Bare ground patches creation [23] and soil compaction [28] by grazing activity, could create microsites with lower water and nutrient availability [29], reducing the establishment of new individuals [25,30].

In mountain Pampean grasslands, grazing by feral horses increases species diversity [31], by favoring specific vegetation groups, such as shrubs, rosette species, annual forbs and unpalatable grasses [32]. Exclusion of herbivory grazing, on the other hand, favors palatable perennial grass species [32,33], which are supposed to be the dominant species in a pristine grassland condition [34]. On the contrary, the forb seed bank is abundant in both, grazed and excluded conditions, while the availability of grass seeds is higher in ungrazed areas [35].

Thus, plant species diversity would reflect seed and safe site availabilities for seedling emergence and survival under grazed and ungrazed conditions [36].

With the absence of herbivores increases competition for resources and litter accumulation [14,21,37], grazing makes more limiting the abiotic conditions at a microsite scale. The information about the effects of these factors could be important for restoration purposes. There are various studies about that in temperate grasslands [12,38,39], but the information for grazed areas is scarce [40].

In the present work, we hypothesize that seedling emergence in ungrazed areas is controlled by biotic factors (competition, litter presence), while under grazing conditions it is controlled by abiotic factors (bare ground, ground cover, soil compaction). Consequently, the objectives of this work were to evaluate the differential effects on seedling emergence of the: (i) above- and below-ground competition, and litter presence in an ungrazed area; and (ii) vegetation cover, soil removal and presence of litter or dung in a grazed area.

Materials and Methods

Study Site

The study area was located in Central East Argentina (38° 02'-38° 04' S and 61° 57'-62° 00' W), at Ernesto Tornquist Provincial Park (ETPP). Piedmont valleys with 3% to 11% slopes and occasional rocky outcrops dominate the area. Climate is sub-humid temperate [41]. Medium annual air temperature is 14°C with an annual precipitation of 800 mm, concentrated during spring and summer. Snow occasionally falls in winter [42].

Soils are classified as Lytic Hapludolls and Argidolls, and are characterized by a high organic content (ca. 7%) in superficial horizons [43,44]. In the absence of grazing, the physiognomy of the vegetation is a grassland with sparse small shrubs. Grass canopy reaches 50 to 60 cm in height, and is dominated by perennial cool-season grasses such as *Piptochaetium hackelii*, *Nassella melanosperma* and *Briza subaristata* [34]. Autumn is the major growth period in the area [44]. Above-ground primary productivity averages 500 g m^{-2} [45]. The system has also been affected by recurrent fires and droughts [46,47], disturbances that may promote grazing exaptations [48]. During the study period the area was subjected to moderate drought conditions during autumn 2008 to winter 2009 (Figure 1).

The area has evolved under grazing pressure of native herbivores [49,50], being the guanaco (*Lama guanicoe*) the most abundant species in the past several thousand years [51]. Unfortunately, there is no documentation of population levels or grazing intensity of guanacos. Additionaly, ETPP has a recent history (around 30 years) of heavy, continuous, year-round grazing by feral horses. Feral horses reached a maximum density of 32.5 horses per km^2 in early 2002 [52]. After that, the population was reduced by mass mortality caused by extreme windy and rainy conditions in November 2002 [53], and in 2006 due to a planned retirement of horses by ETPP authorities [54]. In 2008 the feral horse population was reduced to a minimum density of ca. 7.25 horses km^2, and are slowly recovering thereafter (A. Scorolli, pers. com).

Experimental Design

Grass and forb emergence and survival were compared in different microsites in domestic natural grasslands. We choose two areas: one area was grazed by feral horses, while the other was excluded herbivores grazing. Plant emergence was registered for a period of two years (April 2008 to March 2010). Colonization of microsites is primarily limited by seed availability, thus 50 seeds of *Nassella trichotoma*, collected in the study area, were sown in every microsite to enssure that seed availability was not a limiting factor for this species. This species is a native perennial grass, easily recognizable as seedling. It was scarcely represented in the seed bank in the study area (42.4 seeds m^{-2} in the ungrazed area in April 2008, A. Loydi, pers. obs.), although there were established plants in the aboveground vegetation.

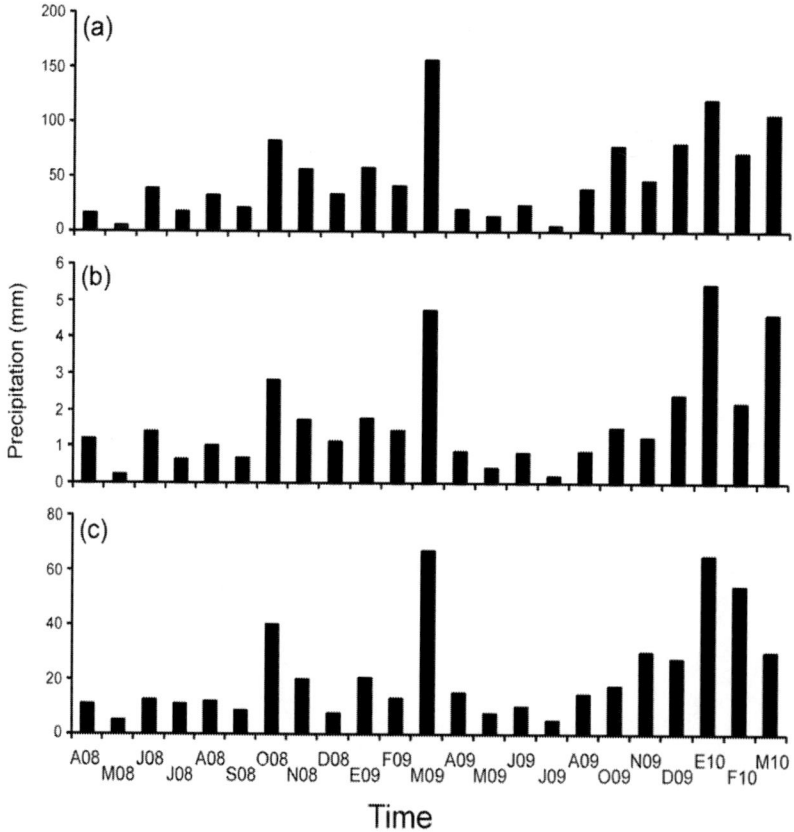

Figure 1. Total (a) daily (b) and daily maximum (c) precipitation during the study period.

To avoid the effect of these individuals, all plots in the study areas were placed away (>2 m) from every existing individual of *N. trichotoma*. Before setting the experiment, a germination essay with *N. trichotoma* was made. Petri dishes (n= 10) were used, each with 50 seeds of *N. trichotoma*. Seeds showed a germination power of 30 ± 2.5% (mean ± S.E.); thereby ca. 15 seedlings per plot were expected. Differences in emergence of *N. trichotoma* in every microsite were expected to depend only on microsite conditions, not on seed availability, as could be the case from spontaneous seedling emergence.

Every microsite consisted at a circular plot of 10 cm diameter surrounded by natural vegetation. They were installed in March 2008, resulting in 23 monthly sampling dates until March 2010. In all sampling dates, all seedlings were counted and marked with color wires, without being removed. All plots were randomly disposed within either area of 1 ha open to grazing by feral horses or a 28 ha exclosure of 15-year-old. In the exclosure, we established microsites with different levels of competition and litter, resulting in six treatments. Half of them were in bare ground (BG), and the other half were ground covered with litter (L). Within each of these two groups, three different competition levels were settled: microsites with above- (A) and below-ground (B) competition (BG_{A+B} and L_{A+B}), microsites with only below-ground competition (BG_B and L_B), and microsites without above- or below-ground competition (BG and L). Bare ground treatments were achieved by removing litter from plots, without disturbing the soil. For the litter treatments, natural litter was removed from the plots, which were covered with plant material collected from standing vegetation without any seed content. This plant material was collected in the field, and it was air dried under laboratory conditions. Previous to application, it was cut into 2 cm long pieces and 2 g per plot were used, forming a homogeneous layer of 1 cm thick. This represents 250 g m^{-2}, which is the natural litter content in ungrazed areas at the study site [44]. Competition treatments were achieved modifying the vegetation surrounding the plots. Treatments without above-ground competition were accomplished through periodical herbaceous vegetation clipping at ground level within a circular area of 60 cm diameter centered in the plots. Treatments without below-ground competition were achieved by buring steel tubes (10 cm diameter, 30 cm deep) in the soil. *Nassella trichotoma* seeds were sown over the ground in all plots. Immediately after that, only in the litter treatments, litter was added. Each treatment was replicated eight times, totaling 48 plots. In every treatment, one plot did not receive *N. trichotoma* seeds, functioning as a control plot for the natural emergence of *N. trichotoma*. In the grazed area, six treatments were established. One of these consisted of natural microsites with established vegetation (V). Two treatments included bare-ground (BG), two others ground covered by litter (L); in one of each of these treatments it was also applied superficial soil removal (BGR or LR). One last treatment includes bare ground microsites covered with horse dung (D). Microsites with natural vegetation (V) were selected on places covered by grass, and far away (>1m) from shrubs and rocks. Litter, when present, was removed from the bare ground microsites. Litter treatments were accomplished like in the excluded area, trying to mimic the possible litter accumulation in the eventual case of herbivore reduction or complete exclusion. Soil removal was made in the first 2 cm of soil with a garden shovel, without turning it up; the removal made by feral horse trampling and small animals in the study area (i.e. rodents, rabbits, armadillos) was simulated. The dung cover treatment was achieved using disaggregated, air-dried horse dung; it formed a layer of 1 cm thick and 3 g weight (ca. 380 g m^{-2}). As seed content in the horse dung was unknown, seedlings that were clearly rooted in dung were not counted. In every case, 15 plots were established, but due to the activity of feral horses and others animals, some plots were destroyed during the study. At the end of the experiment, there were 14 plots for treatments with bare ground (BG) and established vegetation (V), 13 for bare ground with soil removal (BGR), and 11 for treatments with soil cover with litter (L), litter and soil removal (LR) and soil covered with dung (D). This made a total of 74 from the original 90 plots. Every treatment included at least one plot without the addition of *N. trichotoma*.

Statistical Analysis

Cumulative emergence of seedlings of *N. trichotoma*, grasses, forbs, and total natural seedling emergence (i.e., grasses + forbs), were evaluated through one-way ANOVA. In every case, data were rank transformed prior to analyses [55]. Post-hoc analyses were made with Tukey's test with a 5% global error. Grass data did not include *N. trichotoma* emergence, which was analyzed separately. A priori contrasts were also made to compare emergence in the different treatment groups. In the excluded area, we compared seedling mean emergence between treatments with and without above- and below- ground competition, independently of ground cover. Treatments with and without litter cover were also compared, independently of their competition level. In the grazed areas, contrasts were made to detect differences in seedling emergence between treatments with and without litter addition, and treatments with and without soil removal. All analyses were made using the InfoStat statistical software [56] . We also describe emergence and survival of grass and forb seedlings during the study period.

Results

Ungrazed Area

During the study period, we registered 521 seedlings (12% *N. trichotoma*, 57% grasses and 31% forbs). The highest emergence peaked was in March 2009 with 112 seedlings. *Nassella trichotoma* seedlings were 10% of the expected amount (i.e., 63 of 630). It was registered only one *N. trichotoma* seedling in the control plots; thus the effect of natural emergence of this species was not considered anymore in further analyses. Total seedling emergence varied across competition levels (Figure 2), being higher in treatments without competition compared to those with above- and below-ground competition ($F_{(5, 42)} = 3.08$, $p < 0.05$). Treatments without above-ground competition showed intermediate emergence values. *Nassella trichotoma* and grass emergence were higher in treatments without competition ($F_{(5, 36)} = 3.16$, $p < 0.05$ and $F_{(5, 42)} = 4.43$, $p < 0.01$, respectively); forb emergence was not affected by any treatment ($F_{(5, 42)} = 0.68$, $p > 0.60$). There was a positive effect of reduced competition on seedling emergence, except for forb species (Contrast 1, Table 1). Instead, litter cover had no effect on seedling emergence (Contrast 2, Table 1). Figure 3 shows monthly emergence and survival of grass and forb seedlings during the study period. Highest grass survival was in bare-ground treatments without above- or below-ground competition. Microsites with litter cover and no competition showed the same seedling survival as microsites without above-ground competition, suggesting that litter cover decrease seedling survival for grass species. Forb survival was higher in microsites covered with litter and without competition, suggesting a positive effect of litter on survival for this group.

Grazed Area

In the grazed area, 1293 seedlings were registered during the study period (3.3 % *N. trichotoma*, 10.4% grasses and 86.3% forbs). The highest emergence was in March 2009 with 550 seedlings. *Nassella trichotoma* seedlings reached 4.34% of the expected amount (43 of 990 sowed seeds); they were not found in control plots, suggesting that natural germination was very low, and will not be considered further. Different microsite treatments resulted in different seedling emergence (Figure 4). Natural emergence of seedlings was higher in treatments with soil removal with and without cover of litter, compared to the rest of the microsites. Treatments with bare-ground and removed soil or soil covered by horse dung showed intermediate emergence values. The same response was found for *N. trichotoma*, grass and forb seedlings.

Table 1. Contrasts between treatments with (AB) and without (WC) above- and below-ground competition (Contrast 1) and treatments with bare-ground (BG) or ground cover by litter (L) (Contrast 2)

		Total		*N. trichotoma*		Grass		Forb	
		$F_{(1, 42)}$	P	$F_{(1, 36)}$	P	$F_{(1, 42)}$	P	$F_{(1, 42)}$	P
Contrast 1	AB vs WC	14.80	$p < 0.01$	7.34	$p < 0.05$	20.89	$p < 0.01$	1.91	$p > 0.15$
Contrast 2	BG vs L	<0.001	$p > 0.95$	2.16	$p > 0.15$	0.59	$p > 0.40$	0.32	$p > 0.55$

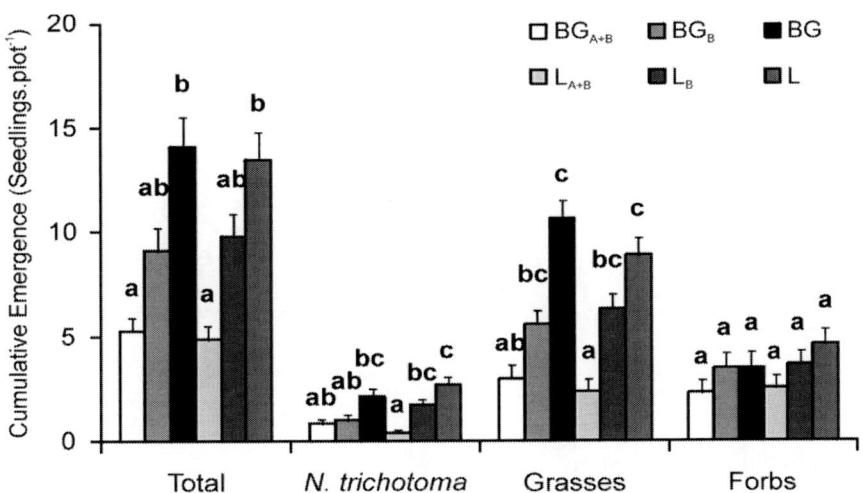

Figure 2. Cumulative seedling emergence of different vegetation groups in ungrazed area at the end of the study period (two years). Different letters indicate statistically different mean values. BGA+B: bare-ground with above- and below-ground competition. BGB: bare-ground only with below-ground competition. BG: bare-ground without above- or below-ground competition. LA+B: litter with above- and below-ground competition. LB: litter only with below-ground competition. L: litter without above- or below-ground competition.

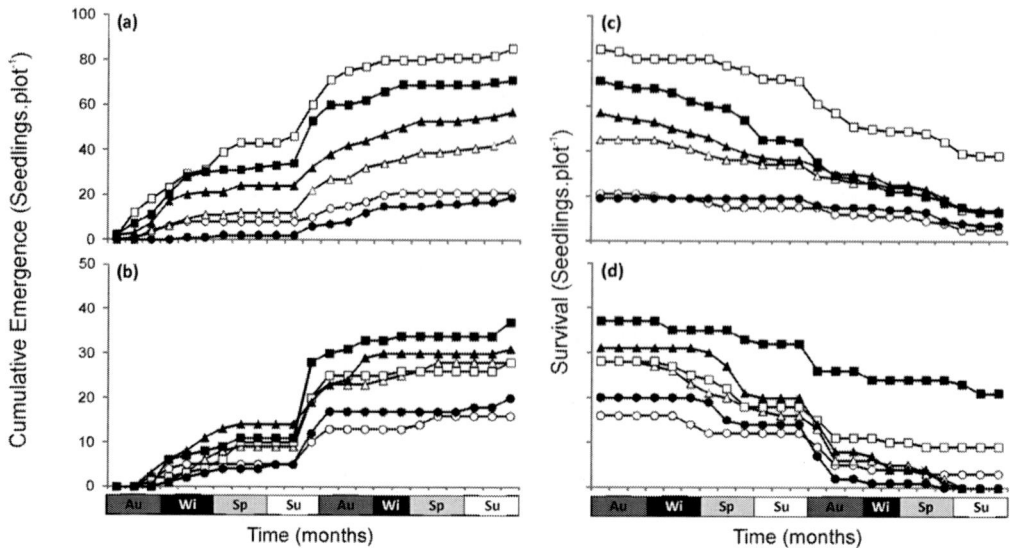

Figure 3. Cumulative emergence and survival of grass (a and c) and forb (b and d) seedlings during the study period. Au: Autumn. Sp: Spring. Su: Summer. Wi: Winter. Filled symbols: treatments with litter application. Empty symbols: treatments without litter application. Treatments: (○●) with above- and below-ground competition, (Δ▲) only with below-ground competition and (■□) without above- or below-ground competition.

Table 2. Contrasts between treatments with bare-ground (BG) or cover by litter (L) (Contrast 1) and treatments with (+R) and without (-R) soil removal (Contrast 2)

		Total		N. trichotoma		Grass		Forb	
		$F_{(1, 68)}$	P	$F_{(1, 49)}$	P	$F_{(1, 68)}$	P	$F_{(1, 68)}$	P
Contrast 1	**BG vs L**	1.15	p> 0.25	0.09	p> 0.75	4.90	p< 0.05	0.86	p> 0.35
Contrast 2	**+R vs -R**	24.50	p< 0.01	16.92	p< 0.01	4.30	p< 0.05	26.25	p< 0.01

Litter application in microsites only increased grass emergence (Contrast 1, Table 2). Soil removal significantly increased emergence of all seedling types (Contrast 2, Table 2). In all cases, seedling mortality was high (>90%) (Figure 5). For grasses, emergence increased during the first spring and mortality of these seedlings occurred during the following summer. While forb emergence peaked at the end of the first summer, it was followed by a great mortality at the beginning of the next autumn.

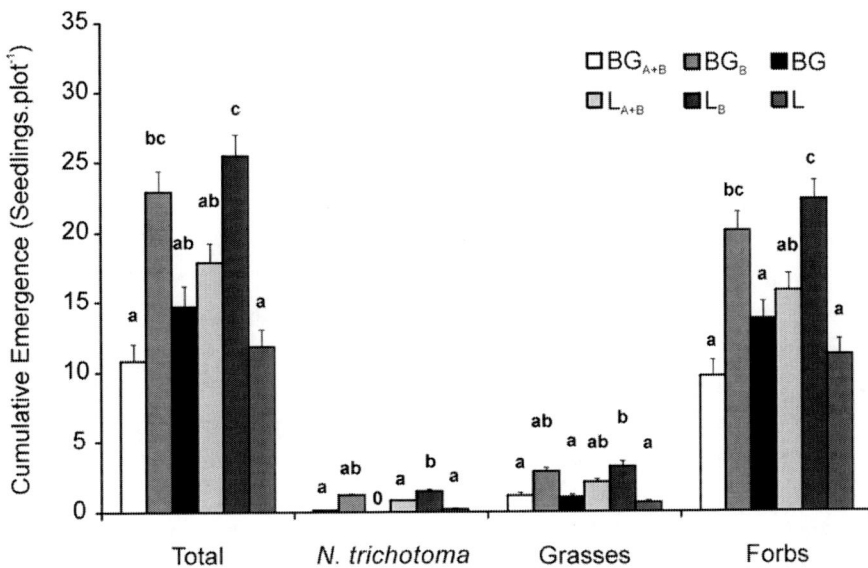

Figure 4. Cumulative seedling emergence of different vegetation groups in the grazed area at the end of the study period (two years). Different letters indicate statistically different mean values. BG: microsites with bare ground. BG-R: microsites with bare ground and soil removal. D: microsites covered with dung. L: microsites covered with litter. L-R: microsites covered with litter and soil removal. V: microsites with natural vegetation.

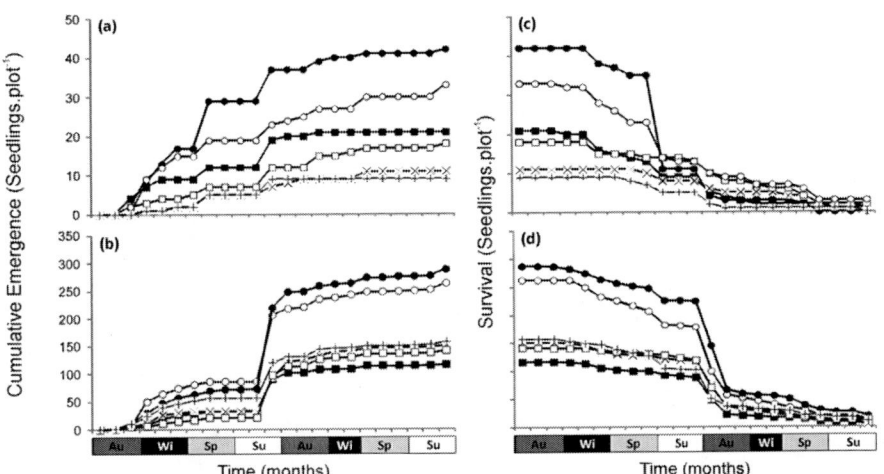

Figure 5. Cumulated emergence and survival of grass (a and c) and forb (b and d) seedlings during the study period in the grazed area. Au: Autumn. Sp: Spring. Su: Summer. Wi: Winter. Filled symbols: treatments with litter application. Empty symbols: treatments without litter application. Treatments: (○●) with soil removal, (■□) only without soil removal, (x) microsites covered with dung and (+) microsites with natural vegetation.

Discusion

The present work shows that microsite conditions are an important factor in determining emergence, and subsequent recruitment of seedlings. The increase in emergence in ungrazed areas is affected by competition with established vegetation, while in grazed areas it depends on soil removal or litter cover.

Results were partially consistent with the hypothesis, which suggests that emergence is controlled by biotic factors in ungrazed areas. Both, grass and *N. trichotoma* emergence increased with the reduction of competition from established vegetation, which was not observed for forb emergence. For grass species, it has been reported that competition with established vegetation reduces seedling emergence [14,57-59]. The same pattern was found for other species [12,60-62]. Forb species seem to have a persistent soil seed bank in the study area and their presence is abundant in excluded areas [35]. Nonetheless, their presence in above-ground vegetation is reduced in the same excluded areas [33]. Under these conditions, forb emergence was considerably lower than that on grasses, regardless of the level of competition or litter application. Probably, these species require disturbances of higher intensity that may stimulate the germination of seeds and emergence of seedlings. Reproduction by seeds in perennial grasslands is important when high intensity disturbances take place [63]. In the study area, there are several opportunistic forb species (i.e., annuals, non-native) that may require this kind of disturbances to complete their life cycle, being unable to germinate or emerge when dominant vegetation has not been affected.

Litter addition in the excluded area did not affect seedling emergence, although it increased forb seedling survival. Generally, the main effects of litter are to improve microclimatic conditions of temperature and humidity [6,64,65], favoring the emergence of seedlings. But in the excluded area, the presence of a dense vegetation stratum might have a similar effect [1,66], protecting the soil surface and reducing the air circulation over it. Probably these effects of above-ground vegetation, in or near the plots, partially explain the lack of grass responses to litter application.

In the grazed area, improvement of microsite conditions increased seedling emergence, which was consistent with the proposed hypothesis. Grass emergence increased with litter application. This might improve microsite conditions, reducing the thermal amplitude and increasing soil moisture [9,67,68]. Litter presence has generally inhibited species with small seeds [68,69]; and most of the dominant grass seeds in the study area are relatively large (>1 mm diameter, A. Loydi, pers. obs.). This might be relevant since land use [70,71] and climate changes [72-74] may induce accumulation of litter in many grasslands around the world, affecting vegetation dynamics. On the other hand, soil removal largely increased the emergence of seedlings of all studied groups. In general, soil removal favors seedling anchorage and improves soil structure, promoting an easier water infiltration, aeration and penetration of roots [25], favoring seedling germination and emergence. Soil compaction by cattle trampling has promoted erosion [29], reduced available water and oxygen [25], and the recruitment of new individuals [30,75]. This suggests that compaction associated with cattle grazing could limit colonization by seeds. Nevertheless, superficial soil removal has provided appropriate microsites for seedling emergence [40,76,77]. In this way, grazing may have a dual effect, favoring seedling emergence by soil removal, and reducing seedling emergence by creating bare-ground compacted soil patches, which showed the lowest emergence values in the study period. Nevertheless, none of these treatments improved seedling survival during the study period. Forb seedlings emerged massively after a rain peak at the end of summer, but died shortly after this germination pulse. This suggests that this vegetation group has germination pulses during benign weather conditions, which is typical for ruderal species [63]. While grass emergence was gradual during winter, emerged seedlings died as a result of the stressful conditions (i.e., high temperature and evapotranspiration) typical of the summer period in the area [45].

Microsite colonization by new individuals depends primarily on seed availability [68]. Sowing of *N. trichotoma* overcome this limitation for seedling emergence of this species. Overall, response of *N. trichotoma* was similar to emergence of natural seedlings (i.e., higher emergence with reduction of competition in excluded areas, and with soil removal treatments in grazed area), which showed that shortage of seeds could be a limiting factor for many species. However, availability of safe sites also showed to have an important role in the establishment of new individuals.

In natural grasslands dominated by species with sexual reproduction, availability of seeds and safe sites are the primary determinants of species regeneration. Grazing may affect negatively seed availability [78-81] and safe sites for seedling establishment [5,82]. However, exclusion of grazing could have a similar effect on vegetation, particularly affecting the availability of safe sites. One possible solution for Pampean grasslands in Ventania mountains might be an intermediate disturbance level that does not limit seed regeneration whether by seed shortage or lack of safe sites for establishment, favoring in this way an increased plant diversity [31].

Acknowledgment

This work was funded by the Consejo Nacional de Investigaciones Científicas y Tecnicas (CONICET) and Universidad Nacional del Sur, Argentina. Roberto A. Distel and Sergio M. Zalba provided helpful comments to a former version of this manuscript.

References

[1]   S. Janeček and J. Lepš, *Acta Oecol.* 28, 141 (2005).

[2]   N. Hölzel, *App. Veg. Sci.* 8, 115 (2005).

[3]   P. J. Grubb, *Biol. Rev.* 52, 107 (1977).

[4]   J. L. Harper, *J. Ecol.* 55, 247 (1967).

[5]   J. E. Kinloch and M. H. Friedel, *J. Arid Environ.* 60, 163 (2005).

[6]   T. W. Donath and R. L. Eckstein, *J. Ecol.* 96, 272 (2008).

[7]   S. Xiong and C. Nilsson, *J. Ecol.* 87, 984 (1999).

[8]   B. Boeken and D. Orenstein, *J Veg. Sci* 12, 825 (2001).

[9]   R. L. Eckstein and T. W. Donath, *J. Ecol.* 93, 807 (2005).

[10]   E. Ruprecht, J. Józsa, T. B. Ölvedi and J. Simon, *J Veg. Sci* 21, 1069 (2010).

[11]   A. G. van der Valk, *Aquat. Bot.* 24, 13 (1986).

[12] B. L. Foster and K. L. Gross, *Ecology* 78, 2091 (1997).

[13] N. Hagenah, H. Munkert, K. Gerhardt and H. Olff, *Plant Ecol.* 201, 553 (2009).

[14] A. S. Moretto and R. A. Distel, *Austral Ecol.* 23, 419 (1998).

[15] T. W. Donath, N. Hölzel and A. Otte, *Biol. Conserv.* 130, 315 (2006).

[16] R. L. Eckstein, *New Phytol.* 166, 525 (2005).

[17] P. Ryser, *J Veg. Sci.* 4, 195 (1993).

[18] S. Xiong, M. E. Johansson, F. M. R. Hughes, A. Hayes, K. S. Richards and C. Nilsson, *J. Ecol.* 91, 976 (2003).

[19] P. E. Hulme, *J. Ecol.* 84, 609 (1996).

[20] L. M. Martin and B. J. Wilsey, *J. Appl. Ecol.* 43, 1098 (2006).

[21] A. Altesor, G. Piñeiro, F. Lezama, R. B. Jackson, M. Sarasola and J. M. Paruelo, *J Veg. Sci* 17, 323 (2006).

[22] M. D. Bertness and R. Callaway, *TREE* 9, 191 (1994).

[23] E. S. Bakker and H. Olff, *J Veg. Sci* 14, 465 (2003).

[24] X. Dai, *J Veg. Sci* 11, 715 (2000).

[25] I. E. Bassett, R. C. Simcock and N. D. Mitchell, *Austral Ecol.* 30, 827 (2005).

[26] M. J. Crawley, in *Seeds: The ecology of regeneration in plant communities. 2nd. ed.*, CABI International, Wallingford, UK 167 (2000).

[27] J. E. Malo and F. Suárez, *Oecologia* 104, 246 (1995).

[28] J. J. Drewry, K. C. Cameron and D. Buchan, *Aust. J. Soil Res.* 46, 237 (2008).

[29] M. Kirby, *Soil Till. Res.* 93, 472 (2007).

[30] D. Alameda and R. Villar, *Soil Till. Res.* 103, 325 (2009).

[31] A. Loydi and R. A. Distel, *Ecol. Austral* 20, 281 (2010).

[32] A. E. de Villalobos and S. M. Zalba, *Acta Oecol.* 36, 514 (2010).

[33] A. Loydi, S. M. Zalba and R. A. Distel, *Plant Ecol. Evol.* (in press).

[34] J. L. Frangi and O. J. Bottino, *Rev. Fac. Agron.* 71, 93 (1995).

[35] A. Loydi, S. M. Zalba and R. A. Distel, *Acta Oecol.* (in press).

[36] H. Olff and M. E. Ritchie, *TREE* 13, 261 (1998).

[37] J. Liira and K. Zobel, *Plant Ecol.* 146, 185 (2000).

[38] K. A. Hovstad and M. Ohlson, *Plant Ecol.* 204, 33 (2009).

[39] I. Špačková and J. Lepš, *Folia Geobot.* 39, 41 (2004).

[40] R. Mayer and B. Erschbamer, *Basic Appl. Ecol.* 12, 10 (2011).

[41] J. Burgos and A. Vidal, *Meteoros* 1, 3 (1951).

[42] J. Burgos, in *Flora de la Provincia de Buenos Aires*, Colec. Cient. INTA, Buenos Aires 33 (1968).

[43] D. Cappannini, C. O. Scoppa and J. Vargas Gil. in *Reunión Geol. Sa. Australes Bonaerenses.* 203 (CIC, Prov. Buenos Aires).

[44] J. L. Frangi, N. E. Sánchez, M. G. Ronco, G. Rovetta and R. Vicari, *Bol. Soc. Arg. Bot.* 19, 203 (1980).

[45] C. A. Pérez and J. L. Frangi, *J Range Manag.* 53, 518 (2000).

[46] M. D. Barrera and J. L. Frangi, *Ecotropicos* 10, 161 (1997).

[47] A. A. Medina, *Bosque* 28, 234 (2007).

[48] M. B. Coughenour, *Annals Missou. Bot. Gar.* 72, 852 (1985).

[49] D. Bilenca and F. Miñarro, *Identificación de Áreas Valiosas de Pastizal (AVP) en las Pampas y Campos de Argentina, Uruguay y sur de Brasil*, Fundación Vida Silvestre Argentina, Buenos Aires (2004).

[50] W. H. Hudson, *The naturalist in La Plata*, M Dent and Sons, London (1929).

[51] W. K. Lauenroth, *Ecol. Austral* 8, 211 (1998).

[52] A. L. Scorolli and A. L. Cazorla, *Wildl. Res.* 37, 207 (2010).

[53] A. L. Scorolli, A. C. Lopez Cazorla and L. A. Telera, *Mastozoo. Neotr.* 13, 255 (2006).

[54] M. N. Smorzeñuk. *Fidelidad al área de actividad de las tropas-harenes de caballos cimarrones en el Parque Tornquist luego de un manejo poblacional*, Universidad Nacional del Sur, (2008).

[55] J. H. Zar, *Biostatistical analisis*, 4th Edition edn, Prentice Hall, Upper Saddle River, New Jersey (1999).

[56] J. A. Di Rienzo, F. Casanoves, M. G. Balzarini, L. Gonzalez, M. Tablada and C. W. Robledo, *InfoStat*, UNC, Córdoba, Argentina (2011).

[57] E. Haugland and M. Tawfig, *Grass Forage Sci.* 56, 193 (2001).

[58] G. Liu, P. Mao, Y. Wang and J. Han, *Ecol. Res.* 23, 197 (2008).

[59] G. X. Liu and J. G. Han, *Rangeland Ecol. Manage.* 60, 624 (2007).

[60] T. E. Adams, P. B. Sands, W. H. Weitkamp and N. K. McDougald, *J Range Manag.* 45, 93 (1992).

[61] D. R. Gordon, J. M. Menke and K. J. Rice, *Oecologia* 74, 533 (1989).

[62] G. N. Harrington, *Ecol.* 72, 1138 (1991).

[63] J. P. Grime, *Plant strategies, vegetation processes, and ecosystem properties*, 2nd Edition edn, John Wiley and Sons, LTD, Chichester, UK (2001).

[64] E. S. Deutsch, E. W. Bork and W. D. Willms, *Plant Ecol.* 209, 135 (2010).

[65] N. L. Fowler, *Am. Midl. Nat.* 115, 131 (1986).

[66] B. Kruk, P. Insausti, A. Razul and R. Benech-Arnold, *J. Appl. Ecol.* 43, 227 (2006).

[67] J. M. Facelli and S. T. A. Pickett, *Bot. Rev.* 57, 1 (1991).

[68] M. Fenner and K. Thompson, *The ecology of seed*, Cambridge University Press, Cambridge, UK (2005).

[69] E. M. Everham, R. W. Myster and E. Van De Genachte, *Am. J. Bot.* 83, 1063 (1996).

[70] D. Moog, P. Poschlod, S. Kahmen and K. F. Schreiber, *App. Veg. Sci.* 5, 99 (2002).

[71] F. Quétier, S. Lavorel, W. Thuiller and I. Davies, *Ecol. Appl.* 17, 2377 (2007).

[72] S. Díaz, S. Lavorel, F. De Bello, F. Quétier, K. Grigulis and T. M. Robson, *PNAS* 104, 20684 (2007).

[73] C. E. Owensby, J. M. Ham, A. K. Knapp and L. M. Auen, *Global Change Biol.* 5, 497 (1999).

[74] W. J. Parton, J. M. O. Scurlock, D. S. Ojima, D. S. Schimel and D. O. Hall, *Global Change Biol.* 1, 13 (1995).

[75] A. K. Skinner, I. D. Lunt, P. Spooner and S. U. E. McIntyre, *Austral Ecol.* 34, 698 (2009).

[76] P. J. Grubb, *Biol. Conserv.* 10, 53 (1976).

[77] R. J. Hobbs and H. A. Mooney, *Oecologia* 67, 342 (1985).

[78] M. B. Bertiller, *J Veg. Sci* 3, 47 (1992).

[79] T. G. O'Connor and G. A. Pickett, *J. Appl. Ecol.* 29, 247 (1992).

[80] M. Ortega, C. Levassor and B. Peco, *J. Biogeogr.* 24, 177 (1997).

[81] M. Sternberg, M. Gutman, A. Perevolotsky and J. Kigel, *J Veg. Sci* 14, 375 (2003).

[82] J. M. Bullock, in *Seeds. The ecology of regeneration in plant communities*, 2nd Edition edn, CABI Publishing, Wallingford, UK 375 (2000).

In: From Seed Germination to Young Plants

Editor: Carlos Alberto Busso

ISBN: 978-1-62618-653-8

© 2013 Nova Science Publishers, Inc.

Chapter 8

Germination Traits of the Native *Hyalis argentea* (Asteraceae)

Julia Camina[*1], *Elian Tourn*[2,3], *Ana Andrada*[2]
and Cecilia Pellegrini[2]

[1] Dept. Biología, Bioquímica y Farmacia,
Universidad Nacional del Sur (UNS), Bahía Blanca, Argentina
[2] LabEA, Dept. Agronomía, UNS, Bahía Blanca, Argentina
[3] CIC (Comisión de Investigaciones Científicas Pcia. Buenos Aires),
La Plata, Argentina

Abstract

The germination requirements of native species are key factors in the ecological restoration of degraded ecosystems. A variety of plant seeds are affected by dormancy. This process provides the advantage of spreading survival during unfavorable years. Therefore, the knowledge of processes and environmental factors that control germination can aid in decision making on how to use species for restoration. This study evaluated the germination responses of *Hyalis argentea* cypselas to different photoperiod conditions.

The existence or not of a germination inhibitor located in the pappus of the cypselas was also investigated. Cypselas were gathered randomly from a plant population in the semiarid Pampa. They were exposed to four germination treatments during 18 days at 25°C under different light regimes: (i) darkness and (ii) long or short photoperiods (this latter treatment using cypselas with-iii- and without-iv- pappus).

The percentage of germination was 14% for long photoperiods and an average of 77% for darkness and short photoperiods, with and without pappus. Our results suggest that *Hyalis argentea* is a non-dormant species that germinates during autumn, taking advantage of soil resources over other species that germinate during the spring in the semiarid Pampa region.

---

[*] E-mail address: juliacamina@yahoo.com.ar.

Introduction

The process of ecological restoration involves recovering or re-establishing an ecosystem that has been degraded, damaged or destroyed. These changes can be caused by either abrupt (damage and destruction) or gradual events that occur over longer time periods (degradation). Both types of processes involve a large loss of ecological diversity and integrity and deterioration of the physical environment that contains it [1]. Although ecosystems have the capacity to recover themselves, this natural phenomenon takes too long. Therefore, the practice of re-vegetation has been proposed as an alternative to the natural process, establishing the necessary conditions for the continued development of the environment. The use of native plants has been proposed to recover the local plant community structure [1-4]. The choice of species for re-vegetation is a critical issue in this practice. Besides knowing the species traits such as the availability of fruits or the feasibility of using nurseries, it is equally important to bear in mind certain characteristics that facilitate their handling (i.e., the root system, tolerance to salinity, etc.). Dalmasso (2007, 2010) [2,3] proposed a species selection index (SSI) for ecosystem restoration that considers all of these attributes for the optimum selection of the species. As part of the restoration project in Malargüe-Mendoza, he determined that the favourable species for restoration of the natural vegetation were those with a SSI higher than 18. Furthermore, he proposed that the usable vegetation had to consist mostly of perennial species. This is because they not only have a wide plant coverage year-long and various reproductive cycles but also ensure a more advanced succession stage and a faster recovery of the area. Another key factor in the restoration process is to know the characteristics of the native flora. It is always useful to be familiar with the phenology of the species that inhabit the area, their reproductive and seed dispersal strategies, the presence or absence of seed banks, and the characteristics necessary for germination and establishment of a target species in the areas to be restored [5].

The dormant stage in plants is the seed. Perennials do not necessarily depend on seeds for persistence and establishment, however in annual plants the seed stage is the connection between generations. Rees (1994) [6] showed that as the longevity of the species increased, natural selection drove against seed dormancy, although selection effects were slight. Consequently, there is not a marked relationship between the life cycle and the capacity of the seed to either germinate or not in a specific environment.

A dormant seed is unable to germinate under suitable conditions of temperature, humidity and oxygen [7, 8]. Dormancy is usually tested by exposing the seeds to various ranges of temperature in the presence or absence of light [9]. Recently, a conceptual model that negatively related the degree or level of dormancy of a species with its capacity to colonize new habitats was proposed [10]. The model suggests that it is possible to obtain a suitable approximation to the germination response of a seed by testing the favourable thermal and osmotic potential ranges, and the sensitivity of the seed to the "dormancy terminating factors" (e.g., light). Seeds not only possess basic requirements of water, oxygen and temperature to germinate, but also depend on factors such as light and/or the presence of nitrates in the soil. Light can stimulate germination and finalize the dormancy stage, and together with temperature could be an indicator of the depth at which the seed is located before germinating [8]. As an adaptive advantage, dormancy spreads the possibility of survival over time during unfavourable years [6], and results in the adaptation of the species to the habitat in which it develops [9-11]. The fact that an Angiosperm persists in a natural environment, implies that it has germinated, established and reproduced successfully [12]. Species belonging to the Asteraceae family are examples of Dicotyledons that germinate in a wide range of conditions (non-dormant species). Although cooling seems to help germination in some genera within this family, this pretreatment is not strictly necessary. This is the case, for example in the genera *Baccharis*, *Chrysanthemum*, *Grindelia*, *Helipterum*, *Heterotheca*, *Senecio*, *Taraxacum*, *Wyethia* [13-20].

Hyalis argentea D. Don ex Hook and Arn. var. *latisquama* Cabrera, called "olivillo", is a native Asteraceae, common in the dunes of central and western Argentina. It is found associated with *Sporobolus rigens* (Trin.) E. Desv., "junquillo", and *Panicum urvilleanum* Kunth, "tupe", making up the characteristic community of the coastal dunes (Figure 1a). It is a shrub up to 1m in height, with long rhizomes, and striated and branching stems. Its leaves are simple, sessile, lanceolated, alternated and trinervated (Figure 1b). The entire surface of the stem is cover by hairs, which give the plant its characteristic green-silver colour. The violet, dimorphic flowers are arranged in capitula at the end of the branches (corymb). Each capitulum is composed of four to five marginal, bilabiated flowers and a central actinomorphic, 5-lobed, tubular flower (Figure 1c). Flowering of this species extends from December to February [4, 21] with a peak between late December and early January. The fruit is a cypsela with a pappus formed by numerous white hairs [22] (Figure 1d). Weight of a thousand seeds is 27.11 ± 0.06 g (mean $\pm$ 1 standard error; J. Camina, unpublished data).

Figure 1. (a) Native population of *Hyalis argentea* in Pehuen-co, Argentina; (b) details of the leaves; (c) a capitulum and (d) a cypsela.

Several studies on *H. argentea* have been done in Argentina. Dalmasso (2010) [3] evaluated its importance as a species for the re-vegetation of degraded and desert environments. The "olivillo" was also proposed as one of the species to use in xeriscaping: an alternative to landscaping, with native species that involves a more efficient use of the drinkable water [4]. Melissopalynological studies have shown the presence of *H. argentea* pollen in honeys coming from the maritime seaboard counties of the south western Buenos Aires, Argentina [23]. The pollen grains of this species characterise the honeys of this region together with other pollen types such as those of *Cyclolepis genistoides* D. Don [21, 24]. This species has been proposed as endangered with low risk in the province of Buenos Aires [25]. The importance of this species has been enhanced from a pharmacological point of view since a series of secondary metabolites that could be involved in the inhibition of the activity of the HIV were found in *H. argentea* var. *argentea* [26]. This work proposes expanding the available knowledge of the fruit characteristics of *H. argentea*, since it seems to be a native species with great potential for developing in different areas. We evaluated the cypselas germination responses to long and short photoperiods, and if there is a physical dormancy effect associated to some inhibitor, possibly located in the dispersion structures (*pappus*) of the fruit.

Materials and Methods

Study Site and Plant Material

Hyalis argentea grows naturally in certain areas of the Bahía Blanca county. It is located within the phytogeographical province of the Espinal, in the Caldén District, characterised by xerophilous forest vegetation. In this region, it is common to find genera such as *Prosopis*, *Geoffroea*, *Condalia*, *Larrea* and grasses like *Stipa*, *Piptochaetium* and *Panicum*. The climate is temperate dry with mean annual temperatures between 15-20°C.

Figure 2. Natural population of *Hyalis argentea* in bloom, in Bahía Blanca, Argentina.

Figure 3. Cypselas: four mature and healthy, and two vanes (arrow).

The mean annual rainfall is roughly 600 mm, subject to a large variation according to the zone. The predominating topography is plain with some low hills up to 1200m [27]. The soil is classified as typical haplocalcic, composed mainly of fine sands over rough and rock pebbles or more ancient sandy silt material; the pH is roughly 7 [28, 29]. The study material was gathered during the fructification period of the species (late February to early March) from plants of a natural population grown at the field of the Agronomy Department (Universidad Nacional del Sur: 38°41′36,12″ S, 62°15′06,17″W; Figure 2). The field was randomly walked over (5230 m²) and around 700 cypselas were gathered from several different individuals. During the collection, special attention was taken to collect fruits only from standing organisms that were mature, non-vane, and healthy (undamaged by herbivores, mechanical factors, etc.) (Figure 3). Samples from a single genic pool were stored in a plastic container at room temperature.

Germination Test

Four treatments were conducted at 25°C for germination tests: (i) short photoperiod (8:16 hours light/darkness - SP); (ii) long photoperiod (16:8 hours light/darkness - LP); (iii) darkness (D), and (iv) the possible physical dormancy within the SP (Masini and Dalmasso, pers. comm.): with this purpose, cypselas with (WP) and without *pappus* (WOP) were used. The choice of a short photoperiod arose from previous observations of the optimum of other common species in the area (Tourn, pers. comm.). The balance of light/darkness hours mirrors the average hours of light/darkness registered in spring (LP) and autumn (SP) [30].

Three plastic Petri dishes lined with cotton and Whatmann paper filter over it were used for each treatment. Each dish had 25 cypselas from which the *pappus* was removed manually (Figure 4a). In the case of WP treatment, cypselas were intact (Figure 4b). The Petri dishes were placed inside a growth chamber built out of medium-density fibre wood (85 cm x 85 cm x 90 cm) and covered with an asphalt membrane. It allowed protecting the wood from humidity and having the aluminium to reflect the light. The temperature of the chamber was kept constant for the duration of the test. A thermostat and a 2,000 W Panasonic heating fan.

The temperature was fixed at 25°C because some authors have reported that the optimum germination, for some Asteraceae species, is achieved in a range between 20 and 30°C [13-17, 20]. Three smaller chambers, made out of polycarbonate and covered with aluminium, were placed transversally inside the growth ones. Inside two of the small chambers, 20 W Osram fluorescent lamps achieved 3,000 lux intensity. This setup allowed fixing the three different photoperiod regimes. The third chamber remained under darkness. The cypselas were checked daily and irrigated with distilled water. At every observation date, the Petri dishes were rotated so that each sample unit received the same conditions. A cypsela was considered germinated when the radicle of the embryo was observed to burst out of the seminal tegument (Figure 4-insert). Each germinated seed was removed from the Petri dishes. Testing lasted two and a half weeks; after five consecutive days with no germination activity the study was concluded. It is remarkable that several observations during the experiment suggested that these cypselas were not viable (e.g. fungus growth, exudates, collapse, etc.) [31].

Figure 4. Petri dishes with cypselas (a) without *pappus* treatment (WOP), and (b) with cypselas with *pappus* treatment (WP). The insert shows the radicle emergence during the germination process.

Measurements and Statistical Analyses

The germination percentage, the time to reach 50% germination (t_{50}), and the initial time (t_i) or moment that germination started were calculated for each treatment. Results were compared using one-way ANOVA; mean separation, after F tests were significant, was made using the Bonferroni test at the 1% level of probability.

Results

The percentage of germination of the cypselas under each light regime showed significant differences ($F_{3, 7} = 54, 09$; $p < 0,01$) between those under LP (17%) and the rest of the treatments (over 74% for SP-WP/WOP and D) (Figure 5). It was also observed an identical germination response of cypselas with and without *pappus*.

Similar results to those of the germination percentage were obtained for the t_{50}. For the SP and D treatments, t_{50} took four days meanwhile for LP it was achieved on the eight day after germination started (Figure 6).

Regarding t_i, with exception of the LP that was out of phase by a week (t_i = day 8), the cypselas of *H. argentea* began to germinate immediately (t_i = day 2) (Figure 6).

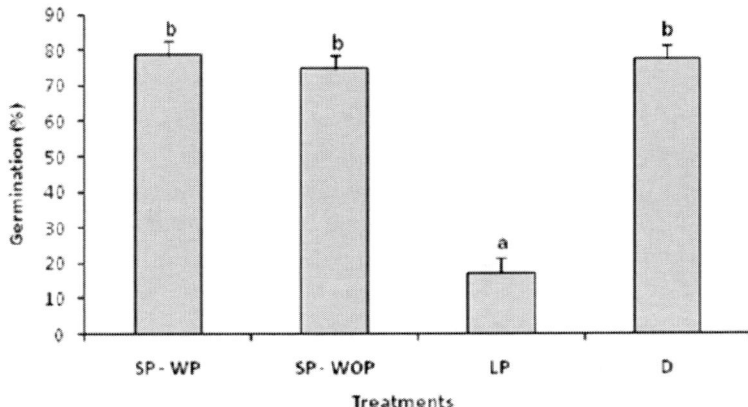

Figure 5. Germination percentage of *Hyalis argentea* cypselas under four treatment conditions (SP = short photoperiod; LP = long photoperiod; D = darkness; WP = with *pappus*; WOP = without *pappus*).

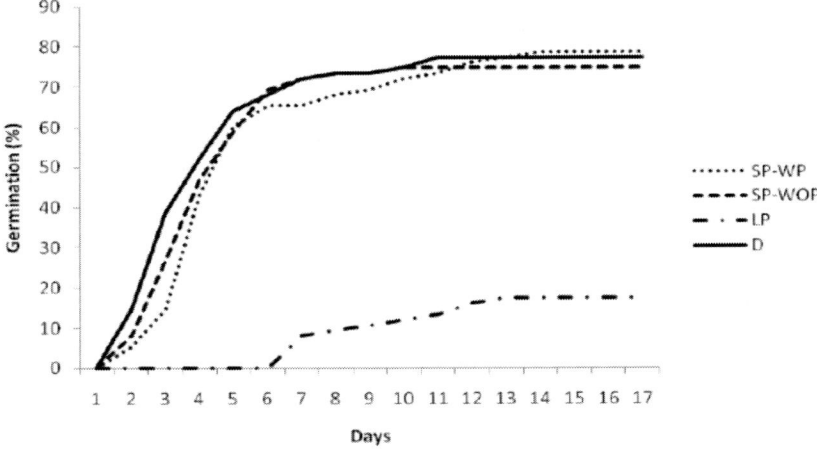

Figure 6. Cumulative germination response of *Hyalis argentea* cypselas under light conditions of spring (LP: long photoperiod) and autumn (SP: short photoperiod with-WP- and without *pappus*–WOP-, and D: darkness).

Discussion

Knowing and understanding the ecophysiological characteristics of the dispersion structures of Angiosperm offer advantages when establishing guidelines for improving degraded environments across different regions of the world. This information also helps to understand how species establish and survive in different ecosystems. The germination traits of any species are often understood as an adaptation of that species to the natural environment where it persists [7, 9, 12]. The possibility of great vegetative reproduction guarantees the persistence and colonization of the species in periods where the development, biological surrounding and climatic characteristics are favourable.

Seed dormancy may be due to an inability to germinate during favourable conditions as a consequence of an incomplete development of the embryo -innate dormancy- and/or the presence of some structures in the seminal cover that prevent the exchange of fluids with the medium -physiological dormancy- [7, 9, 11]. Understanding species dormancy has direct influences on practical applications, like predicting sowing models to avoid weed presence; handling native populations, or using native plants as usable crops [10].

The germination percentage of *Hyalis argentea* was similar to that of other Asteraceae species around the world [12, 14, 15, 19, 20, 32]. Our results are comparable to studies done in other natural populations in Argentina. Populations of *H. argentea* in southern Mendoza, where the species achieved potential germination values similar to those reported here (71%), showed a notably lower and more variable reproductive success than in our study (28%, 35 % and 41%; Dalmasso, pers. comm.). No differences were found between seeds with and without *pappus* in our study, consequently we can state that *Hyalis argentea* cypselas germination is not affected by chemical compounds in the *pappus.* This is certainly an advantage since plant-produced seeds would not have a constraint for germination as long as they are under a short photoperiod and even darkness.

There are few reports that evaluate the t_i parameter, in other words, the time the species takes to germinate, in the Asteraceae. The germination percentage, the germination pattern and potential of over 20 species of the temperate-humid forest in Chile was analysed [17]. The only member of the Compositae was *Baccharis magallanica*. This species presented an immediate germination, and a synchronic germination pattern. Therefore, the authors suggested that this species would not form part of the seed bank in Chile. *Hyalis argentea* presents a very similar strategy to the native species of Chile. This germination pattern is a good strategy to colonize dune environments, where *H. argentea* lives, since a fast germination reduces the chances of physical damage by wind erosion and damaging actions by insects or nocturnal mammals. The fact that the "olivillo" showed immediate germination also indicates that the species lacks innate dormancy. However, it remains the possibility that the species was dormant when it was gathered (February-March), and that dormancy was overcome during storage [5]. Species of the Compositae family require a post-harvest period so that embryo development is finished and summer water stress is avoided [5, 14, 31]. Therefore, we must test the possibility of innate dormancy with studies of germination performed immediately after harvest in this native species of the semiarid Pampa region.

Factors that are related with seasonal changes, e.g., soil temperature, and those that more directly indicate the favourable conditions for germination, like germination initiation factors, may affect seed dormancy [10]. In this chapter we emphasized testing one of the most important factors that affect germination: light. If light is limiting germination, seeds do not germinate until some external factor uncovers them. The underlying mechanism is controlled by a series of phytochromes, which make the individual increasingly sensitive to light, and decreasingly dormant as they activate [8-10]. Similar results in the germination response to light and dark conditions have been found in *Crysothamnus nauseosus* [15] and *Heterotheca subaxillaris* [14]. However, positive responses to light as a regulator of dormancy have been observed in *Taraxacum* sp. [20] and *Boltonia decurrens* [12]. The ecological interpretations that underlie this phenomenon are related to the capacity of identifying areas devoid of vegetation, to the fallow times during tillage, and to the capacity of measuring seed burial depth [10]. The observed response when exposing *H. argentea* to different photoperiods was unexpected, considering that this species lives in environments where marked seasonal conditions are present. Firstly, the species germinated both in the presence of light and in its absence. This would point out that the "olivillo" is insensitive to light, resulting in an advantage when seeds are completely covered or exposed during the movement of the dunes by wind. Secondly, *H. argentea* had its best germination performance during the season of the year when days shorten. Germinating in autumn would allow the seedlings to build a suitably mature and deep root system capable of developing a great capacity to compete and survive during the summer. It is noteworthy that *H. argentea* has large reserves in the cotyledons which will most likely help persistence of the species under unfavourable conditions (Figure 7).

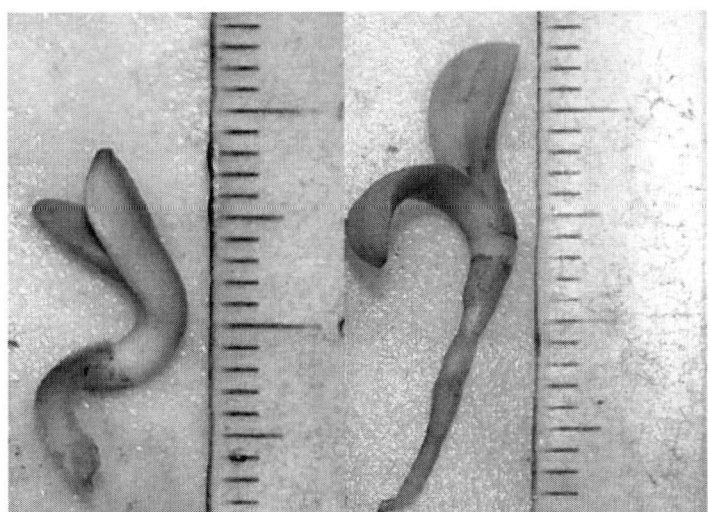

Figure 7. *Hyalis argentea* var. *latisquama* seedlings, with epigeous cotyledons.

The size and weight of the seed are variables that explain the dispersion patterns of the species. Plant species are subject to a constant trade off between producing large amounts of small seeds, which have good chances of long-distance dispersal and the opportunity of arriving into various and/or new habitat types, or few large seeds, with a better capacity for emergence and higher chances of survival [33, 34]. The seed weight has been associated with different processes such as growth rate of the seeds, t_i, reproductive cycle, photoblastism, and dispersal capacity. As a member of the Asteraceae family, *Hyalis argentea* presents an airborne dispersal type. While in the south of Buenos Aires province fruits reach an average weight of 27 mg, they achieve average values of 5 mg in Malargüe (northern Mendoza province). This weight difference could be due to the fact that in Malargüe the conditions are truly adverse for the species, whereas in the vicinity of the sea, environmental conditions are much more stable and favourable, allowing the plant to translocate and accumulate photosynthates in the dispersal structures. Leishman et al. (1995) [33] suggested that the variation in seed weight was not associated to the settlement conditions of the location. Therefore the variation in seed size of the "olivillo" could depend more on the growing conditions of the mother plant than on the settlement conditions of the seedling. In any case, further research should focus in understanding the factors that affect the variability of weight among fruits in this species.

The advantage of culturing a native species for production is that it already persists in the ecosystem, and has a co-evolutionary advantage over other cultured or introduced species. In comparison to *Chenopodium* sp., *Amaranthus* sp. or some species of legumes, the levels of total crude protein in *H. argentea* cypselas (9.87 ± 1.01%, J. Camina, unpublished data) are somewhat low, but the total lipids (39.95 ± 2.67%, J. Camina, unpublished data) are within standard parameters [35-37]. These results point to regard *H. argentea* as a possible complement to livestock diet, since it is also palatable [2, 3, 38]. However, several studies that complement this information are needed before considering the "olivillo" as a forage species for livestock grazing.

Conclusion

Hyalis argentea var. *latisquama* is a non-dormant species, like a large number of Asteraceae around the world. This species germinates during autumn, taking advantage of soil resources that might contribute to its settlement before other species that germinate during spring in the semiarid Pampa region. This type of pattern allows us to better understand the settlement mechanisms of the species of this family around the world. Based on the nutritional composition of the cypselas of this native species, we consider that the "olivillo" might be a good forage resource for livestock grazing. However, further studies are needed to test this hypothesis.

References

[1] Society for Ecological Restoration International (SER), Tucson, Arizona. Available: www.ser.org (2004).

[2] A. D. Dalmasso, Tesis doctoral, Universidad Nacional de Córdoba, Argentina (2007).

[3] A. D. Dalmasso, *Bol. Soc. Arg. Bot.* 45, 149 (2010).

[4] A. D. Dalmasso, R. Candia and C. Ganci, *Bol. Ext. Cient.*, ADIZA-CONICET, Mendoza, Argentina (2008).

[5] L. E. Commander, D. J. Merritt, D .P. Rokich and K. W. Dixon, *J. Arid Environ.* 73, 617 (2009).

[6] M. Rees, *Am. Nat.* 144, 43(1994).

[7] J. L. Harper, *Population biology of plants*, Academic Press, London (1977).

[8] W. E. Finch-Savage and G. Leubner-Metzger, *New Phytol.* 171, 501 (2006).

[9] C.C. Baskin and J. M. Baskin, *Seed. Ecology, Biogeography and Evolution of Dormancy and Germination.* San Diego, Academic Press, (1998).

[10] D. Batlla and R. L. Benech-Arnold, *Plant Mol. Biol.* 73, 3 (2010).

[11] J. A. Figueroa, P. León-Lobos, L. A. Cavieres, H. Pritchard and M. Way, *Fisiología Ecológica en Plantas. Mecanismos y Respuestas a Estrés en los Ecosistemas*, EUV, Valparaíso (2004).

[12] C. C. Baskin and J. M. Baskin, *Am. Midl. Nat.* 147, 16 (2002).

[13] N. T. Mirov, *Ecology* 17, 667 (1936).

[14] J. M. Baskin and C. C. Baskin, *Bull. Torrey Bot. Club* 103, 201(1976).

[15] M. A. Khan, N. SankhUr, D. J. Wehcr and E. D. McArthur, *Great Basin Nat.* 47, 220 (1987).

[16] E. Jurado and M. Westorby, *Aus. J. Ecol.* 17, 341 (1992).

[17] J. A. Figueroa, J. J. Arrnesto and J. F. Hernandez, *Rev. Chil. Hist. Nat.* 69, 243 (1996).

[18] V. L. Negrin and S. M. Zalba, *Bol. Soc. Arg. Bot.* 43, 261 (2008).

[19] A. C. A. Masini, Tesis de grado, Universidad Nacional del Sur, Bahía Blanca, Argentina (2011).

[20] J. Luo and J. Cardina, *Weed Res.* 52, 112 (2012).

[21] A. Forcone and A. Andrada, *Flora melífera de las regiones Pampeana - Austral y Patagónica – Andina*, EdiUNS, Bahía Blanca (2006).

[22] A. Cabrera, *Flora de la provincia de Buenos Aires. Parte VI (Compositae)*, Colección Científica del INTA. Buenos Aires (1963).

[23] A. Valle, A. Andrada, E. Aramayo, M. Gil and S. Lamberto, *Spanish J. Apicult. Res.* 5(2), 172 (2007).

[24] A. Valle, A. Andrada, E. Aramayo and S. A. Lamberto, *Invest. Agr. Prod. Protec. Veg.* 10, 375 (1995).

[25] G. Delucchi, *APRONA Bol. Cient.* 39, 19 (2006).

[26] M. I. Ybarra, S. A. Borkosky, W. A. N. Catalan. C. M. Cerda-Garcia-Rojas and P. Joseph-Nathan, *Phvtochemistry* 44, 479 (1997).

[27] Á. L. Cabrera, *Fitogeografía de la República Argentina. Bol. Soc. Arg. Bot.* 14, 1 (1971).

[28] N. Amiotti, M. C. Blanco and L. F. Sánchez, *Catena* 43, 137 (2001).

[29] C. Saint Pierre, Tesis Magister, Universidad Nacional del Sur, Bahía Blanca, Argentina (2001).

[30] M. S. Sakcali and M. Serin, *EurAsian J. BioSci.* 3, 107 (2009).

[31] E. Imbert, *Can. J. Bot.* 77, 508 (1999)

[32] B. Montes and A. Long, *I Congr. Latinoam. (IV Argentino) Conserv. Biodivers.* Tucumán, Argentina (2010).

[33] M. R. Leishman, M. Westoby and E. Jurado, *J. Ecol.* 83, 517 (1995).

[34] C. Furey, G. Funes, M. R. Cabido and F. Biurrun, *XXV Reunión Arg. Ecología*, Lujan, Argentina (2012).

[35] L. A. Bello Pérez, J. Solorza Feria, M. L. Arenas Ocampo, A. Jiménez Aparicio and M. Velázquez del Valle, *Agrociencia* 35, 459 (2001).

[36] F. Hevia, M. Berti, R. Wilckens and C. Yévenes, *Agrosur*, 30, 24 (2002).

[37] M. Freyre, E. Astrada, C. Blasco, C. Baigorria, V. Rozycki and C. Bernardi, *Cien. Tec. Alim.* 4(1), 41 (2003).

[38] W. A. Muiño, *Ethnobotany Res. Appl.* 8, 219 (2010).

In: From Seed Germination to Young Plants
Editor: Carlos Alberto Busso

ISBN: 978-1-62618-653-8
© 2013 Nova Science Publishers, Inc.

Chapter 9

Physiological Processes of Adaptation to Water Deficit on Seedlings of Three Native Grasses in Semiarid Argentina

Roberto Brevedan[*1], *María Fioretti*[2], *Sandra Baioni*[2]
and Claris Cabeza[†2]

[1] CERZOS (CONICET), Bahía Blanca, Argentina
[2] Departamento de Agronomía, Universidad Nacional del Sur (UNS),
Bahía Blanca, Argentina

Abstract

Continuous grazing with high stocking rates has been used for many years in the semiarid Caldenal. Overstocking and overgrazing led to a rapid deterioration of the natural resources, with desertification of vast areas as a result. Perennial, warm season forage grasses (e.g., *Digitaria californica*, *Pappophorum vaginatum* and *Trichloris crinita*) can improve animal production at the beginning of summer, when cool season grasses became unproductive. Seedling establishment is very sensitive to water deficit, and is a critical phase in pasture production. We evaluated the seedling response of those three species to water deficit using a series of reliable laboratory techniques.

Introduction

The Caldenal is located in the temperate, semiarid region of central Argentina in the ecotone between the cultivated, humid Pampa to the east and the arid Monte to the west. It has a highly variable distribution of rainfall (400-600 mm), which is mainly concentrated in

[*] Email address: ebreveda@criba.edu.ar.
[†] *In memoriam.*

the spring and summer. Temperatures vary between very high (42°C) and very low (-15°C) values. Soils are mostly Mollisols, with a petrocalcic hard pan horizon at a depth of 0.5 to 1.0 m. Grass cover is approximately 70% of the soil surface.

The primary activity in the Caldenal is cattle production on natural vegetation, and there may rarely be some cultivated pastures of weeping lovegrass (*Eragrostis curvula* (Schrad.) Nees) or annual crops (wheat, sorghum, etc.).

The production system is characterized by very low levels of technology and management: continuous grazing on unimproved rangelands, minimal livestock management. The tendency is to over-use and under-invest in the region, and as a consequence the area is degrading rapidly.

Forage from natural rangeland provides a major component of the animal feed resources. Pressure from the increasing animal populations has led to over-stocking and overgrazing. There are changes on vegetation structure due to the selective removal of palatable species [1]. The valuable grasses disappeared and those of medium value diminished, which allowed the multiplication of non-forage herbaceous plants and shrubs. Overgrazing is affecting top soil properties in the region. An increased bulk density and penetration resistance have been reported [2].

Water availability is the dominant factor that regulates biological activity, and soil water content is the single most important factor to determine the seedling establishment success of warm season grasses.

Plants are most vulnerable to adverse environmental conditions, mainly to water deficit, during germination and seedling growth. That is, community structure and ecosystem functioning in the Caldenal depends not only on livestock management but also on other adverse factors. In this work, we considered one of the major stress factors that limit plant growth: water deficit.

Growth and production of rangeland grasses is predominately limited by available water supply. The effects of water deficit on range grasses are particularly pronounced during germination and early seedling growth. Seeds can survive when exposed to wetting and drying cycles, but once germination started, the roots appear and cells become vacuolated resulting in tissues that are much more susceptible to injury from dehydration.

Lack of soil moisture is one of the major sources of mortality in seedlings, when they switch from internal reserves to external sources of support, such as water. Because water is the most limiting factor to grass forage production in semiarid regions, plants must possess water-conserving traits to prolong the water supply.

The problem of establishment is especially acute under semiarid conditions, where the soil surface is wetted only infrequently and irregularly, and the rate of evaporation is high. In such circumstances a seedling must compete with the process of atmospheric drying which rapidly reduces moisture of the surface soil layers.

The seedling phase goes from the time of germination until the plant is autotrophic. However, it is often extended to the point when the plant is considered as established. Grass seedlings are considered established when they have developed adventitious roots of sufficient length and number, and sufficient leaf area to assure adequate water and nutrient supply to sustain growth [3].

Soils in dry regions are usually subjected to rapid drying and crusting. This interferes with grass seedling emergence [4] and increases seedling susceptibility to other stresses in addition to drought.

The establishment phase is critical in the growth cycle of plants, as it determines the density of the stand obtained, and influences the degree of weed infestation and its eventual success.

Basic developmental events include radicle emergence, development of the primary or seminal roots, and shoot extension and emergence throughout the soil. The seminal root system supplies water to the developing seedling for a brief period of time. The permanent adventitious root system forms at the coleoptilar node and soon replaces the seminal system. This transition is a crucial stage in seedling development. The establishment of the adventitious root system requires moist soil during 5 to 7 days [5-7].

Water relations of seedlings during establishment are complex. They involve a transition from air-dry seeds, with a very low water potential (< -100 MPa), to seedlings with vacuolated cells which rarely have a water potential of < -1.0 to -1.5 MPa. Water uptake by seeds is driven by the water potential gradient between the dry seed and the soil.

Water deficits in the top 10 cm of the soil are critical to seedling establishment. Germination of the seed and development of seedling roots are dependent on appropriate precipitation after sowing or an adequate soil water storage. After germination occurs, seedling establishment is dependent on an adequate soil water content, and favorable environmental conditions for root development and penetration into the soil. Still another obstacle to seedling emergence is the tendency of certain soils to form a hard crust upon drying after wetting. Tender seedlings unable to emerge through the crust may lie repressed only a few millimeters below the soil surface. The extreme temperature fluctuations which take place at bare soil surfaces may further limit the chances of seedling survival in arid soils. Species capable of successful production in these regions must possess drought tolerance at the seedling as well as at later stages. It should be stressed the importance to have reliable screening tests to compare the response of different warm season range grasses to water deficit. Attempts to find selection factors for improving yield and survival of plants during drought (e.g., osmotic adjustment, leaf waxes, etc.) often did not produce the expected results. Drought responses of plants are very interactive and a change in one characteristic may be compensated by other characteristics [8]. Various attempts have been made to define specific chemical or physiological characteristics that are indicative of drought resistance. Up to date, we do not know of definitive responses of these reactions. Finding forage species to fill the forage deficit periods, and successfully establishing these species from seeds, are critical needs for producers. The following three warm season, native grasses were included in our study:

Trichloris crinita (Lag.) Parodi

The species belongs to the tribe Chlorideae. Perennial, rhizomatous and stoloniferous with flattened tillers forming large clumps. It has a wide area of distribution in semiarid and arid zones: it is common in southern United States, Mexico, Paraguay, Bolivia and Argentina. It is a desired grass by livestock, and considered good forage for their protein content and nutritional value [9, 10].

It sprouts in September and blooms from November to February. Low temperatures and frosts during winter (April, May) make it enter into a dormancy stage. Seed dispersal occurs from January to August. It has a high animal preference in the green stage. It is considered as excellent forage because of its high nutritional value and high dry matter production. It makes it a keystone species of pasture management where it has the potential to be dominant [11]. However, it presents difficulties for introduction into cultures as a result of having large deciduous caryopses [12]. This summer growing perennial grass is well adapted to water deficiency conditions, and is tolerant to grazing by domestic animals. It seems a suitable species for using in processes of land rehabilitation [13].

Pappophorum vaginatum Buckley

The species belongs to the tribe Pappophoreae. Perennial, caespitose with erect or decumbent shoots, lives on sandy-stony, calcareous or alkaline soils, being a natural forage often eaten by animals [9]. It is in the south of Brazil and Bolivia, Uruguay; in Argentina, it extends from northern and central Argentina to northern and central Patagonia. It is also a component of plant communities in the southern United States and Mexico.

Digitaria californica (Benth.) Henr.

Digitaria is a morphologically diverse genus distributed in variable habitats in tropical and warm temperate regions, and estimated to comprise from 170 to 230 species [14].

The species belongs to the tribe Paniceae. Perennial, tufted, with short rhizomes 20-70 cm tall. It was described for the southern and western United States, and grows in tropical and subtropical regions of Mexico, and Central and South America. In Argentina, it is located in the NE and in the Andean provinces, from Jujuy to Mendoza, Buenos Aires, La Pampa and Río Negro. It is located at high altitudes and in sandy soils at sea level [9].

This summer growing perennial grass forms a clump of stems reaching up to one meter height. The branching root system can reach up to one meter soil depth. The leaves are usually short and narrow. The inflorescence is a dense, narrow panicle containing pairs of haired spikelets. It usually sprouts in September, flowers in November and disperses seeds from December to May.

Although it rarely becomes dominant, it is found in all the vegetation from the forests of black locust and white quebracho (in the plains of La Rioja) to the forest of Southern Caldén in San Luis. Animal preference for this species is high, especially in the dry winter. Its absence or scarcity in the plant community is an indication of its disproportionate use. It presents difficulties for introduction into the cultivation because of having large hairy, deciduous caryopses [9]. This species grows in a number of habitat types, including desert scrubs and shrublands, shrubsteppes, and savannas. It tolerates varying precipitation amounts and survives easily under drought conditions, becoming dormant at times, and regrowing quickly when rains return. Much of its growth occurs in the summer, after the spring and summer rain cycles.

Objective

To evaluate the seedling response to water deficit of three major warm season native grasses of the Caldenal: *Trichloris crinita* (Tcr), *Pappophorum vaginatum* (Pva) and *Digitaria californica* (Dca).

Seedling Emergence, Soil Texture and Moisture Level

The soil matric potential can be recognized as being of fundamental importance to seed germination. This is mainly because of three factors: (i) A direct effect on the energy required for the absorption of water by the seed; (ii) An indirect effect on the control of the contact area between the soil and the seed (hydraulic conductivity), and (iii) An indirect mechanical effect on the mechanical strength of the soil and the seed compression [15].

Often it has been demonstrated that seeds germinate in soil with a water chemical potential of 1500 J kg^{-1}, but the germination rate is low. These effects are associated with the degree of contact between both seed and soil, which is affected by the size of the soil aggregates and the soil water content. If the average diameter of soil aggregates exceeds 20% of the seed, the water absorption can retard germination even with high water contents in the soil [16].

Objective

To analyze the effects of soil texture and moisture level on seedling emergence.

Materials and Methods

Soil from three rangeland sites were selected according to textural differences as sandy (S), sandy loam (SL) and loamy (L). Their water retention values, obtained using a pressure plate (Soil Moisture Equip.), are indicated in Table 1.

Table 1. Soil water retention (%) in sandy (S), sandy loam (SL) and loamy (L) soils at different tensions (atm)

Soil	Tension (atm)			
	15	11	7	0.3
S	3.81	3.87	4.01	6.60
SL	5.73	5.74	6.88	11.83
L	8.26	8.32	9.60	16.96

The procedure used to wet the soil was as follows: soil was first dried for 24 hours at 105°C. Thereafter, the dry soil was distributed in an even layer on a sheet of 1 m^2 polyethylene. The calculated volume of water for each soil moisture level was added with a hand sprayer. Moistened soil was stored in polyethylene bags, and closed for 24 hours. Soil moisture content was determined gravimetrically. Plastic pots with lids were used. In each pot, 75 g of soil were placed from which 25 g were used to cover the caryopses. An alternating temperature of 25°C for 9 hours and 15°C for 15 h was used. During the experience, pots remained closed without additional water contributions. After 15 days, the number of emerged seedlings was recorded. A completely randomized design was used with six replicates of 25 caryopses each. ANOVA was applied after transforming the data by arcsine √x, being x the variable under study. Tukey's test was applied (p = 0.05) to compare treatment means.

Results

Initial moisture values obtained for each treatment are listed in Table 2. In *Trifolium repens* L., germination decreased as matric potential decreased. However, germination was correlated with soil texture. Germination percentages increased at finer soil texture [17]. Overall emergence under high water content in the soil did not allow bringing out the texture effect (Table 3). Thus in the studied species the values obtained in the three soils were not different among themselves. The medium moisture level resulted in responses that failed to establish a clear pattern of response. The low humidity level, which imposed the most limiting moisture condition, highlighted the effect of the finest texture in the species germination.

Table 2. Initial soil moisture (%) for sandy (S), sandy loam (SL) and loamy (L) soils at different tensions (atm)

Initial water content (%)			
	H0	H1	H2
S	6.3 e*	4.5 f	4.1 f
SL	11.6 b	8.4 d	6.8 e
L	14.7 a	10.8 b	9.1 c

*Values followed by the same letter are not statistically different (p=0.05) according to the Tukey's test.

Table 3. Seedling emergency percentage (%) for sandy (S), sandy loam and loamy (L) soils with different moisture levels in three grass species

	S			SL			L		
	H0	H1	H2	H0	H1	H2	H0	H1	H2
Tcr	87 a*	46 d	24 e	85 a	67 c	68 c	76 b	77 b	69 c
Pva	91 a	94 a	50 d	86 ab	83 b	63 c	58 cd	89 ab	76 b
Dca	78 b	91 a	38 e	64 c	59 cd	51 d	68 c	84 ab	56 d

*Values in a row followed by the same letter are not statistically different (p=0.05) according to Tukey's test.

Seedling Emergence and Low Soil Temperature

Seedling vigor is one of the most important factors for the establishment of pastures.

The rate and total quantity of emerged seedlings are important components in the reseeding of pastures [18]. Chilling injury results from exposure of seeds to low temperatures during the early stages of imbibition.

There is evidence that low temperatures increase the loss of solutes from seeds; this most likely indicates alterations in cell membranes [19]. It has been suggested that there is a good correlation between germination tests conducted under cool conditions and field germination [20].

Objective

To analyze the effects of low soil temperatures on seedling emergence.

Materials and Methods

One hundred grams of saturated soil was added to plastic pots with lids. After planting the caryopses, pots were placed in a growth chamber at 10°C for 7 days. At the end of that time period, they were transferred to a regime of alternating temperatures: 25°C for 9 hours and 15°C for 15 h. After 15 days, the number of emerged seedlings per pot was recorded. During the experiment, pots remained closed without additional water contribution. A completely randomized design was used with eight replicates of 25 naked caryopses each. ANOVA was applied after transforming the data by arcsine $\sqrt{x}$. Tukey's test was used to compare treatment means (p = 0.05).

Results

Two species groups were obtained (Table 4). The first group included Pva and Tcr with higher emergence levels after the cold treatment. In the second group, values were significantly lower in Dca than in Pva and Tcr.

Table 4. Seedling emergence percentage (%) under low temperatures in three grass species

	Emergency (%)
Tcr	72 a*
Pva	77 a*
Dca	19 b*

*Values followed by the same letter are not statistically different (p=0.05) according to the Tukey's test.

Seedling Survival under Soil Water Deficit

Seed germination is relatively insensitive to water stress. As a seedling develops, it becomes sensitive to desiccation, and irreversible changes occur with drying.

This susceptibility to loss of water may be due directly to some metabolic changes, but the increase in vacuolization of cells under development, could make them more sensitive to drying, and decreased cellular activity may result as a consequence of increased protoplasmic disruption [21].

Seedling survival, affected by environmental conditions, has limited usefulness.

This is because of practical problems for its measurement. A large number of seedlings and a wide range of environmental conditions are needed to determine correctly the relationship between drought and seedling death [22].

Objective

To determine the survival of seedlings under soil water deficit conditions.

Materials and Methods

Planting was conducted in 200 cm^3 pots filled with 150 g of air dry, sandy loam soil.

Between 20 and 30 caryopses of each species were placed in each pot. After watering, pots were placed in a growth chamber with 9 hours of light at 25°C and 15 hours of darkness at 15 °C.

Pots were watered daily. Fifteen days after sowing, plants were thinned to 6 plants per pot. Thirty days after sowing, irrigation was suspended for 12 days.

Thereafter, irrigation was restarted for 7 more days, when surviving plants were counted. A completely randomized design was used with sixteen replicates for each species.

ANOVA was applied after transforming the data by arcsine $\sqrt{x}$. The Tukey's test was applied to compare treatment means (p = 0.05).

Results

Seedling survival was greater in Tcr than in Pva and Dca (Table 5).

Table 5. Seedling survival percentage (%) with water deficit in three grass species

	Survival (%)
Tcr	83.3 a*
Pva	66.7 b
Dca	66.4 b

*Values followed by the same letter are not statistically different (p=0.05) according to the Tukey's test.

Seedling Growth under Water Deficit

Ecologists have reported that natural selection has favored those plants with physiological adaptations that allow their survival and growth based on their most efficient use of available water in natural grasslands under semiarid conditions. Water deficit affects virtually all physiological and biochemical processes in plants and in many cases also affect their anatomy and morphology [23]. The first effect of water deficit on plants is a delay in stem and leaf growth probably as a result of a reduction in their turgidity [23-25].

In seedling establishment, a good seedling vigor, which involves a fast growth of stems and roots, it is of equal or greater importance than germination rate. The rapid elongation of the seedling root is considered a prerequisite for successful establishment in pasture reseeding [26], especially in semiarid and arid environments where soil dries out quickly after rainfall [27-29].

A single cycle of water deficit should be easier to interpret than recurrent cycles. The nature of plant response should be compared under both conditions: during drought and after release from water deficit [30].

Objective

To evaluate the effects of water stress on shoot and root growth on seedlings exposed to a single water deficit period.

Materials and Methods

Pot Trial

The experiment was conducted in the greenhouse. Sowing was carried out in pots filled with 1.2 kg of soil. Thirty days after planting, the experiment was thinned leaving 20 plants per pot.

Two treatments were imposed: 1) control (C), in which irrigation was applied daily throughout the test period, and 2) water deficit (D), which was similar to the control during the first 60 days, and then watering was stopped during the next 10 days.

After this period, the test was finished. At the beginning and end of the 10-day-period, plant height and number of tillers per plant were registered in six plants per pot.

A factorial experiment in a completely randomized design was used with six replicates per treatment. ANOVA was applied after transforming the data by arcsine $\sqrt{x}$. To compare treatment means, the Tukey's test was applied ($p = 0.05$).

Test Tubes

The experiment was conducted in the greenhouse. Sowing was done in PVC tubes (6 cm diameter, 50 cm height) filled with fine sand. Fifteen days after sowing, seedlings were thinned to 4 per tube. At the same time watering began with nutrient solution at the rate of 20 cm^3 per tube during 3 days. The solution composition was 1M $Ca(NO_3)_2.4H_2O$, 1 M KNO_3, 1 M $MgSO_4.7H_2O$, 1 M KH_2PO_4, 0.03 M Fe-EDTA.

We used 7, 7, 2.8, 1.4 and 7 ml per liter of solution of these chemicals, respectively. Treatments were as follows: (i) Control (C), daily watering during the test period; (ii) water deficiency D1, suspension of irrigation after 26 days from sowing for 7 days, and (iii) water deficiency D2, irrigation was suspended after 19 days from sowing for 15 days. After 33 days, leaf water potential, and root and shoot dry weights (DW) were determined in each treatment.

The material was dried at 70°C for 24 hours. Water potential was determined on the last developed leaf using a pressure chamber. A factorial experiment in a completely randomized design was used with six replicates per treatment. ANOVA was applied after transforming the data by arcsine $\sqrt{x}$. The Tukey's test was used to compare treatment means (p = 0.05).

Results

Pot Test

Ten days after irrigation was suspended, no significant differences were observed in plant height relative to the control (Table 6). The number of tillers per plant was not affected by the drought period.

Plant height and number of tillers of Pva and Tcr were not affected after suspending irrigation during 10 days. It would take several cycles or a longer dry cycle to observe growth differences. In Dca plant height and number of tillers per plant were affected by water deficiency.

Table 6. Height and number of tillers per plant for control (C) and water deficit (D) treatments in three grass species

	Height		Tillers	
	(cm)		(tillers plant^{-1})	
	C	D	C	D
Tcr	24.20 a*	26.10 a	4.84 a	4.88 a
Pva	28.76 a	24.64 a	4.00 a	3.80 a
Dca	31.04 a	27.72 b	6.96 a	4.40 b

*Values for each parameter between treatments followed by the same letter are not statistically different (p=0.05) according to the Tukey's test.

Test Tubes

Treatment D1 and the control showed similar water potentials, and shoot and root weights (data not shown).

In contrast, significant differences were found for these parameters in the control (-0.13 MPa) and the D2 water deficit treatment (-0.28 MPa), regardless of the species considered. The highest shoot and root dry weights were shown by *D. californica* (Table 7). This species and *P. vaginatum* showed the highest root dry weights.

Suspension of the irrigation period during 7 days did not affect neither the water status nor the growth of the species under study.

After 15 days of water shortage, despite significant differences in growth of shoots and root it was not possible to establish differences in the relation shoot:root.

The observed differences were solely due to characteristics of each species. Shoot:root dry weight ratio was higher for Tcr (2.9) and Dca (2.7) than for Pva (1.3).

Table 7. Shoot (Sh) and root (R) dry weights (mg plant^{-1}) for water deficit (D2) treatment in the three grass species

	Sh	R
Tcr	41.87 b*	14.42 b
Pva	47.86 b	31.39 a
Dca	87.07 a	32.37 a

*Values between shoot and root followed by the same letter are not statistically different (p=0.05) according to the Tukey's test.

Seedling Water Status and Recovery

Water potential has been the most widely used measure of water deficit during the last decades. There is no doubt that the concept of water potential is well founded, and that the water potential gradient is of fundamental importance for water transport. However, water potential *per se* is not critical for physiological functions. The absolute water content is perhaps the most direct measure of a tissue water status. Uses of water deficit indicators might result confusing because of differences in size and dry matter content of the tissues. The use of relative water content avoids these difficulties, but requires an accurate measure of tissue water content at saturation or full turgor [31].

The stomata closure in leaves exposed to the sun results in an increase of foliar temperature, if other relevant factors as wind speed and vapor pressure remain relatively constant [32]. This increase in temperature can be detected with an infrared thermometer. The canopy temperature specifically responds to changes in plant water potential [33].

Objective

To evaluate the effects of water deficit, followed by a recovery period, on plant water status indicator parameters.

Materials and Methods

The experiment was conducted in the greenhouse. Sowing was done in pots filled with 2.3 kg of soil. Thirty days after planting the experiment was thinned leaving 20 plants per pot. There were three treatments: (1) Control (C) irrigation was applied daily throughout the test period; (2) Water deficit (D) similar to the control during 13 days, then watering was stopped for 10 consecutive days; and (3) Recovery (R) the same as D but after 10 days of suspended irrigation, it was resumed for a 3-day period. The following parameters were determined: water potential, relative water content and canopy temperature at the end of each treatment. Thereafter, irrigation was re-started and recovery of those parameters was determined after 3 days. Dry weight of the pots was recorded daily throughout the water deficit cycle (data not shown).

Leaf water potential was measured using a pressure chamber on the last expanded leaf. The leaf relative water content (RWC) was determined on fully expanded leaves using the Slavik technique [34]. Canopy temperature was determined with an infrared thermometer (Barnes Instatherm). Greenhouse air temperature and relative humidity were recorded using a thermohygrograph during the study.

A factorial experiment in a completely randomized design with 18 replicates was conducted. Ten replicates were used for measurements of leaf water potential, and 10 for determinations of relative water content and 15 for leaf temperature in each pot. ANOVA was applied after transforming the RWC data by arcsine √x. To compare treatment means, the Tukey's test was applied (p = 0.05).

Results

Water potentials were significantly lower in the water deficiency treatment (D) than in the control (C) in all three species (Table 8), the largest decline was observed in Tcr. Relative water content significantly decreased compared with the control in all three species (Table 8).

Once the irrigation was restarted, water potential and RWC increased in all three species, and values were equal to their controls after 3 days (Table 8).

Suspension of irrigation had a significant increase in canopy temperature (Table 9).

A dry period of 10 days was enough to significantly change all studied water status parameters. Differences among species in response to the observed variables were most often not detected. Although not quantified, it is interesting to note that while Tcr folded leaf blades in response to water deficit, Dca and Pva showed a marked curling of them. Finally, leaf water status was reestablished after a 3-day-recovery period. Also, it is interesting to note that 3-4 hours were sufficient to produce leaf unfolding after reestablishing irrigation.

Table 8. Water potential (MPa) and relative water content (RWC) (%) before and after a recovery period from plant water deficit in three grass species

	Before recovery				After recovery			
	Water potential		RWC		Water potential		RWC	
	(MPa)		(%)		(MPa)		(%)	
	C	D	C	D	C	R	C	R
Tcr	-0.27 a*	-1.55 c	96 a	33 bc	-0.30 b	-0.28 b	97 a	97 a
Pva	-0.29 a	-1.06 bc	93 a	39 b	-0.26 b	-0.15 a	96 a	96 a
Dca	-0.27 a	-0.77 b	92 a	23 c	-0.13 a	-0.17 a	97 a	98 a

*Values between treatments followed by the same letter are not statistically different (p=0.05) according to Tukey's test.

Table 9. Canopy temperature (°C) affected by water deficit in three grass species

	Canopy temperature (°C)	
	C	D
Tcr	26.6 b*	28.2 a
Pva	27.2 b	29.2 a
Dca	27.4 b	29.0 a

*Values followed by the same letter are not statistically different (p=0.05) according to the Tukey's test.

Seedling Membrane Integrity under Water Deficit

Tissue drought tolerance can be evidenced and measured through either physiological or metabolic functions [35]. A valid and functional test of drought tolerance could be related to the integrated response of the plant to a low level of organization (e.g. tissue growth) or to a unique attribute associated with basic cellular aspects or to a tissue response to stress [36].

Membrane stability is one of the most important components of drought tolerance [21]. The level of drought damage on cell membranes can be estimated by measuring the loss of electrolytes from the cells [36].

A nonlethal water stress applied to higher plants tolerant to desiccation might result in changes of cell ultrastructure. When water stress is removed, structures go back to normal in a few hours. Plants under lethal levels of water deficit are unable to reverse those changes producing an increase in the fragmentation of organelles and membrane structures [21].

Objective

To evaluate the effect of a water deficit period, and a subsequent recovery from it, on membrane stability.

Materials and Methods

The experiment was conducted in the greenhouse. Sowing was done in pots filled with 2.3 kg of soil. Thirty days after planting the experiment was thinned leaving 20 plants per pot. There were three treatments: (i) Control (C) irrigation was applied daily throughout the test period; (ii) Water deficit (D) similar to the control during 13 days, then watering was stopped for 10 consecutive days; and (iii) Recovery (R) the same as D but after 10 days of suspended irrigation, it was resumed for a 3-day period. The following parameters were determinated: water potential, relative water content and canopy temperature at the end of each treatment. Thereafter, irrigation was re-started and recovery of those parameters was determined after 3 days.

After 10 days of irrigation withdrawal, it was determined the cell sap electrical conductivity, and irrigation re-started thereafter. Three days later, that parameter was reassessed. In both cases the technique of Sullivan was used [37]. Damage percentage was defined as:

$$\text{Damage \%} = [(T_1/T_2) / (C_1/C_2)] \times 100$$

where T and C were the treated and control treatments, respectively. The subscripts 1 and 2 are used to identify the initial and final conductivity values, respectively [36]. Membrane integrity percentage was determined as: *100% - damage %*.

Measurements were made on ten fully expanded leaves per replicate. The material was collected in the morning between 5 and 6 A.M. There were ten repetitions per treatment.

A factorial experiment in complete randomized designed was applied with an arcsine √x transformation of the data. There were 6 replicates per treatment. For comparison of treatment means, it was used the Tukey's test ($p = 0.05$).

Results

Membrane structures in Pva and Dca were affected significantly by the degree of water deficit reached after 10 days of irrigation suspension (Table 10). A rehydration period of 3 days allowed restoring the maximum values of membrane integrity (data not shown). Tcr was not affected by the treatment.

Table 10. Membrane integrity percentage (%) in three grass species

	Membrane integrity percentage (%)	
	C	D
Tcr	100 a*	99 a
Pva	098 a	82 b
Dca	099 a	86 b

*Values followed by the same letter are not statistically different (p=0.05) according to the Tukey's test.

Seedling Nitrate Reductase Activity under Water Deficit

The absorption rate of nitrates is generally low in desert environments because of the low levels of soil available nitrogen. In these environments, mineralization and nitrification are typically low because these processes are dependent on humidity. Furthermore, during drought, leaf tissue nitrate levels can be toxic to animals, because synthesis, activity and stability of the nitrate reductase decrease with water deficit [38].

A high capacity for recovery of the nitrate reductase activity once water deficit is over might be advantageous. This is because it will reduce tissue nitrates to sustain protein synthesis. Also, it is highly likely that energy conservation is an essential requirement for survival during stress; therefore, a rapid loss of nitrate reductase activity could be a metabolic advantage under those conditions [39].

Objective

To evaluate the effects of a drought period, followed by a recovery period, on the nitrate reductase activity.

Materials and Methods

The experiment was conducted in the greenhouse. Planting took place in polyethylene pots filled with 2.3 kg of soil. Thirty days after planting, plants were thinned leaving 20 plants per pot.

The three treatments imposed were: (1) Control (C), irrigation was applied daily throughout the test period; (2) water deficit (D), similar to the control during 13 days, then watering was stopped for 10 consecutive days, and (3) Recovery (R), the same as D but after 10 days of suspended irrigation, it was resumed for a 3-day-period.

The nitrate reductase activity was determined on 0.5 g of fresh leaf material. It was placed in beakers with 10 cm^3 of incubation medium: 0.1 M KH_2PO_4, 0.2 M KNO_3 and 0.1% Triton X-100 (w/v), pH 7.0. Samples were incubated in a N_2 atmosphere for 30 minutes in darkness and continuous agitation. The nitrite content in the incubation medium was determined using a 1 cm^3 aliquot; then 2.5 cm^3 of 1% sulphanilamide in 2.8 N HCl were added, followed by 2.5 cm^3 of a solution of N-(1-naphthyl) dihydrochloride etielendiamino 0.02% (w/v) in ethanol 94% (v/v). Absorbance was read at 540 nm. Water potential was measured on the last expanded leaf using a pressure chamber. Samples for enzyme activity and water potential measurements were taken at noon.

A factorial experiment in a completely randomized design was used. There were six replicates per treatment. ANOVA was applied after transforming the RWC data by arcsine $\sqrt{x}$. The Tukey's test was applied to compare treatment means (p = 0.05).

Results

After 10 days of irrigation withdrawal, there was a significant reduction in leaf water potential in all three species (Table 11). The activity of the enzyme nitrate reductase was significantly lower on Pva and Tcr, while Dca was not affected (Table 11). After 3 days of restarting the irrigation, water potentials increased in Pva, Tcr and Dca, such as these values were similar to those in the controls (Table 11). The activity of the nitrate reductase was lower in the recovery than in the control treatment in Pva and Tcr, but not in Dca.

Ray and Sisson [38] reported that activity values of nitrate reductase were between 1.60 and 0.55 mg NO_2^-.g DW h^{-1} under leaf water potentials in the order of -2.7 and -0.3 MPa, respectively, in *Panicum coloratum* L.

The activity values obtained for the controls of Tcr (33.48 µg NO_2^-.100 mg DW h^{-1}) and Pva (48.52 µg NO_2^-.100 mg DW h^{-1}) were within the range of values found in *Lolium perenne* cv S24 [40].

Table 11. Water potential and nitrate reductase activity before and after a recovery period in three grass species

| | Before recovery | | | | After recovery | | | |
| | Water potential (MPa) | | NR activity (μg NO_2^-.100 mg DW h^{-1}) | | Water potential (MPa) | | NR activity (μg NO_2^-.100 mg DW h^{-1}) | |
	C	D	C	D	C	R	C	R
Tcr	-0.24 a*	-2.85 d	33.48 b	0.50 c	-0.29 a	-0.32 a	32.06 b	19.96 c
Pva	-0.25 a	-1.00 b	48.52 a	29.74 b	-0.36 a	-0.25 a	47.08 a	24.48 bc
Dca	-0.26 a	-2.34 c	9.94 c	6.50 c	-0.31 a	-0.32 a	12.04 d	10.73 d

*Values between treatments followed by the same letter are not statistically different (p=0.05) according to the Tukey's test.

Recoveries of 99.4% were observed in barley after 48 hours of rehydration. In maize seedlings there was a partial or total recovery after 24 hours [39]. In *T. crinita* and *P. vaginatum*, activity values were increased after recovery, but they still resulted significantly lower than values in the controls. In contrast, in Dca there was an increase in the enzyme activity values after recovery.

Seedling Acid Phosphatase under Water Deficit

Water stress can lead to a disruption of the solute compartmentalization within leaf cells in sensitive plants. The released acid hydrolases, previously latent, can have a profound destructive action, and most of the metabolic effects of water stress could be explained by such destruction [41]. The increased activity of hydrolytic enzymes can be assigned to both an increase in their solubilization and total activity [42].

Cytochemical and ultrastructural studies on stressed tissues indicate that activation of acid and alkaline lipases are responsible for the alteration of membranes and their decompartmentalization. Drought-resistant plants did not show such activation; it is then possible that they have more resistant compartments [41].

Objective

To evaluate the effect of a drought period followed by a period of recovery on the activity of acid phosphatase.

Materials and Methods

The experiment was conducted in the greenhouse. Sowing was done in pots filled with 2.3 kg of soil. Thirty days after planting the experiment was thinned leaving 20 plants per pot. There were three treatments: (1) Control (C) irrigation was applied daily throughout the test period; (2) Water deficit (D) similar to the control during 13 days, then watering was stopped for 10 consecutive days; and (3) Recovery (R) the same as D but after 10 days of suspended irrigation, it was resumed for a 3-day period. The following parameters were determinated: water potential, relative water content and canopy temperature at the end of each treatment. Thereafter, irrigation was re-started and recovery of those parameters was determined after 3 days.

The acid phosphatase activity was determined at the end of each treatment.

The phosphatase activity was determined on 0.5 g of fresh leaf tissue. Samples were homogenized in a mortar with a Tris-HCl buffer solution (pH 7.5), 0.3 M sucrose, 0.1% Triton-X100 (w/v) and 1 mM 2-mercaptoethanol. Thereafter, they were filtered and centrifuged, and 0.5 cm^3 were taken from the supernatant. This was incubated at 30°C in 0.5 cm^3 of p-nitro-phenyl-phosphate; 40 μmol cm^{-3} of 0.2 M acetate buffer (pH 5); mannitol 0.3 M, and 5 mM Na metabisulphite. Incubation was completed after the addition of 1 cm^3 of tampon solution buffer pH 10.7. Absorbance was read at 400 nm. Enzyme activity was expressed in μM paranitrophenol per 100 mg dry weight per hour [43].

A factorial experiment in a completely randomized design was used with six replicates per treatment. ANOVA was applied after transforming the data by arcsine $\sqrt{x}$. The Tukey's test was applied to compare treatment means (p = 0.05).

Results

After 10 days from the suspension of irrigation there was a significant increase in acid phosphatase activity in Pva and Tcr, but not in Dca (Table 12). Acid phosphatase activity under water stress increased 2 and 3 times, respectively, in Pva and Tcr. The level of acid phosphatase activity in Dca was not affected by the degree of water stress reached. In Pva the activity increased significantly as a result of water deficiency, resulting three days enough to restore activity values similar to those of the control, that was not the case for Tcr.

The phosphatase hydrolytic activity increased from 2 to 4 times relative to controls in leaf tissues of cotton stressed plants sensitive to drought [41].

During recovery, plant responses of the studied species were similar to those reported in wheat, where the acid phosphatase activity reached values similar to the watered controls after a rehydration period of 24 hours [44].

Table 12. Acid phosphatase (AP) activity (μM paranitrophenol.100 mg DW h^{-1})
before and after a recovery period in three grass species

	Before recovery		After recovery	
	C	D	C	R
Tcr	33.72 d*	107.28 c	27.12 b	140.40 b
Pva	179.10 b	384.84 a	335.46 a	367.92 a
Dca	20.58 d	36.06 d	40.14 b	32.76 b

*Values between treatments followed by the same letter are not statistically different (p=0.05) according to the Tukey's test.

Seedling Osmotic Adjustment

The metabolic activity of plants is linked to the maintenance of turgor pressure. Under drought conditions, loss of cell water causes an increase in the concentration of solutes in the cytoplasm. This results in a decreased cellular and tissue osmotic potentials that will affect the maintenance of a positive turgor. This is a purely passive mechanism.

Certain species, and cultivars of different species, are able to decrease their osmotic potential because of the migration of osmotic solutes from organs such as roots, stems and leaf blades. This process of osmotic adjustment, which is active, is a character of resistance [45].

In arid and semiarid regions, water deficiency occurs for varying periods and with different frequencies. As a result, it is important to know the degree of osmotic adjustment during stress, and its loss after such stress is removed. This information, together with data on other changes in water relation characteristics, will allow to gain a better understanding of the acclimation responses of stressed plants [46].

Objective

To evaluate the occurrence of osmotic adjustment after plants of the studied species are exposed to several cycles of water stress.

Materials and Methods

Seeds were sown in polyethylene bags filled with 2.3 kg of soil. Thirty days after planting, they were thinned leaving 10 plants per bag. Three treatments were imposed: (i) Control (C), irrigation was applied daily throughout the test period; (ii) Water deficit (D), similar to the control during 12 days; then plants were exposed to four cycles of water stress of 7 days each, and (iii) Recovery (R), same as D but plants were irrigated during two days after exposure to four cycles of water deficit.

The following parameters were determined at the end of each treatment: relative water content (RWC), osmotic and water potentials, and turgor pressure. RWC was determined at noon on 5 fully expanded leaves [34]. Osmotic potential was determined on leaf tissue collected at noon. Material was frozen in liquid nitrogen and kept in a freezer until measurement; determinations were made with an osmometer Wescor 5100B. Water potential measurements were made at noon on the last expanded leaf using a pressure chamber. Turgor pressure was calculated as the difference between the water and osmotic potentials.

A factorial experiment in a completely randomized design was used. There were six replicates for each treatment. RWC data were transformed by arcsine $\sqrt{x}$. For comparison of treatment means, it was used the Tukey's test ($p = 0.05$).

Results

At the end of the fourth cycle of water stress, the three species showed a significant decrease in RWC compared to control plants (Table 13). After 2 days of rehydration, however, RWC was similar between treatments The osmotic potential measured on plants after rewatering was significantly lower on water-stressed than control plants in Pva and Tcr. At the same time, Dca plants showed similar values between treatments.

Table 13. Relative water content percentage (%) before and after a recovery period in three grass species

	Before recovery		After recovery	
	C	D	C	R
Tcr	97.40 a*	63.48 c	98.26 a	98.88 a
Pva	97.68 a	88.14 b	98.40 a	98.77 a
Dca	96.38 a	60.61 c	97.91 a	97.13 a

*Values between treatments followed by the same letter are not statistically different (p=0.05) according to the Tukey's test.

Values of -1.25 and -1.44 MPa of osmotic potential have been reported in rice leaves of control plants of 31 days of age. Values of osmotic potentials between -1.6 and -1.8 MPa were determined in rice plants subjected to three cycles of drought, between 4 to 6 days, with a recovery period of 2 days prior to measurement [47]. The values obtained in this experiment were close to those observed in rice. The maximum decrease in osmotic potential under maximum turgor pressure in *Cenchrus ciliaris*, subjected to a drying cycle of five weeks, has been -0.66 MPa. The values obtained in Tcr (-0.45 MPa) and Pva (-0.47 MPa) were somewhat higher than those determined in *C. ciliaris*, but were within the range of values observed in *Centrosema* (-0.16 to -0.56 MPa) [48].

Turgor pressures were similar in the recovery treatment than in the controls in all study species (Table 14).

As a result of successive cycles of water stress, Pva and Tcr most likely acclimated to water stress by a significant decrease of the osmotic potential. At the same time, however, Dca did not show this resistance mechanism. Values of turgor pressure in Tcr and Pva were similar to those found in rice, both in the controls (1.25 to 1.35 MPa) and in rice plants subjected previously to successive cycles of water stress (1.54 to 1.65 MPa) [47].

Table 14. Osmotic and turgence potential after a recovery in three grass species

	Osmotic potential (MPa)		Turgor pressure	
	C	R	C	R
Tcr	-1.51 b*	-1.96 a	1.32 a	1.61 a
Pva	-1.50 b	-1.97 a	1.20 a	1.53 a
Dca	-1.23 b	-1.53 b	1.05 a	1.14 a

*Values between treatments within each species followed by the same letter are not statistically different (p=0.05) according to the Tukey's test.

General Discussion

Seedlings of Dca achieved the highest growth in both shoots and roots under water stress conditions among all study species. Furthermore, Pva was capable of duplicating its root dry weight at the same time, which will confer it an advantage at the establishment time. The greater growth in Dca was associated with higher water potentials. Meanwhile, Tcr, which presented the lowest water potentials, was the species most affected by drought in its growth. The three species increased leaf temperatures to the same level; at the stress level reached, there was no correspondence between leaf temperatures and water potentials.

These species also showed a high capacity to withstand desiccation without having a permanent damage in their membrane structures. The response of water status and membrane stability during the recovery from water deficit was rapid and complete.

The indicators of plant water status (water potential, RWC) used in this study showed that the plants usually recovered when drought stress was removed. However, the recovery of normal physiological functioning often is delayed [49]. Thus, the recovery rate will be an indication of the level of damage caused by water deficit and the varying strength of the material [50]. Enzymatic activity of Dca was unaffected by water deficit. At a given water deficit condition, there was not neither a significant reduction in the activity of the nitrate reductase nor a significant increase in the acid phosphatase activity. In contrast, water deficit caused a marked reduction in the activity of the nitrate reductase and an increase in the acid phosphatase activity in Tcr and Pva.

The two study enzymes responded similarly to water deficit, but they showed a different response during the recovery period to such stress. While the nitrate reductase did not regain its normal activity level, the recovery was higher in the acid phosphatase. Thus, we might assume that the level of damage on the nitrate reductase was more important than that on the acid phosphatase.

Osmotic adjustment results in an increased ability to maintain turgor pressure, and presumably mediated processes that could be maintained, despite reductions in rice plant water potential [47]. As a result of successive cycles of water stress, Pva and Tcr showed a significant decrease in osmotic potential, as well as a clear tendency toward higher values of turgor pressure than Dca. *Digitaria californica* did not revealed this mechanism of resistance.

References

[1]   E. A. Cano, H. D. Esterlich, M. Montes, B. Fernández and E. Morici, *Rev. Fac. Agron. UNL Pampa* 3, 11 (1988).

[2]   M. B. Villamil, N. M. Amiotti and N. Peineman, *Soil Sci.* 166(7), 441 (2001).

[3]   D. D. Wolf, J. A. Balasko and R. E. Ries, In: L.E. Moser, D. R. Buxton and M. D. Casler (Eds.), *Cool-season Forage Grasses*, p. 71-85, Agron. Monog. 34. ASA, Madison, Wi. (1996).

[4]   A. V. French and S. Clarke, *Trop. Grassl.* 27, 387 (1993).

[5]   A. M. Wilson and D. D. Briske, *J. Range Manage.* 32, 209 (1979).

[6]   P. R. Newman and L. E. Moser, *Crop Sci.* 28, 148 (1988).

[7]   P. R. Newman and L. E. Moser, *Agron. J.* 80, 383 (1988).

[8]   A. B. Frank, S. Bittman and D. A. Johnson, In: L.E. Moser, D. R. Buxton and M. D. Casler (Eds.), *Cool-season Forage Grasses*, p. 127, Agron. Monog. 34. ASA, Madison, Wi. (1996).

[9]   M. N. Correa, *Flora Patagónica*. Tomo VIII. Parte III. Gramineae. INTA, Buenos Aires (1978).

[10] INTA and UNLPam, *Inventario Integrado de los Recursos Naturales de la Provincia de La Pampa*, Buenos Aires (1980).

[11] J. L. Saenz, EEA San Luis, INTA, *Informativo Rural* 21, 6 (1985).

[12] G. Covas, *Agrarius* 2, 44 (1983).

[13] J. B. Cavagnaro and S. O. Trione, *J. Arid Environ.* 68, 337 (2007).

[14] W. D. Pitman, C. G. Chambliss and J. B. Hacker, In: L.E. Moser, D. R. Buxton and M. D. Casler (Eds.), *Cool-season Forage Grasses*, p. 715–744, Agron. Monog. 34. ASA, Madison, Wi. (1996).

[15] N. Collis-George and J. B. Hector, *Aust. J. Soil Res.* 4, 145 (1966).

[16] F. L. Milthorpe and J. Moorby, *Introducción a la Fisiología de los Cultivos*. Ed. Hemisferio Sur, Buenos Aires (1982).

[17] J. A. Young and R. A. Evans, *Agron. J.* 62, 743 (1970).

[18] D. L. Kittock and A. G. Law, *Agron. J.* 60, 286 (1968).

[19] J. D. Bewley and M. Black, *Physiology and biochemistry of seeds in relation to germination,* Vol. 1, Springer-Verlag, New York (1982).

[20] W. N. Rice, *Proc. Assoc. Offic. Seed Anal.* 50, 118 (1960).

[21] J. D. Bewley, *Annu. Rev. Plant Physiol.* 30, 195 (1979).

[22] A. M. Wilson and J. A. Sarles, *Agron. J.* 70, 231 (1979).

[23] R .W. Brown, In: R. E. Sosebee (Ed.), *Rangeland Plant Physiology,* pp. 97-140, Soc. Range Manage., Denver (1977).

[24] A. S. Crafts, In: T.T. Kozlowski (Ed.), *Water deficits and plant growth,* pp. 85-133, Vol. 2. Academic Press, New York (1968).

[25] T .C. Hsiao, *Annu. Rev. Plant Physiol.* 24, 519 (1973).

[26] A. P. Plummer, *J. Am. Soc. Agron.* 35, 19 (1943).

[27] C. J. Asher and P. G. Ozzane, *Plant Soil* 24, 423 (1966).

[28] N. H. Tadmor and Y. Cohen, *Crop Sci.* 8, 416 (1968).

[29] J. R. Mc William, R. J. Clements and P. M. Dowling, *Aust. J. Agric. Res.* 21, 19 (1970).

[30] C. T. Gates, In: T. Kozlowski (Ed.), *Water deficit and plant growth*, pp. 135-190, Vol. 2. T. Academic Press, New York (1968).

[31] K. J. Bradford and T. C. Hsiao, In: O. L. Lange, P. S. Nobel, C. B. Osmond and H. Ziegler (Eds.), *Physiological Plant Ecology. II Water Relations and Carbon Assimilation,* pp. 263-324, Heidelberg (1982).

[32] W. L. Ehrler and C. H. M. van Bavel, *Agron. J.* 59, 243 (1967).

[33] Ehrler, W. L., S. B. Idso, R. D. Jackson and R. J. Reginato, *Agron. J.* 70, 251 (1978).

[34] B. Slavik, *Methods of studying plant water relations,* Springer-Verlag, New York (1974).

[35] A. Blum, In: H. Mussell and R. C. Staples (Eds.), *Stress physiology in crop plants*, pp. 429-445, John Wiley and Sons Inc, New York (1979).

[36] A. Blum and A. Ebercon, *Crop Sci.* 21, 43 (1981).

[37] C. Y. Sullivan, In: N. G. P. Rao and L. R. House (Eds.), *Sorghum in Seventies*, pp. 247-264, Oxford and India House, New Delhi (1972).

[38] I. M. Ray and W. B. Sisson, *J. Range Manage.* 39, 531 (1986).

[39] S. K. Sinha and J. D. Nicholas, In: L. G. Paleg and D. Aspinal (Eds.), *The Physiology and Biochemistry of Drought Resistance in Plants,* pp. 145-169, Academic Press (1981).

[40] S. M. Smith and D. B. James; *Plant Soil* 68, 223 (1982).

[41] J. B. Vieira da Silva, In: O. L. Lange, L. Kappen and E. D. Schulze (Eds.), *Ecological studies. Analysis and Synthesis*, pp. 207-224, Vol. 19, Water and Plant Life, Springer Verlag, Berlin (1976.)

[42] J. B. Vieira da Silva, *Physiol. Veg.* 8, 413 (1970).

[43] K. Linhart and K. Walter, In: H. U. Bergmeyer (Ed.), *Methods of Enzymatic Analysis,* pp. 783-785, Academic Press, New York (1963).

[44] O. A. Jonas, M. C. Pereyra, A. D. Golberg and J. F. Ledent, *Cereal Res. Communic.* 18(4), 299 (1990).

[45] C. Renard, *Tropicultura* 1, 128 (1983).

[46] J. R. Wilson and M. M. Ludlow, *Aust. J. Plant Physiol.* 10, 15 (1983).

[47] J. M. Cutler, K. M. Shahan and P. L. Steponkus, *Crop Sci.* 20, 307 (1980).

[48] M. M. Ludlow, A. C. P. Chu, R. J. Clements and R. G. Kerslake, *Aust. J. Plant Physiol.* 10, 119 (1983).

[49] W. S. Iljin, *Annu. Rev. Plant Physiol.* 8, 257 (1957).

[50] A. T. Pham Thi and J. Vieira da Silva, *Plant Sci. Let.* 11, 121 (1978).

In: From Seed Germination to Young Plants
Editor: Carlos Alberto Busso

ISBN: 978-1-62618-653-8
© 2013 Nova Science Publishers, Inc.

Chapter 10

Plant Traits Contributing to the Performance of Native and Introduced Rangeland Grasses in Arid Argentina

*Yanina Torres[1], Carlos Busso[*1], Oscar Montenegro[2], Hugo Giorgetti[2], Gustavo Rodríguez[2] and Leticia Ithurrart[1]*

[1] Dept. Agronomía, Universidad Nacional del Sur
and CERZOS (CONICET), Bahía Blanca, Argentina
[2] Chacra Experimental Patagones,
Ministerio de Asuntos Agrarios, Argentina

Abstract

Rangelands from central Argentina characterize by insufficient warm-season, palatable perennial grasses to domestic livestock. Two palatable and water-stress tolerant cultivars (''Magnar'' and ''Trailhead'') of *Leymus cinereus* were introduced into Argentina with the purpose of increasing the warm-season forage offer. Some mechanisms involved in determining defoliation tolerance and competitive ability, and subsequent dry matter production, were investigated in the study genotypes. Studies were conducted within an exclosure to domestic livestock. Two severe defoliation treatments were applied in 2006/07 and 2007/08. Only *Pappophorum vaginatum*, the native genotype, produced and dispersed seeds. Natural water stress during the second study year advanced phenology of all three genotypes. This advancement, together with the high production of total, reproductive and daughter tillers in *P. vaginatum*, even under defoliation, might contribute to explain its greater abundance than the other native, warm-season perennial grasses in rangelands of central, arid Argentina. Although plant survival and forage production were lower ($p < 0.05$) on the introduced cultivars than in the native

[*] E-mail address: cebusso@criba.edu.ar.

genotype, *Leymus* genotypes showed greater (p<0.05) total (green + dry) leaf numbers/cm^2, total green sheath lenght/cm^2, and arbuscular mycorrhiza (AM) colonization percentages than *P. vaginatum*. Root length densities, total (green + dry) leaf length and leaf blade/sheath ratios were greater (p<0.05) or similar (p>0.05), but not lower, in the introduced than in the native genotype. Future research is needed to substantially improve plant establishment of *L. cinereus*, genotype which can be seeded by conventional drilling.

Introduction

Cattle production industry in 75% of continental Argentina, characterized by arid and semiarid areas, is based upon grazing of native vegetation [1]. Rangelands of central Argentina are characterized by the scarcity of warm-season, native perennial grass genotypes, palatable to domestic livestock [2]. In rangelands at the south of Buenos Aires Province, Argentina, *Pappophorum vaginatum* Buckey is the more abundant native, warm-season, C$_4$, palatable perennial grass [3].

Grazed grasses can be repeatedly defoliated to various intensities during their life cycle. Tolerance to defoliation is given by a rapid leaf area reestablishment after such event is produced [4]. Perennial grasses can reestablish photosynthetic tissues lost after grazing through production of new leaf blades and sheaths. These can grow from (i) undefoliated tillers, (ii) defoliated tillers which have their intercalary and/or apical meristems intact, and/or (iii) from axillary bud activation and subsequent new tiller production and growth [5].

The mechanisms involved in determining tolerance to defoliation and competitive ability in plant species include components of aerial and root growth [4]. For example, the association degree of roots with fungi which form the arbuscular mycorrhiza (AM) mutualism, root length density (cm root / cm^3 soil), and aerial and root biomass production [6]. Any disturbance, such as defoliation, which reduces those components, could limit regrowth capacity in perennial grasses [7].

Biomass production after defoliation of *P. vaginatum* plants exposed to various disturbances in previous years has been reported [8]. However, they evaluated this variable between prolonged measurement periods and/or did not compare simultaneously the response of defoliated and undefoliated plants. Research on other warm-season species in Argentina [*Pappophorum pappiferum* (Lam.) Kuntze, *Schizachyrium plumigerum* (Ekman) Parodi, *Bothriochloa springfieldii* (Gould) Parodi, *Digitaria californica* (Benth.) Henrard, *P. caespitosum* Fries, *Trichloris crinita* (Lag.) Parodi, *Setaria leucopila* (Scribn. and Merr.) K. Schum., *Diplachne dubia* (Kunth) Scribn.] have studied the effects of the origin location or various defoliation frequencies and intensities of these species on their persistence, nutritive value, phenology, height, inflorescence number, and dry matter production [9-12]. One important constraint in most of these works is that they did not mention if meristems in active growth remained on the plants after various defoliation treatments were applied. This is very important since the decreasing order of growth speed is intercalary meristerms > apical meristem > axillary buds [4]. Defoliation often produces an immediate root growth reduction [4]. Some grazing tolerant and competitive plant species have preferentially assigned carbon to aerial sinks (i.e., shoots) after defoliation [13]. This mechanism allows to rapidly reestablish the photosynthetic surface area lost to defoliation and the return to the root/shoot equilibrium [4]. However, plant responses to defoliation have been contradictory [14], and other studies report either lack of effects [7] or even greater root growth after defoliation [15]. Greater root length densities and arbuscular mycorrhiza (AM) root colonization have been associated with an increased nutrient acquisition in perennial grasses [16], mainly in poorly productive environments where soil resource competition is intense [17]. Plant species with lower root length densities have been shown to be more dependent on AM root colonization for soil resource acquisiton such as phosphorus [18]. Even though it could be expected that as much root length density as AM root colonization would contribute to tissue nutrient enrichment, both mechanisms appear to be alternative strategies [19]. Importance of the root system and its colonization by fungi forming AM has not yet been studied in *P. vaginatum* to explain potential differences in its competitive ability and defoliation tolerance. In native C_3 perennial grasses, root proliferation and root length density, and percentage AM root colonization were similar in defoliated and undefoliated plants, after a first defoliation which did not remove the apical meristems and a second defoliation 30 days later which removed them [20]. Several works in C_3 perennial grasses, native to central Argentina, have reported overcompensation in annual forage production when defoliations were applied early than when they did not occur during the growing season [21]. This has also been reported in other perennial grasses and herbaceous species [22]. Plant phenology can also be affected by the timing of defoliation [5]. Defoliation during late phenological stages (i.e., under high air temperatures and low soil moisture contents) has advanced the phenology of *Stipa tenuis* Phil. and *Piptochaetium napostaense* Speg. [21]. Research conducted over several years on the phenology of *P. vaginatum* [23] was realized under field exclosures to domestic animals, with no information on wheather defoliation might affect or not plant phenology of this species.

Plant production is partially determined by the photoassimilate distribution among various organs [24]. Studies on dry matter partitioning among various plant organs are scarce in general [25] and lacking in *P. vaginatum*. These studies are important because they can provide information of differences in dry matter distribution in aerial organs between species, with direct consequences in animal preference on the offered forage [26].

Needs of increasing cattle forage production during the warm season in rangelands of central Argentina is critical. This can be achieved via perennial grass species introduction into the region [27]. Severe water stress during the growing season is a constraint the introduced species might be exposed to [3]. However, results on productive performance and plant survival in native versus introduced species are scarce [28] and contradictory: there had been either increments or no differences or decreases in forage production and survival of introduced plant species [29-30]. Since native species of any given environment are adapted to the local conditions of such environment, it is expected a greater plant establishment and persistence, and a greater forage production, in the native than in the introduced species. Failures of introducing plant species in arid rangelands have been attributed to the lack of adequate precipitations [31]. These authors emphasized that appropriate management of native vegetation would be the best approach to achieve a good plant cover of palatable forage for livestock in those rangelands. Other authors [2] compared the productive performances of perennial grass genotypes coming from U.S. states other than Nevada [*Achnatherum hymenoides* (Roemer and J.A. Schultes.) Barkworth cvs. Paloma, Nezpar and Rimrock], with a local perennial grass ecotype [*Leymus cinereus* (Scribn. and Merr.) A. Löve cv. Gund] to restore degraded rangelands in Nevada. Introduced genotypes established successfully in the experimental plots, which were irrigated daily and excluded from grazing by domestic and wild animals. Measured parameters which contributed to biomass production were much greater in the local ecotype than in the introduced genotypes. Soil physicochemical characteristics are an important factor to consider in attempts of species introductions into degraded areas in addition to climate [32]. There are several examples where species introduction has contributed to increase forage supply in cultivated and rangeland areas of Argentina and other countries. For example, introduction of several perennial grass species: *Agropyron desertorum* (Fischer ex Link) Shultes from Eurasia into the Western U.S. rangelands; *Eragrostis curvula* (Schrader) Nees and *Tetrachne dregei* Nees, from South Africa into the cultivated area of the semiarid Pampas region, and into the central rangelands of La Pampa Province, Argentina, respectively, and *Cenchrus ciliaris* L. from Africa into Catamarca Province, Argentina [33-35]. *Leymus cinereus*, a native rangeland perennial grass from the western mountain region of the U.S., was introduced to evaluate the possibility of increasing the palatable forage supply for domestic livestock in rangelands of central Argentina during the warm-season. Commercial cultivars introduced of this species were 'Magnar' and 'Trailhead'. *Leymus cinereus*, a C_3 grass, is a sturdy, very drought tolerant species that often propagates by short rhizomes. It is adapted to grow on a variety of sites in dry climates during summer, also excellent to form a good soil structure and a good forage resource for domestic livestock [36]. 'Magnar' is a very sturdy and productive cultivar, with great longevity and drought tolerance [37]. It is potentially a forage species of great value [38]. 'Trailhead' is more drought tolerant than 'Magnar', and can suvive in areas with just 150 mm annual precipitation [37]. Several studies have been made on the morphophysiology, demography, plant growth and responses to grazing in *L. cinereus* [38-41]. However, research of this nature has not yet been conducted in Argentina. This work provided valuable information on the possibility to introduce *L. cinereus* cvs. 'Magnar' and/or 'Trailhead' in temperate-semiarid and arid rangelands of central Argentina. In addition, it contributed to determine some response mechanisms to defoliation, and some defoliation tolerance and competitive ability mechanisms in the study genotypes. Objectives of this study were (i) to evaluate components which contribute to determine (a) total annual forage production per plant [total tiller number per plant; tiller height, and numbers and total length of green and dry leaves, and new daughter tiller production], and (b) some competitive ability (root length density and percentage root colonization by AM) and defoliation tolerance traits (production of new tillers); (ii) to quantify the effects of early and midway defoliations during the growing season versus undefoliated controls on the parameters described under item (i); (iii) to measure leaf (blades versus sheaths) dry matter partitioning in *P. vaginatum* and *L. cinereus* cvs. 'Magnar' and 'Trailhead', and (iv) to determine plant survival at the end of the study on defoliated and undefoliated plants of all three genotypes.

Materials and Methods

Study Site

Studies were conducted during 2006, 2007 and early 2008 within an exclosure to domestic livestock in the Chacra Experimental de Patagones, Buenos Aires, Argentina

(40°39'49,7"S, 62°53'6,4"O; 40 m.a.s.l.), within the Phytogeographical Province of the Monte [42].

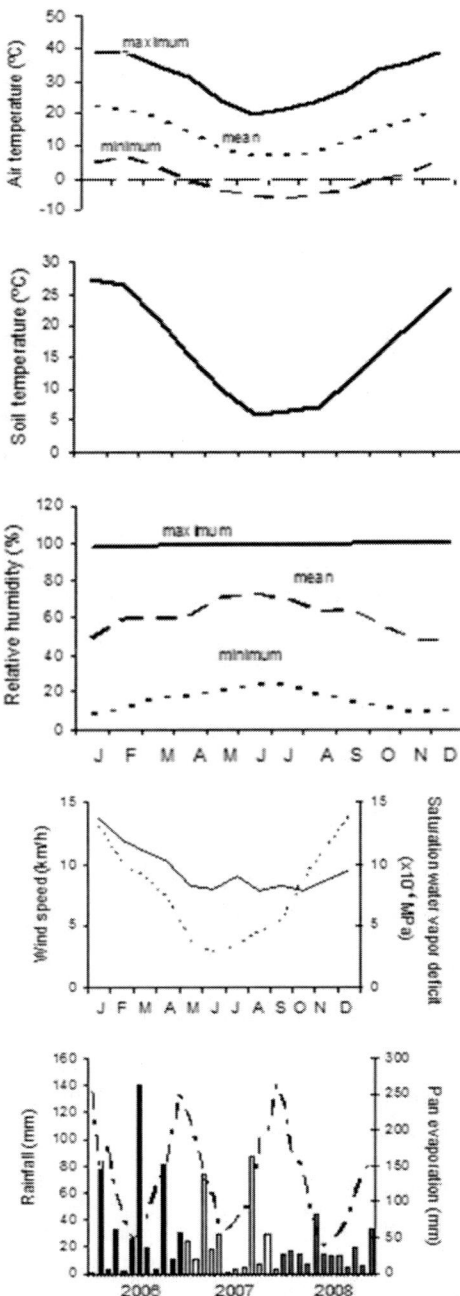

Figure 1. Absolute monthly maximum and minimum, and mean monthly air temperatures; mean monthly soil temperatures at 0-20 cm soil depth; absolute monthly maximum and minimum, and mean monthly relative humidities, mean monthly wind speed and saturation water vapour deficit, and mean monthly pan evaporation and monthly rainfall during 2006, 2007 and 2008 at a meteorological station located at the study site.

Climate is temperate semiarid, with higher precipitations during the spring and fall seasons [23]. Several climatic parameters measured at the study site are shown in Figure 1.

Soil is a typical Haplocalcid. Average pH is 7, and depth is not a constraint in the soil profile. The plant community is characterized by an open, shrubby stratum which includes different-quality, herbaceous species for cattle production [3]. Dominance of a particular grass or shrubby species in the study region is partially dependent on grazing history and fire frequency and intensity [43].

Experimental Design and Defoliation Treatments

At the end of 2006, 48 intraspecific plots (3 genotypes x 2 defoliation treatments x 8 plots (replicates)/treatment x 12 plants/plot = 576 plants) were established at the field. We used transplants, obtained from seeds, of *P. vaginatum,* and *L. cinereus* cvs 'Magnar' and 'Trailhead'. Plant distance among plants in horizontal and vertical lines was 30 cm (from center to center in each plant base). In this way, competitive relationships among plants were most likely similar. Similar approaches for transplant disposition in plots have been reported in other studies [44]. During the two study years, plants were exposed to natural rainfall (Figure 1). Randomly chosen plants of all three genotypes were defoliated in 2006 to 5 cm stubble height during the winter, plant dormancy period. In this way, only aerial plant growth produced during the following growing season was used for sampling. During mid- (22 November) to late-spring (19 December) 2006, and mid-spring 2007 (5 and 11 November), half of the plants was defoliated twice (during the vegetative and immediately after differentiation of the growth apex from vegetative to reproductive) leaving 5 cm stubble height. The other half of the plants remained undefoliated (controls) under natural, rainfed conditions. Tillers of all three genotypes were periodically disected and observed under a binocular microscope to determine both the developmental stage and height of the apical meristem.

As a result, actively growing meristems (intercalary and leaf primordia in the gowth apex) remained on the plants after the defoliation events. Despite literature reports an early internode elongation in *L. cinereus* [41] this did not occur in our study. Height of the apical meristem from the tiller basal area was 5.8 ± 0.9 mm (n=15) in 2006, and 2.7 ± 0.4 mm (n=15) in 2007 on undefoliated tillers of all three genotypes at the timing of defoliation.

Measurements

Plant Phenology

During each growing cycle (2006/2007 and 2007/2008), phenology was determined on a monthly basis on randomly-chosen plants of all three genotypes. This was conducted on 24 undisturbed plants (8 plants/genotype x 3 genotypes). Study phenological stages were: (a) vegetative, (b) boot stage, (c) early panicle exposure, (d) panicle exposed, (e) flowering (anthesis), (f) immature grain, (g) mature grain, (h) seed dispersal, and (i) either plant dormancy or death.

Components of Leaf Area Production

Tiller growth and demography were determined monthly. So, 48 plants randomly chosen were marked with this purpose. On each one of these plants, one progenitor tiller was permanently marked with wire cables (3 genotypes x 2 defoliation treatments x 8 plants/defoliation treatment x 1 tiller/plant). In these plants were determined: (a) plant circumference, to calculate plant basal area and express the results on a plant or unit basal area basis, (b) the number of total and reproductive tillers per plant and of daughter tillers per progenitor tiller, and (c) tiller growth and demography: (i) height, (ii) number of total leaves (green+dry) and (iii) total, dry and green leaf length [of blades + sheaths (green + dry)] [45].

Root Length Density and Arbuscular Mycorrhiza (AM)

At the beginning of the study, 48 different plants than those used for plant demography were permanently marked. At the time of each defoliation and three weeks afterwards, a soil + root sample was extracted from each plant using an auger (volume = 181.5 cm^3). Each sample was used to determine (1) root length density [total root length (dead + alive) /auger volume; cm root/cm^3 soil], (2) percentage root colonization by fungi forming arbuscular mycorrhiza, and (3) soil available phosphorus. Roots were obtained from the soil by manual washing, using a 35-mesh screen. They were then placed between 2 glass plates, avoiding root superposition, and scanned. Obtained images were processed using the software ROOTEDGE 2.3b [46] to obtain root length. Root length density was thereafter calculated. These roots were fixed in FAA (formaldehyde, glacial acetic acid, ethanol) until AM were determined. Percentage root colonization by AM was obtained [47]. Pieces of 1.5 cm root length were placed in flasks containing 10% KOH to clear the root cell cytoplasm. Thereafter, they were heated to 90°C during 15 minutes. Roots were washed in distilled water, and hyphae, vesicules, and/or arbuscules of mycorrhiza were stained using Trypan blue. This treatment was conducted during 20 minutes to 90°C. Finally, roots were washed in distilled water and kept in the refrigerator into lactoglycerol. Stained roots were mounted on glass microscope slides (10 root segments per slide), and the number of intersections containing hyphae, vesicules and/or arbuscules was counted when 3 runs along each of three glass microscope slides/sample were conducted under a microscope (100-400X). Percentage root colonization by AM was obtained from the number of colonized points (CP) with respect to the total number of observed points (OP) as follows: percentage root colonization by AM = CP/PO*100. Available soil P was determinated [48].

Aboveground Biomass Production

Aboveground biomasses were harvested at the time of defoliation (springs 2006 and 2007) on defoliated plants, and at the end of the growing season (falls 2007 and 2008) in all study plants. This material was dried to 70°C during 72 h and weighed. Biomass determinations were made to a plant scale level (the same scale used for growth, phenology and demography measurements).

Aboveground biomass was separated into blades, sheaths, stems, panicles and seeds, and each plant part was weighed. This allowed studying dry matter partitioning on plants of the study genotypes, and calculating the blade/sheath ratio.

Plant Survival

At the end of fall 2007 and 2008, number of surviving plants (SP; plants which regrow and continue growing) was counted in each of the 48 experimental plots. This allowed obtaining percentage plant survival (PS) at a plot scale as:

$$PS = SP / IN \times 100,$$

where IN = initial plant number/plot (=12).

Statistical Analyses

We used a completely randomized design. Data were analyzed using the statistical software INFOSTAT [49]. Values of percentage mycorrhizal colonization, total dry matter and root length density were transformed to arcsen$\sqrt{x}$, ln (x+1) and $\sqrt{x}$, respectively, to comply with the normality and homocedasticity assumptions of variance. Untransformed values are shown in illustrations. Demography and growth variables were analyzed using repeated-measures two-way ANOVA. Whenever the interaction between any factor and time was not significant, data from all study dates were averaged. When the interaction was significant, each date was analyzed separately. The remaining variables were analyzed using two-way ANOVA, except for the reproductive tiller production which was analyzed utilizing one-way ANOVA. Mean separation was conducted using protected LSD, with a significance level of 0.05. Some results were expressed on a cm^2 basis for comparative purposes between the study genotypes.

Results

Phenology

At the same time plants of *L. cinereus* were all regrowing at the beginning of the growing season in October (spring) 2006, 12% of *P. vaginatum* plants still remained dormant (Figure 2). In September 2007, however, all plants of all three genotypes had regrown (6.8 mm during July-August). Only the native species completed the reproductive phenological stage during the study. In October 2007, 87.5% of *P. vaginatum* tillers were in boot stage.

During the same month, but in 2006, only 18.75% of its marked tillers were at that stage (Figure 2).

In November 2006 (92.0 mm during October-November: Figure 1), percentage of tillers which were at the boot stage (18.75%) and started to expose panicles (18.75%) was not greater than 40%; at the same time in 2007 (36.5 mm during October-November), all marked tillers were at the reproductive stage. Seed dispersal in 2007 was initiated in January (30.3 mm, 12.5% of all marked tillers), while in the following growing cycle (2007/08), it started in December (2.5 mm) on 87.5% of all marked tillers. Finally, in January 2008 (17.5 mm during December 2007-January 2008), at least 62.5% of *P. vaginatum* tillers were either dormant or dead. This phenological stage was not recorded at the same time in 2006/07 (54.3 mm during

December 2006 and January 2007). In February 2007, 56.5% of tillers in 'Magnar' and 25% of those in 'Trailhead' were either dormant or dead. At the same time the following year, however, 33% of tillers in 'Magnar' and 71% of those in 'Trailhead' were at those stages of phenology, while the remaining stayed at the vegetative phenological stage.

Only the native grass species produced panicles in both study cycles. Defoliated and undefoliated plants had a similar (p>0.05) number of reproductive tillers in 2006/07 (Figure 3A). In 2007/08, however, reproductive tiller production was 64% greater (p<0.05) on undefoliated than on defoliated plants during spring and summer (Figure 3B).

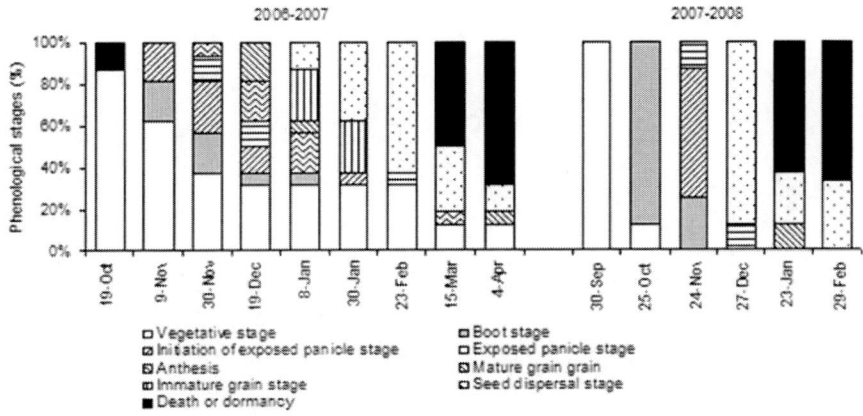

Figure 2. Phenological stages (%) determined on undefoliated plants of *Pappophorum vaginatum* during two growing cycles (n=8). Each histogram corresponds to a sampling date.

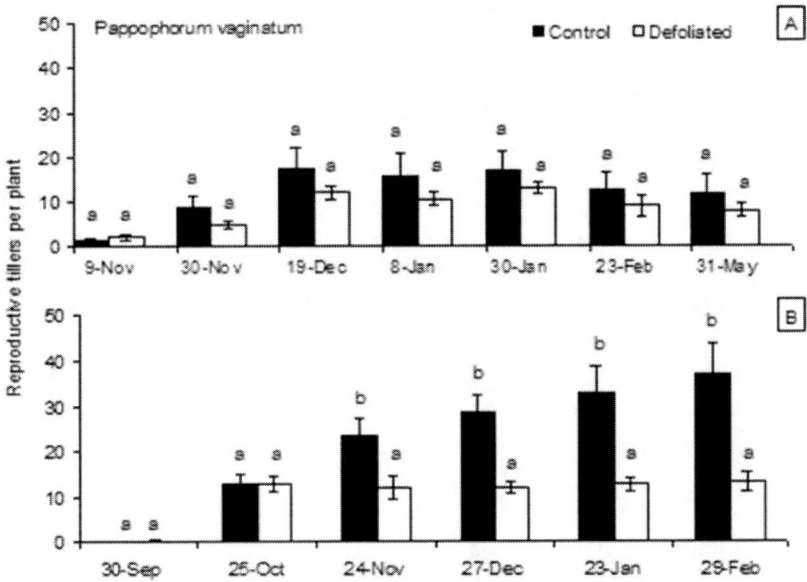

Figure 3. Number of reproductive tillers/plant on plants of *Pappophorum vaginatum* exposed to two defoliation treatments [(defoliated-undefoliated (control)] during the growing season of 2006-2007 (A) and 2007-2008 (B). Each histogram is the mean ± 1 standard error of n=8. Different letters on histograms indicate significant differences (p<0.05) between treatments.

Leaf Area Production Components

Basal Area

At the end of the 2006/07 and 2007/08 (May, fall) growing cycles, basal area was at least 56% and 39% greater (p<0.05), respectively, in the native than in the introduced genotypes (Figure 4).

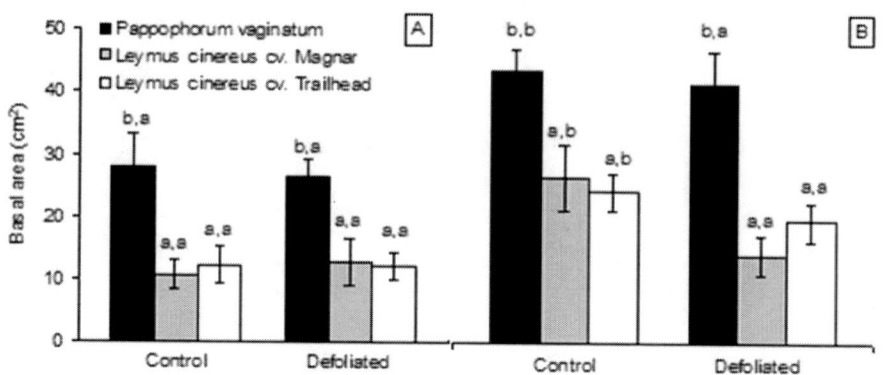

Figure 4. Plant basal area (cm^2) of three genotypes exposed to two defoliation treatments (Control, Defoliated) during the growing period of 2006-2007 (A) and 2007-2008 (B). Histograms are the mean ± 1 standard error of n=8, and correspond to May of each study growing cycle. Different letters on histograms indicate significant differences (p<0.05) among genotypes (first letter) or between treatments (second letter).

Tiller Number

Tiller number on defoliated and undefoliated plants was greater (p<0.05) on the native than on the *L. cinereus* genotypes during the whole 2006/07 study cycle (data not shown). At the end of this growing season, *P. vaginatum* had at least 79% more of tillers/plant. During the second study cycle, until November 2007, plants of *P. vaginatum* had a greater (p<0.05) tiller number than those of *L. cinereus*; this parameter was greater (p<0.05) on undefoliated than on defoliated individuals. This parameter kept being greater (p<0.05) in the native than in the introduced genotypes during December and January; however, no differences (p>0.05) were found between treatments. At the end of the study, there were again differences (p<0.05) between treatments (control > defoliated) in addition to those found between genotypes: tiller number/plant was at least 76% greater in the native than in the introduced genotypes. Between the beginning and the end of the study, tiller number in both treatments increased at least 23% in the native species, and decreased at least 30% in the introduced genotypes.

Pappophorum vaginatum showed a greater (p<0.05; Table 1) daughter tiller production/parent tiller than both *L. cinereus* genotypes in both treatments and years. The only exception occurred in October 2007/08, when *P. vaginatum* and 'Trailhead' showed a similar (p>0.05) number of daughters per parent tiller. At the end of the growing cycles in both years, 100% of daughters were dead on parent tillers of both *L. cinereus* genotypes. Defoliated and undefoliated parent tillers had a similar (p>0.05) production of daughters in all three genotypes during both study years (Table 1). However, regrowth of both *L. cinereus* genotypes, mainly ''Magnar'', occurred 3 weeks earlier than in the native species at least. For example, on 15 September 2009, green tiller number was 60±9.7, 6, and dry weight was 2.72±0.61g, 6 (mean±1 S.E., n) in plants of *L. cinereus* cv. ''Magnar''. At the same time, tiller initiation from axillary buds had not started in plants of the native species.

Table 1. Number of daughter tillers/parent tiller on plants of three genotypes exposed to two defoliation treatments (control-defoliated) during the growing seasons of 2006-2007 and 2007-2008

		P. vaginatum		*L. cinereus* cv. Magnar		*L. cinereus* cv. Trailhead	
		Control	**Defoliated**	**Control**	**Defoliated**	**Control**	**Defoliated**
2006-2007	19-Sep	0.2 ± 0.1 a,a	0.2 ± 0.1 a,a	0.2 ± 0.1 a,a	0.1 ± 0.1 a,a	0.1 ± 0.1 a,a	0.0 ± 0.0 a,a
	9-Nov	0.6 ± 0.3 b,a	0.4 ± 0.1 b,a	0.2 ± 0.1 a,a	0.1 ± 0.1 a,a	0.1 ± 0.1 a,a	0.0 ± 0.0 a,a
	30-Nov	0.8 ± 0.3 b,a	0.7 ± 0.3 b,a	0.2 ± 0.1 a,a	0.2 ± 0.1 a,a	0.2 ± 0.1 a,a	0.0 ± 0.0 a,a
	19-Dec	0.1 ± 0.2 b,a	1.3 ± 0.2 b,a	0.2 ± 0.1 a,a	0.2 ± 0.1 a,a	0.4 ± 0.2 a,a	0.0 ± 0.0 a,a
	8-Jan	0.1 ± 0.3 b,a	0.9 ± 0.1 b,a	0.1 ± 0.1 a,a	0.2 ± 0.0 a,a	0.2 ± 0.1 a,a	0.0 ± 0.0 a,a
	30-Jan	0.8 ± 0.2 b,a	0.7 ± 0.1 b,a	0.0 ± 0.0 a,a	0.1 ± 0.1 a,a	0.0 ± 0.0 a,a	0.0 ± 0.0 a,a
	23-Feb	0.8 ± 0.3 b,a	0.6 ± 0.2 b,a	0.0 ± 0.0 a,a	0.0 ± 0.0 a,a	0.0 ± 0.0 a,a	0.0 ± 0.0 a,a
2007-2008	30-Sep	0.4 ± 0.2 b,a	0.6 ± 0.2 b,a	0.2 ± 0.2 a,a	0.0 ± 0.0 a,a	0.0 ± 0.0 a,a	0.1 ± 0.1 a,a
	25-Oct	0.7 ± 0.4 b,a	1.5 ± 0.33 b,a	0.5 ± 0.3 ab,a	0.9 ± 0.1 ab,a	0.1 ± 0.1 a,a	0.2 ± 0.2 a,a
	24-Nov	0.9 ± 0.4 b,a	1.0 ± 0.3 b,a	0.4 ± 0.2 a,a	0.2 ± 0.2 a,a	0.4 ± 0.2 a,a	0.1 ± 0.1 a,a
	27-Dec	0.5 ± 0.2 b,a	1.2 ± 0.3 b,a	0.1 ± 0.1 a,a	0.2 ± 0.2 a,a	0.4 ± 0.2 a,a	0.1 ± 0.1 a,a
	23-Jan	0.6 ± 0.3 b,a	0.6 ± 0.3 b,a	0.1 ± 0.1 a,a	0.0 ± 0.0 a,a	0.1 ± 0.1 a,a	0.0 ± 0.0 a,a
	29-Feb	1.3 ± 0.3 b,a	0.4 ± 0.3 b,a	0.0 ± 0.0 a,a	0.0 ± 0.0 a,a	0.0 ± 0.0 a,a	0.0 ± 0.0 a,a

Values are the mean ± 1 standard error of n=8. Different letters within the same date indicate significant differences (p<0.05) among genotypes (first letter) or treatments (second letter).

Tiller Growth and Demography

Height

This variable was higher (p<0.05) on undefoliated than defoliated plants of all three genotypes by the end of November 2006 (Figure 5). From December until the end of the growing season, defoliated and undefoliated plants of *P. vaginatum* showed a greater (p<0.05) height than cultivars of *L. cinereus*. Height differences (≥10%) between defoliated and undefoliated plants were maintained from the time of defoliation until the end of the growing season. Differences (p<0.05) between treatments were maintained during the second study year (2007/2008) until October, time when tiller height was equal (p>0.05) in both treatments. In November of the second study year (after the first defoliation), height was greater on undefoliated (p<0.05) than defoliated plants; however, no differences (p>0.05) were found among genotypes. From December (after the second defoliation) until the end of the study, differences (p<0.05) were maintained between control and defoliated plants (i.e., control>defoliated), and height was at least 30.6% greater in *P. vaginatum* than in the introduced genotypes.

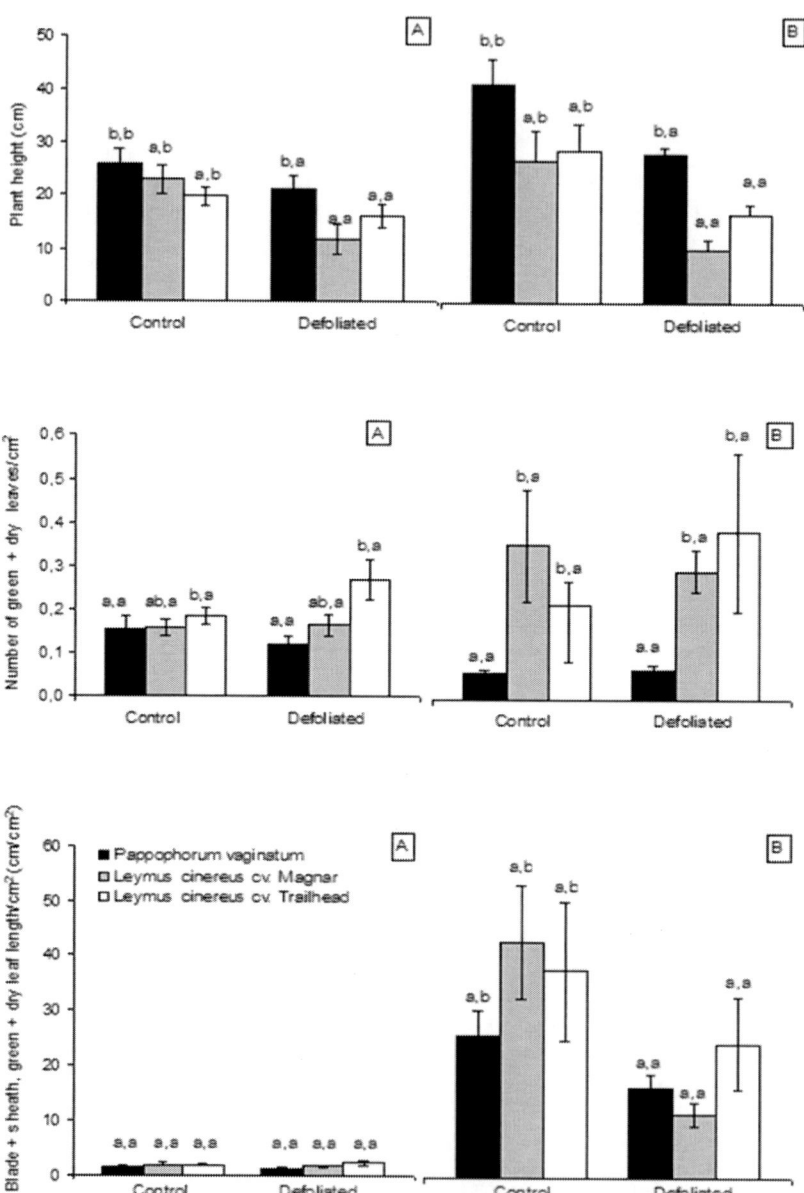

Figure 5. Height; number of (green + dry) leaves cm^{-2}, and leaf length of [blades + sheaths (green + dry)] cm^{-2} on plants of three genotypes exposed to two defoliation treatments (Control, Defoliated) during the growing periods of 2006-2007 (A) and 2007-2008 (B). Defoliations were made on 22 November and 19 December 2006/2007, and 5 and 11 November 2007/2008. Histograms are the mean ± 1 standard error of n=8. Different letters on histograms indicate significant differences (p<0.05) among genotypes (first letter) or between treatments (second letter).

Leaf Number

During the first study year, total leaf production (green + dry) cm^{-2} was 44% greater (p<0.05; Figure 5) in 'Trailhead' than in the native species. No differences (p>0.05) were found between treatments during both the first and second year. Both introduced genotypes produced a total leaf number cm^{-2} at least 80% greater (p<0.05) than *P. vaginatum* during the second year. Between 2006/07 and 2007/2008, leaf number decreased at least 53.3% in *P. vaginatum*; at the same time it increased at least 29% in 'Magnar' and 'Trailhead'.

Leaf Length

During 2006/07, both genotypes and treatments showed similar (p>0.05) total green blade lenght cm^{-2} (Figure 5). From October onward in 2007/08, total green blade lenght cm^{-2} was greater (p<0.05) in both genotypes of *L. cinereus* than in *P. vaginatum*, with no difference (p>0.05) between treatments. Differences (p<0.05) among genotypes were maintained by November, and values were greater (p<0.05) on undefoliated than defoliated plants as a result of the defoliation/s event/s. Total green blade length cm^{-2} was greater (p<0.05) in 'Magnar' than in *P. vaginatum* in December 2007 and January 2008; this variable kept being greater (p<0.05) in control than defoliated plants. At the end of the growing season, no differences (p>0.05) were detected neither among genotypes nor treatments in total green leaf length cm^{-2}. Total green sheath length cm^{-2} was similar (p>0.05) among genotypes and treatments during 2006/07. In the second study year, treatments did not differ significantly (p>0.05); at the same time, however, that variable was at least 68% greater (p<0.05) in 'Magnar' than in *P. vaginatum*.

Total length of green blade + sheaths cm^{-2} did not differ (p>0.05) neither among genotypes nor treatments during the first study cycle. In the following growing season, this variable was greater (p<0.05) in the introduced genotypes than in the native species from October onwards. In November, after defoliation, differences among genotypes were maintained; also, this study parameter was greater (p<0.05) in controls than defoliated plants. 'Magnar' had a greater (p<0.05) total length of green blade + sheaths cm^{-2} than *P. vaginatum* in December 2007 and January 2008.

There were no differences (p>0.05) neither among genotypes nor treatments in this parameter by February 2008. There were no differences neither among genotypes nor treatments during 2006/07 for total leaf (blades + sheaths, green + dry) cm^{-2} (p< 0.05; Figure 5).

Nevertheless, this variable was greater (p<0.05) on undefoliated than defoliated plants during the second study year (Figure 5). From one growing season to the next one, total leaf (blades + sheaths, green + dry) cm^{-2} increased at least 86% on defoliated and undefoliated plants of all three genotypes.

Root Length Density

At the time of the first defoliation in 2006/07, there was a significant (p<0.05) interaction between treatments and genotypes (Figure 6A, B). Root length density was greater (p<0.05) in 'Magnar' than the other two genotypes in both control and defoliated plants (Figure 6A). Root length density in the native species was three times greater (p<0.05) on defoliated than on undefoliated plants (Figure 6 B). However, defoliated and undefoliated plants did not differ (p>0.05) in root length density in the introduced genotypes. At the time of the second defoliation, root length densities were similar (p>0.05) among genotypes and between treatments (Figure 6 C).

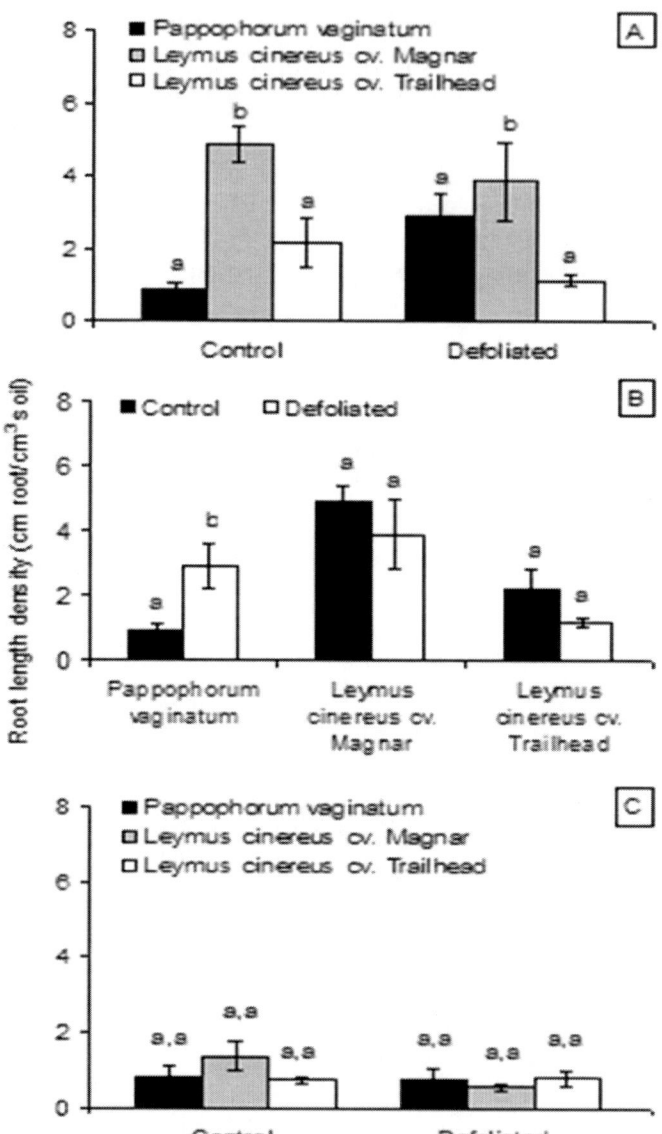

Figure 6. Root length density on plants of three genotypes exposed to two defoliation treatments (Control, Defoliated) during the growing season of 2006-2007. Each histogram is the mean ± 1 standard error of n=8. Different letters on histograms indicate significant differences (p<0,05) among genotypes (A), treatments (B) and genotypes and treatments (C). Results at the time of the first (A and B) or second (C) defoliation in 2006 are shown. The significant (p<0.05) interaction genotype (A) x treatment (B) at the time of the first defoliation determined that results were presented separately. Because of this, different letters within each treatment in A indicate differences (p<0.05) among genotypes, and different letters within each genotype in B indicate differences (p<0.05) between treatments.

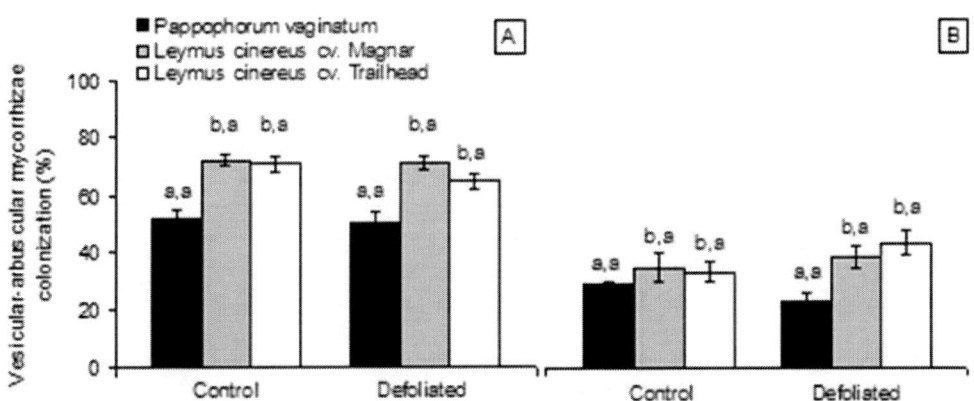

Figure 7. Percentage root colonization by arbuscular mycorrhiza on plants of three genotypes exposed to two defoliation treatments (Control, Defoliated) during the growing periods of 2006-2007 (A) and 2007-2008 (B). Each histogram is the mean ± 1 standard error of n=8. Different letters on histograms indicate significant differences (p<0.05) among genotypes (first letter) or between treatments (second letter).

Neither genotypes nor treatments differ (p>0.05) in root length density during 2007/08. Root length density decreased at least 75% in *P. vaginatum* and 'Trailhead', and at least 83% in 'Magnar', from 2006/07 to 2007/08 for both control and defoliated plants.

AM Colonization

During 2006/07, there were no differences (p>0.05) between treatments, but differences were significant among genotypes with at least 26.4% greater (p<0.05) AM colonization values in 'Magnar' and 'Trailhead' than in *P. vaginatum* (Figure 7A). Similar results were obtained during the second year (Figure 7B), with a difference ((p<0.05) of at least 25.2% between the introduced genotypes and *P. vaginatum*. Between successive growing cycles, AM colonization decreased at least 45.3% in *P. vaginatum*, 46.7% in 'Magnar' and 32.73% in 'Trailhead'.

Soil Available Phosphorus

There were no differences (p>0.05) in soil available phosphorus under plants neither among genotypes nor between treatments during 2006/07 (Table 2). During the next growing season, soil phosphorus availability was greater (p<0.05) under plants of *P. vaginatum* in both treatments than on those of the introduced genotypes. From 2006/07 to 2007/08, soil phosphorus availability increased at least 67.3% under plants of *P. vaginatum*, and 63.3 and 55% under those of 'Magnar' and 'Trailhead', respectively (Table 2).

Table 2. Available soil phosphorus (ppm) below the canopy of three genotypes exposed to two defoliation treatments (control-defoliated) during the growing seasons of 2006-2007 and 2007-2008

	2006-2007		2007-2008	
	Control	**Defoliated**	**Control**	**Defoliated**
P. vaginatum	8.65 ± 1.86 a,a	7.43 ± 0.49 a,a	26..44 ± 2..29 b,a	25.43 ± 1.45 b,a
L. cinereus cv. *Magnar*	7.33 ± 1.56 a,a	6.98 ± 1.37 a,a	23..15 ± 1..54 a,a	19.02 ± 2.53 a,a
L. cinereus cv. *Trailhead*	9.39 ± 0.95 a,a	9.72 ± 1.04 a,a	21..27 ± 1.45 a,a	21.61 ± 2.63 a,a

Values are the mean ± 1 standard error of n=8. Different letters within the same period indicate significant differences (p<0.05) among genotypes (first letter) or treatments (second letter).

Aboveground Biomass Production

During the first study growing season, *P. vaginatum* produced a greater (p<0.05) dry matter production/plant than both introduced genotypes (Figure 8A); additionally, this production was greater on defoliated (p<0.05) than on undefoliated controls (Figure 8A). Genotype and treatment interacted (p<0.05) the subsequent growing season (Figure 8B, C). Dry matter production was greater (p<0.05, Figure 8B) on control and defoliated plants of *P. vaginatum* than on those of both *L. cinereus* genotypes. However, while annual dry matter production was greater (p<0.05) on defoliated than on undefoliated plants of *P. vaginatum*, it was similar (p>0.05) between treatments in *L. cinereus* cvs. 'Magnar' and 'Trailhead' (Figure 8C). From 2006/07 to 2007/08, dry matter production of undefoliated plants increased 56.2% in *P. vaginatum*, 47.4% in 'Magnar' and 72.9% in 'Trailhead'. On defoliated plants, dry matter production increased 31.36% in *P. vaginatum* but decreased 38.9% in 'Magnar' and 1.7% in 'Trailhead'.

Blade: Sheath Ratio

No differences (p>0.05) were found neither among genotypes nor treatments during 2006/07 (Figure 9A). During 2007/08, the interaction genotypes x treatments was significant (p<0.05). The blade/sheath ratio was greater (p<0.05) in the *L. cinereus* genotypes than in *P. vaginatum* on defoliated plants (Figure 9B). In control plants, there were no differences (p>0.05) among genotypes in such ratio (Figure 9B). Blade/sheath ratios were greater (p<0.05) on defoliated than on undefoliated plants on the introduced genotypes (Figure 9C). Both treatments showed similar (p>0.05) blade/sheath ratios in the native species. From the 2006/07 to the 2007/08 growing season, blade/sheath ratios on defoliated and undefoliated plants decreased at least 78.8% in *P. vaginatum*, 20.4% in 'Magnar' and 8.3% in 'Trailhead'. During the first study growing cycle, leaves made the major contribution to aboveground plant biomass in all three genotypes on control (45% in *P. vaginatum*, 84% in 'Magnar' and 89% in 'Trailhead') and defoliated plants (40% in *P. vaginatum,* 71% in Magnar and 72% in 'Trailhead'). In the native species, the second component after leaves with greater biomass values was reproductive structures. Since introduced genotypes did not produce reproductive structures, the second organ after leaves was the sheaths. During the second year, distibution of biomass among plant organs was more variable. In *P. vaginatum*, leaves (29%) or reproductive structures (48%) made the greatest contribution to total aerial plant biomas on undefoliated or defoliated plants, respectively. Results in *L. cinereus* were similar to those in the previous year, with greatest values on leaves on undefoliated (64% in 'Magnar' and 78% in 'Trailhead') and defoliated plants (87% in 'Magnar' and 89% in 'Trailhead'). From one growing cycle to the next, biomass allocation increased to reproductive structures on defoliated plants of *P. vaginatum*. In control plants of this species, biomass allocation to leaves decreased, and it was more distributed in the remaining organs. In both *L. cinereus* genotypes, biomass allocation to leaves decreased in controls, and it increased on defoliated plants.

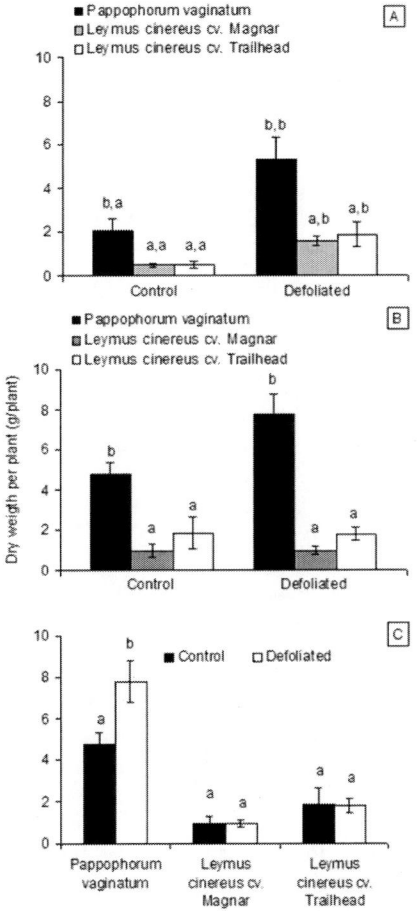

Figure 8. Plant dry weight of three genotypes exposed to two defoliation treatments (Control, Defoliated) at the end of the growing seasons of 2006-2007 (A) and 2007-2008 (B and C). Each histogram is the mean ± 1 standard error of n=8. Different letters on histograms in A indicate significant differences (p<0.05) among genotypes (first letter) or between treatments (second letter). In 2007-2008, the significant interaction (p<0.05) among genotype (B) and treatment (C) determined that results are presented separately. Because of this, different letters within each treatment in B indicate differences (p<0.05) among genotypes and different letters within each genotype in C indicate differences (p<0.05) between treatments.

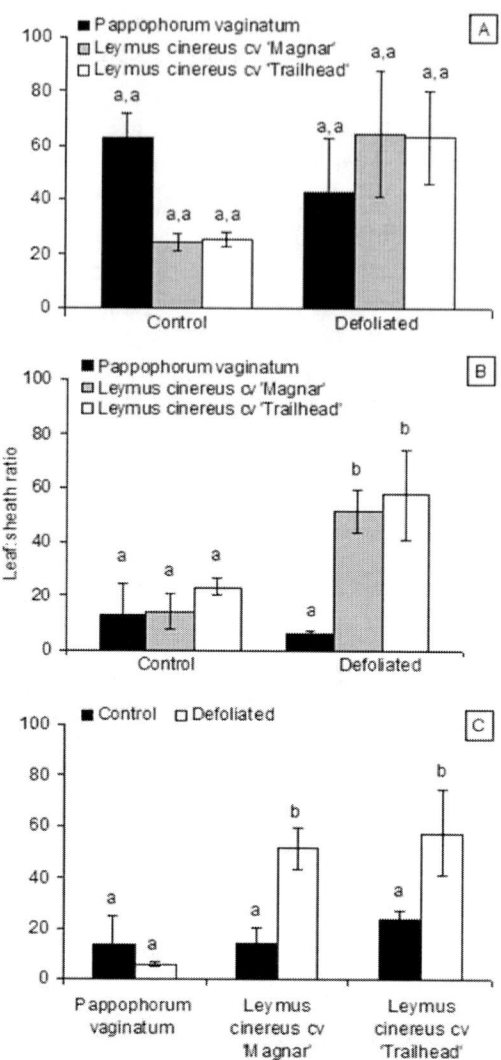

Figure 9. Leaf blade/sheath ratio on plants of three genotypes exposed to two defoliation treatments (Control, Defoliated) during the growing seasons of 2006-2007 (A) and 2007-2008 (B and C). Each histogram is the mean ± 1 standard error of n=8. Different letters on histograms in A indicate significant differences (p<0.05) among genotypes (first letter) or treatments (second letter). In 2007-2008, the significant interaction (p<0.05) among genotypes (B) and treatments (C) determined that results were presented separately. Because of this, different letters within each treatment in B indicate differences (p<0.05) among genotypes, and different letters within each genotype in C indicate differences (p<0.05) between treatments.

Plant Survival

Plant survival was greater (p<0.05) for the native species than for the introduced genotypes at the end of the first growing cycle (Figure 10). This difference (p<0.05) was maintained until the study was completed in 2008, with 92.5% of plants alive in *P. vaginatum*, 19.17% in 'Magnar' and 23.3% in 'Trailhead'.

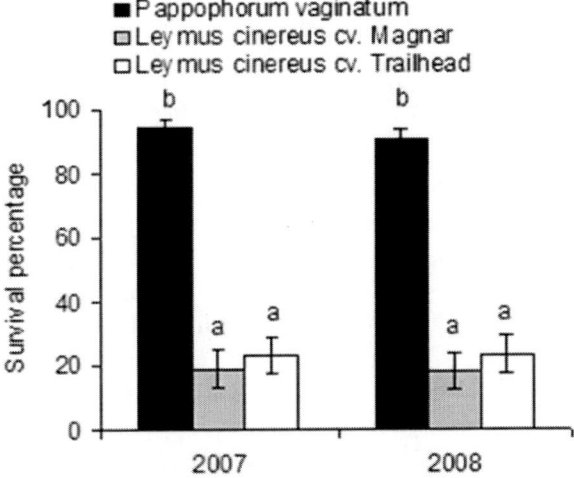

Figure 10. Percentage survival of plants of three genotypes at the end of growing periods of 2007 and 2008. Each histogram is the mean ± 1 standard error of n=8. Different letters on the histograms within each year indicate significant differences (p<0.05) among genotypes.

Discussion

In arid and semiarid ecosystems, which dynamics is mainly dependent upon water availability, plant growth, and thereafter, phenological development, is controlled by precipitation [50]. Water stress has produced an advancement of phenological stages on herbaceous species in many instances [5]. In our study, the lower annual rainfall in 2007 than in 2006, might have caused an advancement of phenological stages on all three genotypes in 2007 compared with the previous year. In addition, only *P. vaginatum* completed its reproductive cycle, reaching the seed dispersal phenological stage. It is possible that this response contributes to a greater opportunity for seed germination and seedling establishment, at a time (December-February) when rainfall represents more than 25.6% of the long-term total annual amount (1981-2005: 434.6 mm). This advancement of the reproductive phenological stages, together with the high amount of reproductive tillers/plant (even under defoliated conditions) could contribute to explain the much greater abundance of *P. vaginatum*, in comparison to the remaining C_4, native species in the region. Similar results for the phenology of this species were reported by Giorgetti et al. (2000) [23]. Fruit morphology in *P. vaginatum* is characterized by a caryopsis of 1.7 to 2 mm length, with 13 to 15 apical awns of about 4 mm length. These fruits have a very good germination percentage, and are easily transported by wind to long distances from the mother plant [51-52]. Plants of *P. vaginatum* are also grazing tolerant (i.e, Figure 8, this study) and the tussock expands via production of new tillers from axillary buds (asexual reproduction) [51]. One interesting question arises from these considerations: what reproduction form (either sexual or asexual or both) is relatively more important to explain the widespread distribution of *P. vaginatum* in arid and semiarid areas of Argentina? Additional research is needed to answer this important ecological question.

Regarding the introduced genotypes, water stress appeared to contribute to plant dormancy, an avoidance mechanism to water stress [53]. The fact that the vegetative period in the introduced genotypes was long-lasting, delaying appearance of later phenological stages, has already been shown [54]. This would be the result that plants of *L. cinereus* require from 2 to 5 years to establish completely [41]. They do not produce inflorescences until reaching a height of at least 90 cm [55].

Defoliation of *P. vaginatum* reduced the number of reproductive tillers per plant during the second study growing season. Our results agree with those reported by Anderson and Frank (2003) [56] on the effects of clipping individual plants at different times during the growing season on grass reproductive biomass. For individual plants, of the 19 studies that included an early defoliation treatment, 16% reported an increase in reproductive biomass, 42% reported no effect, and 42% reported a negative effect. For the 122 individual plant studies with a single defoliation of intermediate timing, reproductive biomass increased in 10%, was unaffected in 39%, and decreased in 51% cases. In the 29 individual plant studies that included a late defoliation treatment, reproductive biomass was not stimulated in any case, but was unaffected in 17%, and negatively affected in 83%. Graminoids from individual plant studies contributed roughly equally to this result; 55% of the graminoid studies, reported a decrease subsequent to defoliation. They also reported the effects of clipping or grazing at various times during the growing season on reproductive biomass of C_4 grasses: in at least 57% of 27 individual plant studies there were negative effects of defoliation on reproductive biomass. They emphasized that historically grazed grass populations had equal or lower reproductive to vegetative biomass ratios than conspecifics from historically ungrazed sites. The grazing of leaf material before internode expansion does not result in the removal of apical meristems and thus regrowth can ensue if soil water is adequate, whereas grazing after internode elongation may remove terminal meristems of developing floral buds [57]. Thus, timing and intensity of grazing in natural systems determines the outcome of grazing on reproductive tiller number [58].

The greater basal area in *P. vaginatum* than in the introduced genotypes was the result of its increase in the tiller number per plant; this was due to its greater daughter tiller production/parent tiller than in the two cultivars of *L. cinereus*. Daughter tillers contribute to the axillary bud bank increase of plants and to immediate carbon gain; they became photosynthetically active at the time parent tillers are already photosynthetically independent or they are senescent [59-60]. Despite daughter tiller production occurred in all three genotypes and in both treatments, tiller mortality was 100% in the introduced cultivars, which compromised their plant survival.

Shoot growth at the beginning of the growing season was about 3 weeks earlier in *L. cinereus* cv. 'Magnar' than in *P. vaginatum*. This response in cv. 'Magnar' would contribute to an earlier occupancy of the more favorable soil parcels, which in turn should improve competitive effectiveness. Favorable soil microsites may be important areas of root competition [61]. One characteristics of *L. cinereus* cv. 'Magnar' may have contributed to its earlier shoot regrowth in early spring. This species had a greater (2006/07) or similar (2007/08), but not lower, rooting density than *P. vaginatum*. This would increase the probability that roots would be in proximity to newly created fertile sites and be able to proliferate more rapidly in these areas. Even in September (early spring), soil temperatures in this area are cool [long-term (1981-2000) mean ± 1 s.e., 9.7 ± 0.23°C] and may limit root growth rates. Although the tendency to proliferate roots in fertile soil microsites may be a common characteristic of many species, those species with sufficient root density to encounter, detect, and grow rapidly in fertile sites will likely have an advantage in exploiting the patchy distribution of nutrients in soils.

The greater height reported for the native species was because reproductive tillers were included in the measurements. The introduced genotypes did not produce reproductive tillers during the two-year study. Tiller heights found in plants of 'Magnar' and 'Trailhead' were greater than those found on plants grown under greenhouse conditions [62].

The introduced genotypes produced a greater amount of total leaves (green + dry) per unit surface area than the native species. It was reported that *L. cinereus* plants rarely produce leaves near the soil surface once they are clipped, possibly due to inactivation of leaf meristems [63]. This disagrees with results found by other authors, who have observed an increased leaf number after defoliation, associated with a greater proportion of photo-asimilates allocated to new leaf production rather than support structures [64].

'Magnar' showed greater green blade + sheath length cm^{-2} than *P. vaginatum*. However, no significant differences were found among genotypes in total (blade + sheath, green + dry) leaf length cm^{-2}. Additionally, total leaf length cm^{-2} was lower on defoliated than on undefoliated plants in all genotypes during the second study year. This indicates that even though meristems in active growth were not removed at the time of defoliations, defoliated plants were unable to compensate total leaf length cm^{-2} in comparison to undefoliated controls. Our results are similar to authors [65] who showed that defoliated plants of the cool-season perennial grasses *Stipa clarazii* Ball. [Syn: *Nassella clarazii* (Ball) Barkworth; 66] and *S. tenuis* [syn.: *N. tenuis* (Phil.) Barkworth, 67] had lower total green leaf length than undefoliated plants under rainfed and irrigated conditions for *S. clarazii*, and irrigated conditions for *S. tenuis*. *Stipa tenuis* also showed a lower green leaf length per tiller during winter when it was defoliated at the vegetative and early, late and post-internode elongation phenological stages under rainfed conditions [21]. However, the increase in total leaf length cm^{-2} from the first to the second study year, despite of the water stress imposed during 2007/08 as a result of lack of rains, would suggest resistance to water stress in all three genotypes.

Root length densities at the time of the first defoliation treatment would be due to differences inherent to the species and not to the defoliation treatment. High root length densities on defoliated and undefoliated plants of 'Magnar' will most likely contribuye to its competitive capacity. After the second defoliation, a decrease was shown in this variable on defoliated and undefoliated plants. Root length density decreased from the 2006/07 to the 2007/08 growing season in both treatments. This was very likely the result, at least partially, of the drought conditions to which plants were exposed during the measurement period in 2007/08. It was also reported decreases in root length density in grass species under dry and warm field conditions [68].

Values found for AM colonization are similar to those reported for other C_3 and C_4 perennial grasses [69-70]. Both cultivars of *L. cinereus*, which did not show differences between them regarding treatments, show greater values than *P. vaginatum* in both study years. These greater values in *L. cinereus* most likely contributed to its competitive ability. All three genotypes showed a decrease in colonization percentage by AM between one year and the following in both treatments. Soil moisture is an important determinant of available soil nutrient supply, and affects water flow throughout the plant and photosynthetic C gain [68]. Plant water relations can be higher in plants colonized by AM than in those without AM under soil water stress conditions [71]. However, available information on the influence of soil water status on defoliated or undefoliated mycorrhizal plants is yet unclear. Water stress can reduce root growth and formation of arbuscular mycorrhiza in various grass species, which is partially associated to stress intensity [68]. It was observed a decrease in percentage colonization by AM in plants of *Schizachyrium scoparium* (Michx.) Nash when soil water content also decreased [72].

Results obtained in relation to root length density and percentage colonization by AM in *L. cinereus* do not support reports which indicate that low root length density plant species are more dependent upon AM root colonization for phosphorus and water acquisition [18]. However, our results showed that both parameters decreased under water stress. Some studies have also suggested that C_3 plants would be slightly dependent on the symbiotic relationship with AM, even in environments with low nutrient availability [73]. In our study, however, the C_3 introduced genotypes showed greater AM colonization percentages than the C_4 *P. vaginatum* during both study years. This suggests a greater AM colonization dependency in the C_3 than in the C_4 perennial grass species.

Either simulated or natural herbivory can reduce AM root colonization levels if the host plant reduces C allocation to mycorrhizal fungi. For example, it was found that AM colonization levels decreased after events of either simulated or natural herbivory in 23 out of 37 studies (60%) [74]. Other studies reported that severe defoliation did modify neither root length density nor AM colonization percentages in the C_4 perennial grasses *Bouteloua curtipendula* (Michx.) Torr., *Aristida purpurea* Nutt. and *Eriochloa sericea* (Scheele) Munro ex Vasey in comparison to undefoliated controls [67]. In our study, defoliation did not appear to affect AM colonization levels, but the measured decrease from the first to the second year could be attributed to the low precipitations registered during 2007/2008. In addition, between years, there was an increase in the aboveground dry matter production/plant in the native species, to the expense of a reduced root growth (estimated as root length density) on defoliated and undefoliated plants. In spite of increasing the photosynthetic material, and then the C source for mycorrhizal fungi, AM root colonization was reduced in the native species. It is likely that under water stress conditions, *P. vaginatum* may increase its fitness investing resources in seed production rather than allocating resources to maintain its symbiotic relationships. A successful seed production, and subsequent germination and seedling establishment, might contribute to maintain its abundance in the community.

Soil available phosphorus concentration increased from the first to the second study year. However, values were greater under the canopy of the native than under that of the introduced genotypes. Biological material decomposition is a fundamental process in ecosystem functioning; it has a close relationship with the supply of nutrients for plant growth. Decomposition dynamics is closely related to quality of the biological material, environmental conditions and decomposer community available in the soil [75]. After sprouting in early 2009, C:N ratios were 14.66 ± 4.75 and 55.27 ± 6.63 (mean ± 1 s.e., n=6) on roots of *L. cinereus* and *P. vaginatum*, respectively. The greater quality [i.e., lower C:N ratio in the introduced than in the native genotype: p<0.001] of the root material underneath the canopy of *L. cinereus* than under that of *P. vaginatum* contributed to obtain greater soil available phosphorus concentrations under the native species canopy. This is because soil available phosphorus would be more available for root uptake in the introduced than in the native genotype.

The leaf blade:sheath relationship, together with the annual dry matter production, provide a qualitative and quantitative estimate of the forage offered to domestic herbivory consumption, because of its plant part selectivity at grazing time. Plants of *P. vaginatum* showed a greater annual dry matter production than those of the introduced genotypes during both study years. This was mainly due to the greater plant sizes of *P. vaginatum*. Defoliated plants of the native and introduced genotypes overcompensated annual dry matter production of the undefoliated controls in the first study year. At the second year, overcompensation in annual dry matter production was only shown by defoliated plants of *P. vaginatum* compared with values on controls. At the same time, exact compensation in annual dry matter production was shown by defoliated plants of both *L. cinereus* genotypes. The various compensation responses in dry matter production reported in perennial grasses has been attributed to several physiological mechanisms [4]. One of them is timing of defoliation: plants defoliated early in the growing season have more time to recuperate from the defoliation event, and actively growing meristems (i.e., intercalary and apical meristem with leaf primordia) remain in the plants after defoliation [76]. Even more, while plants of *P. vaginatum* increased annual dry matter production from the first to the second study year, those of the introduced genotypes maintained annual dry matter production in both years, despite the reduction in precipitation from the first (Dec 2006 to April 2007= 155.9 mm) to the second (Dec 2007 to April 2008= 55 mm) study year, which suggest high resistance to water stress in both study species. This response in defoliated plants has been widely documented in the literature for various perennial grass species [23, 59].

During the first year, leaf blade:sheath ratios were similar among genotypes and between treatments. Defoliated plants of both *L. cinereus* genotypes showed greater leaf blade:sheath ratios than those of the native species during the second study year. At this time, leaf blade:sheath ratios were greater on defoliated than on undefoliated plants of both introduced genotypes. These results agree with those of who reported plant biomass partitioning in *L. cinereus*, grown under field conditions [54]. Nutritive value is greater in leaf blades than leaf sheaths [77]. Thereafter, during the second year, defoliated plants of both *L. cinereus* genotypes would be offering a greater quantity of plant material of a greater nutritive value to grazers. However, the native species was the most productive as a result of its greater basal area in both years. Defoliated plants of *P. vaginatum* achieved a similar leaf blade:sheath ratio than controls in both years of study; at the same time, they increased C allocation to the formation of reproductive structures. This would allow *P. vaginatum* continued seed dispersal in the plant community, enhancing seed germination and new seedling establishment when conditions are favorable during the year for these physiological processes.

The native genotype showed a greater plant survival than the introduced genotypes. Similar results have been obtained for other species of *Pappophorum* [9, 11] under grazing conditions. Survival values determined for the introduced genotypes, after 2 years of establishment, are lower than those reported on plants which grew under field or greenhouse conditions [62, 78]. The tendency of the introduced genotypes to reduce its vigor and new tiller production when facing a water stress event (September 2008 to May 2009 = 131.5 mm precipitation) could compromise its persistance in the plant community. This is because vegetative growth (i.e., tillering) is generally the dominant reproductive form in semiarid rangelands [4].

Survival of trasplants was 74 % lower in *L. cinereus* than in *P. vaginatum*, and plants of the native species were of a greater size than those of the intoduced genotypes. However, established plants of the introduced genotypes had a greater number of total leaves cm^{-2}; greater length of green sheaths cm^{-2}; greater leaf blade:sheath ratio, and greater percentage root colonization by AM. These results foster research focused in substantially improving plant establishment of *L. cinereus*, a constraint that has also been recognized in various studies conducted in the USA [37]. Substantial improvement in *L. cinereus* establishment after field seedings would contribute to inclusion of a warm-season, rhizomatous species of good physiological (root length density, percentage colonization by AM) and morphological (leaf production, high leaf blade:sheath ratio) traits in rangelands of central Argentina. This is because seeds of this genotype, especially cv. 'Magnar' (which sprouted at least 3 weeks earlier in the growing season than the native species), can be seeded by conventional drilling, while those of *P. vaginatum* are very hairy which impede its seeding.

Conclusion

Improvements in seedling establishment of the introduced species *L. cinereus* cvs. 'Magnar' and 'Trailhead' will allow to substantially increase the warm-season, palatable forage offer to livestock. This is because this species can be seeded by conventional drilling, while the warm-season, native perennial grass *P. vaginatum* can not. If seedling establishment of *L. cinereus* cvs. 'Magnar' and 'Trailhead' can be substantially improved, a new puzzle to solve, forage offer to domestic livestock will be partially solved during the warm season, and cattle production will be greatly increased as a result.

References

[1] O. A. Fernández and C. A. Busso, in: O. Arnalds and S. Archer (Eds.), *Agric. Res. Inst. Rep.*, Reykjavik, Iceland, pp. 41-60 (1999).

[2] C. A. Busso, H. D. Giorgetti, O. A. Montenegro and G. D. Rodríguez, *Pyhton, Int. J. Exp. Bot.* 53, 9-27 (2004).

[3] H. D. Giorgetti, O. A. Montenegro, G. D. Rodríguez, C. A. Busso, T. Montani, M. A. Burgos, A. C. Flemmer, M. B. Toribio and S. S. Horvitz, *J. Arid Envir.* 36, 623-637 (1997).

[4] D. D. Briske and J. H. Richards, in: D. J. Bedunah and R. E. Sosebee (Eds.), *Soc. Range Manage.*, Denver, USA, pp. 635-710 (1995).

[5] C. A. Busso and J. H. Richards, *Rev. Fac. Agron.* 10, 127-138 (1989).

[6] B. B. Casper and R. B. Jackson, *Ann. Rev. Ecol. Syst.* 28, 545-570 (1997).

[7] C. Saint Pierre, C. A. Busso, O. A. Montenegro, G. D. Rodríguez, H. D. Giorgetti, T. Montani and O. A. Bravo, *Can. J. Plant Sci.* 84, 195-204 (2004).

[8] H. D. Giorgetti, O. A. Montenegro, G. D. Rodríguez and C. A. Busso, *Rev. Arg. Prod. Animal* 26, 199-200 (2006).

[9] J. B. Cavagnaro and A. D. Dalmasso, *Deserta* 7, 203-218 (1983).

[10] A. D. Dalmasso, *Multequina* 3, 9-34 (1994).

[11] M. J. L. Privitello, E. G. Gabutti and R. U. Harrison, *Rev. Arg. Prod. Animal* 18, 111-115 (1998).

[12] R. E. Quiroga, L. J. Blanco and E. L. Orionte, *Rev. Arg. Prod. Animal* 24, CD-ROM (2004).

[13] J. H. Richards, *Oecologia* 64, 21-25 (1984).

[14] J. S. Murphy and D. D. Briske, *J. Range Manage.* 45, 419-429 (1992).

[15] G. F. Becker, C. A. Busso, T. Montani, M. A. Burgos, A. C. Flemmer and M. B. Toribio, *J. Arid Envir.* 35, 269-283 (1997).

[16] D. Tilman and D. Wedin, *Ecology* 72, 685-700 (1991).

[17] M. M. Caldwell, in: M. M. Caldwell and R. W. Pearcy (Eds.), Academic Press, San Diego, USA, pp. 325-347 (1994).

[18] R. T. Koide and M. Li, *Oecologia* 85, 403-412 (1991).

[19] S. K. Kothari, H. Marschner and E. George, *New Phytol.* 116, 303-311 (1990).

[20] C. Saint Pierre, C. A. Busso, O. A. Montenegro, G. D. Rodríguez, H. D. Giorgetti, T. Montani and O. A. Bravo, *Plant Ecol.* 165, 161-169 (2002).

[21] G. F. Becker, C. A. Busso, T. Montani, R. E. Brevedan, A. Orchansky, M. A. Burgos and A. C. Flemmer, *J. Arid Envir.* 35: 251-268 (1997).

[22] W. G. Gold and M. M. Caldwell, *Oecologia* 80, 289-296 (1989).

[23] H. D. Giorgetti, Z. Manuel, O. A. Montenegro, G. D. Rodríguez and C. A. Busso, *Phyton, Int. J. Exp. Bot.* 69, 91-108 (2000).

[24] N. Monsi and Y. Murata, in: I. Setlik (Ed.) Wageningen, The Nederlands, pp. 115-129 (1970).

[25] L. F. M. Marcelis, *J. Exp. Bot.* 47: 1281-1291 (1996).

[26] R. S. Nowak, C. L. Nowak and J. E. Anderson, *Great Basin Natur.* 53, 222-236 (1993).

[27] D. L. Anderson, *Ecol. Arg.* 4, 9-11 (1980).

[28] B. J. Wilsey and H. W. Polley, *Oecología* 150, 300-309 (2006).

[29] J. G. Ehrenfeld, *Ecosystems* 6, 503-523 (2003).

[30] M. Vilá and J. Weiner, *Oikos* 105, 229-239 (2004).

[31] A. T. Bleak, N. C. Frischknecht, A. Perry Plummer and R. E. Eckert Jr., *J. Range Manage.* 18: 59-65 (1966).

[32] M. C. C. De Graaf, P. J. M. Verbeek, R. Bobbink and J. G. M. Roelofs, *Acta Bot. Neerl.* 47, 89-111 (1998).

[33] M. M. Caldwell, J.H. Richards, D. A. Johnson, R. S. Nowak and R. S. Dzurec, *Oecologia* 50, 14-24 (1981).

[34] O. A. Fernández, R. E. Brevedan and A. O. Gargano, *El pasto llorón. Su biología y manejo.* CERZOS-UNS (Eds.), Bahía Blanca, Argentina (1991).

[35] M. A. Ruiz, A. D. Goldberg and O. Martínez, *Phyton, Int. J. Exp. Bot.* 77, 7-20 (2008).

[36] J. A. Young, R. A. Evans and P. T. in: R. Elston (Ed.), Nevada Survey, Reno, pp. 187-215 (1975).

[37] Granite Seed Company, *The Granite Seed Catalog.* Available: http://graniteseed.com (2003-2004).

[38] R. A. Evans and J. A. Young, *J. Range Manage.* 36, 95-398 (1983).

[39] B. A. Roundy, J. A. Young and R. A. Evans, *Agric. Ecosyst. Envir.* 25, 2-3 (1989).

[40] B. A. D. Hetrick, G. W. T. Wilson and T. C. Todd, *Can. J. Bot.* 68, 461-467 (1990).

[41] D. G. Ogle, L. St. John, L. Holzworth, S. R. Winslow and T. A. Jones, *Basin Wildrye.* NRCS Plant Guide. USDA (2002).

[42] A. L. Cabrera, in: E. F. Ferreira Sobral (Ed.), ACME, Buenos Aires, Argentina (1976).

[43] R. A. Distel and R. M. Bóo, in: E.N. West (Ed.), *Proc. Vth Int. Rangeland Cong.*, *Soc. Range Manage.*, Salt Lake City, USA, pp. 117-118 (1996).

[44] A. C. Flemmer, C. A. Busso, O. A. Fernández and T. Montani, *Can. J. Plant Sci.* 82: 539-547 (2002).

[45] C. A. Busso and J. H. Richards, *J. Arid Envir.* 29: 239-251 (1995).

[46] T. C. Kaspar and R. P. Ewing, *Agron. J.* 89, 932-940 (1997).

[47] M. Giovannetti and B. Mosse, *New Phytol.* 84: 489-499 (1980).

[48] S. R. Olsen and L. E. Sommers, in: A. L. Page, R. H. Miller and D. R. Keeny (Eds.), Madison, Wisconsin, *Am. Soc. Agron.* pp. 403-430 (1982).

[49] Grupo Infostat, *Infostat,* FCA, Univ. Nac. Córdoba, Argentina (2008).

[50] W. Yuan, G. Zhou, Y. Wang, X. Han and Y. Wang, *Ecol. Res.* 22: 784-791 (2007).

[51] A. L. Cabrera and E. M. Zardini, *Manual de la Flora de los Alrededores de Buenos Aires*. ACME S.A.C.I., Buenos Aires, Argentina (1978).

[52] E. Cano, *Pastizales Naturales de La Pampa. Descripción de las Especies más Importantes*. Convenio AACEA-Prov. La Pampa, Argentina (1988).

[53] R. W. Brown, in: D. J. Bedunah and R. E. Sosebee (Eds.), *Soc. Range Manage.*, Denver, USA, pp. 291-413 (1995).

[54] C. A. Busso, B. L. Perryman and H. A. Glimp, in: *89th Ann. Meet. Ecol. Soc. Am.*, Portland, Oregon, pp. 77 (2004).

[55] J. R. Stroh, *Agron. J.* 63: 512-513 (1971).

[56] M. T. Anderson and D. A. Frank, *J. Range Manage.* 56, 501-516 (2003).

[57] O. R. Jewiss, *J. Brit. Grassl. Soc.* 27, 65-82 (1972).

[58] T. G. O'Connor and G. A. Pickett, *J. Appl. Ecol.* 29, 247-260 (1992).

[59] B. E. Olson and J. H. Richards, *Oecologia* 76, 1-6 (1988).

[60] B. E. Olson and J. H. Richards, *Oikos* 51: 374-382 (1988).

[61] A. H. Fitter, *Plant Soil* 45, 177-189 (1976).

[62] L. J. Marty *Development of Acid/Heavy Metal-Tolerant Cultivars Project,* Bridger Plant Materials Center. Available: http://plant-materials.nrcs.usda.gov/pubs/mtpmcrnleci 4b.pdf (2001).

[63] L. J. Perry and S. R. Chapman, *J. Range Manage.* 28, 271-274 (1975).

[64] D. D. Briske and J. H. Richards, in: M. Varra, W. A. Laycock and R. D. Pieper (Eds.), *Soc. Range Manage.,* Denver, USA, pp. 147-176 (1994).

[65] C. A. Busso, R. E. Brevedan, A. C. Flemmer and A. I. Bolletta, in: A. Hemantaranjan (Ed.), *Adv. Plant Physiol.* V, 341-395. Scientific Publishers, Jodhpur, Available: http://angelfire.com/ak5/adv_pp /index.htm (2003).

[66] M. A. Torres, *Nasella (Gramineae) del Noroeste de la Argentina*, CIC (Ed.), 13, 5-46 (1997).

[67] M. E. Barkworth and M. A. Torres, *Taxon* 50, 439-468 (2001).

[68] M. F. Allen, J. H. Richards and C. A. Busso, *Biol. Fert. Soils* 8, 285-289 (1989).

[69] C. A. Busso, D. D. Briske and V. Olalde-Portugal, *Oikos* 93, 332-342 (2001).

[70] C. Saint Pierre and C. A. Busso, *Phyton, Int. J. Exp. Bot.* 75, 21-30 (2006).

[71] E. B. Allen and M. F. Allen, *New Phytol.* 104, 559-571 (1986).

[72] L. J. Cerligione, A. E. Liberta and R. C. Anderson, *Can. J. Bot.* 66, 757-761 (1988).

[73] B. A. D. Hetrick, G. W. T. Wilson and D. C. Hartnett, *Can. J. Bot.* 67, 2608-2615 (1989).

[74] C. A. Gehring and T. G. Whitman, *Trends Ecol. Evol.* 9, 251-256 (2004).

[75] C. E. Prescott, *For. Ecol. Manage.* 220, 66-74 (2005).

[76] D. W. Hilbert, D. M. Swift, J. K. Detling and M. I. Dyer, *Oecologia* 51, 14-18 (1981).

[77] D. P. Poppi, D. J. Minson and J. H. Ternouth, *Austr. J. Agric. Res.* 32, 99-108 (1981).

[78] D. J. Tilley, *Basin Wildrye,* Adv. Eval. Prelim. Rep., Range Conserv. (Plants) (2005).

In: From Seed Germination to Young Plants
Editor: Carlos Alberto Busso

ISBN: 978-1-62618-653-8
© 2013 Nova Science Publishers, Inc.

Chapter 11

Seedling Dynamics in an Environmental Gradient of Andean Patagonia, Argentina

Pablo Martín López Bernal, María Florencia Urretavizcaya and Guillermo E. Defossé [*]

Patagonian Andes Forest Research and Extension Center (CIEFAP-CONICET)
and University of Patagonia, Esquel, Chubut, Argentina

Abstract

In this chapter, we analyze the most important environmental and biotic factors that influence seedling dynamics of three of the most conspicuous species grown in a forest-steppe gradient of Patagonia in Argentina. In a short distance of about 80 km in this gradient, vegetation abruptly changes from pure *Nothofagus pumilio* forests (grown from the timberline down to altitudes of about 1,200 m a.s.l., with 2,000 to 3,000 mm yr^{-1} of precipitation), to mixed *Austrocedrus chilensis-Nothofagus dombeyi* forests and pure *A. chilensis* forests towards the east, and isolated *A. chilensis* forests in the ecotone with the steppe (at altitudes from 900 to 500 m a.s.l. and 1,200 to 500 mm yr^{-1} of precipitation). Vegettion then changes to a grassland steppe zone, located at about 400 to 600 m a.s.l., and in which precipitation barely reaches 300 to 400 mm yr^{-1}. Although precipitation values greatly differ, all areas share the same Mediterranean climate, with rains concentrated during late fall and winter followed by a marked dry period during spring and summer. The most important native species in this gradient are the trees, *Nothofagus pumilio* (lenga), in the higher slopes of the Andes, *Austrocedrus chilensis* (cypress or ciprés de la cordillera) at the mid slope and in the forest-steppe ecotone, and the grass *Festuca pallecens* (coirón blanco), in the steppe zone. All three species reproduce by seed, and form transient soil seed banks. In the case of *Festuca pallescens* this bank is annually replenished, while for *A. chilensis* and *N. pumilio* seed production occurs during favorable periods that may vary from two to several years. Among the environmental factors that affect seedling dynamics, water stress during the summer dry period appears

---

[*] E-mail: gdefosse@ciefap.org.ar.

to be of paramount importance for *Festuca* and *A. chilensis* seedlings, and less important for *N. pumilio*. For the latter species, and after germination, light availability appears to be crucial for further sapling growth. Right after germination, young *Festuca* or *A. chilensis* seedlings appear to grow better under protected microsites provided by either an adult *Festuca* plant or a nurse shrub in the case of *A. chilensis*. This facilitation effect seems to change as seedlings become older, shifting this interaction to competition as secondary succession progresses. Apart from the environmental conditions, seedling dynamics of all three species are also affected by natural (fires, earthquakes, avalanches, falling out of senescent trees in the case of *N. pumilio*) and anthropogenic disturbances (fire, grazing, browsing, and logging). Although the inferences made here about seedling dynamics may be valid for the whole area of distribution of *A. chilensis* and *F. pallescens*, for *N. pumilio*, by instance, will be constrained to continental Patagonia (excluding Tierra del Fuego), in which Mediterranean climatic conditions prevail.

Introduction

Patagonia is a vast territory (about 1 million km^2) that covers the southern tip of South America, going from the Atlantic Ocean to the East in Argentina, to the Pacific Ocean to the West in Chile. Its physiography include plateaus, valleys, rivers, lakes and glaciers, and in its western part it is crossed, north to south, by the Andean cordillera.

The climate of Patagonia has been modeled by the proximity of the two oceans (moderating extreme temperatures), and has also been greatly influenced by the westerly winds that blow year around from the southern Pacific Ocean [1]). Clouds formed in the southern Pacific are driven toward the East-North-East by these winds, discharging most of their moisture (in the form of either rain or snow) on the western slopes and higher peaks of the Andes in Chile.

This precipitation rapidly decreases as the air masses reach the eastern Andean slopes and then the steppe zone in Argentine Patagonia [1-3]. This sharp precipitation gradient, together with either natural (fire, earthquakes, volcanism, landslides, and wind through) and/or anthropogenic disturbances (fire, logging, grazing, afforestation with exotics), have helped shape the structure of the different plant communities we seen today in Patagonia [1, 4-8].

In this chapter, we will focus on seedling dynamics of three of the most conspicuous species grown in this environmental gradient in north-central Patagonia in Argentina, and how this dynamics has been affected by different natural and anthropogenic disturbances.

The species are, from west to east, the trees *Nothofagus pumilio* (Poepp. and Endl.) Krasser (locally known as *lenga*), grown in the higher slopes of the Andes, *Austrocedrus chilensis* (D. Don) Pic.Serm.et Bizzarri (cypress or *ciprés de la cordillera*), located in mid-slopes and in the forest-steppe ecotone, and the grass, *Festuca pallecens* (St.Yves) Parodi (locally known as *coirón blanco*), which appears in the steppe zone.

Nothofagus pumilio forests are distributed, North to South, along the whole Andean Region of Argentine Patagonia, while *Austrocedrus chilensis* forests are only found in the North-Central Patagonian Andean region [9, 10]. *Festuca pallecens,* by contrast, is the most important grass species of the Sub-Andean Floristic District of the Patagonian Phyto-geographic Province [11, 12], occupying a narrow and discontinuous belt in the piedmont of the Andes in western Patagonia, widening toward the east as it approaches the Atlantic Ocean in Santa Cruz province.

Most of the data, analyses and the inferences made here about seedling dynamics of these three species may be valid for their whole distribution area in continental Patagonia (excluding the island of Tierra del Fuego), although it is worth to note that they were based on studies carried out in different areas of Chubut and Río Negro provinces, as it is shown in Figure 1.

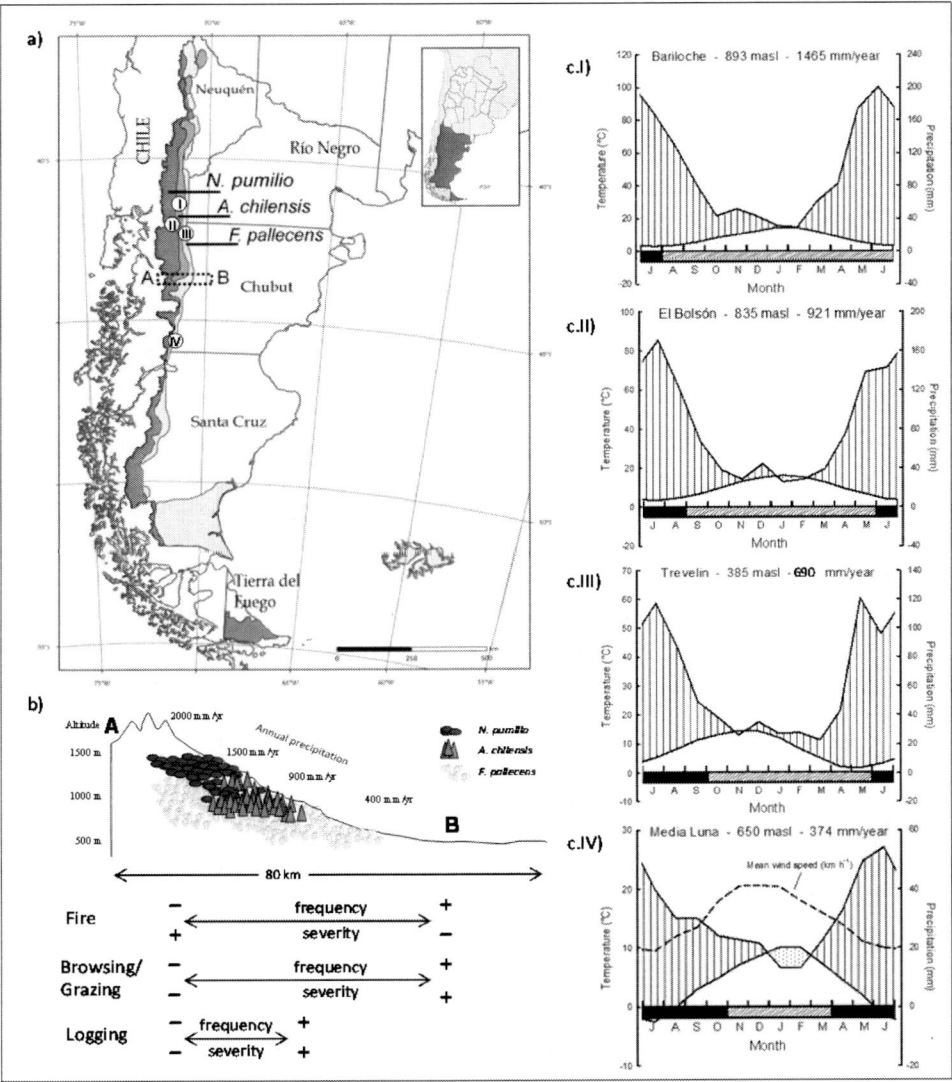

Figure 1. a) The Patagonian territory of Argentina (upper right). Enlarged, the distribution of *Nothofagus pumilio, Austrocedrus chilensis* and *Festuca pallescens* in the eastern slopes of the Andes, across the five Patagonian provinces. Each circle represents intensive study sites, while the A-B is a crosscut of the Andes in parallel 44 SL; b) Schematic representation in the A-B cut of the distribution of the three species in the forest-steppe gradient, and the regime of the associated disturbances; c) Climatic diagrams of the study sites where most data were drawn (in the format of Walter and Lieth, 1967, *Klimadiagramm-Weltatlas*, VEB Gustav Fischer, Vienna).

Although from the Andean peaks to the steppe zone there is a huge difference in total annual precipitation, the whole Patagonian region in the Argentine territory shows a Mediterranean type of climate, with rains concentrated during winter and early spring, followed by a marked dry period during summer and early fall (Figure 1.c).

This characteristic makes the whole region very susceptible to fire disturbance during the summer drought [7, 13, 14].

Description of the Ecology and Seedling Dynamics of the Species Studied

Festuca pallescens (coirón blanco)

Ecological Description and Use

Festuca pallescens, is the dominant grass species in the Sub-Andean Floristic District of the Patagonian Phytogeographic Province, reaching from 60 to 95% of its vegetation cover [11, 12, 15, 16]. The western limit of this District is not abrupt, and many times its vegetational traits intermix with those related to *A. chilensis* forests in northwestern and central Patagonia, and with *N. pumilio* forests to the south of the 44° S latitude parallel. Most vegetation of this District, from northern to southern Patagonia, has been exposed to disturbance by year-long sheep and cattle grazing since the beginning of the last century [6, 12, 17-20]. This grazing pressure, however, have not been homogeneous, and a great variety of grassland conditions occur today. These conditions may vary from lightly grazed areas with spatial homogeneity and high diversity and cover, to others heavily grazed areas with sparse vegetation cover and low specific diversity [16, 19, 21, 22]. Continuous grazing is then the main human disturbance affecting vegetation of this District, followed by episodic fires (either natural or human caused). It is interesting to note that even in homogeneous paddocks with uniform stocking rates, differences in topography can shift grazing pressure from plain areas to the warmer and wind protected north facing slopes. In these slopes, overgrazing has produced a reduction in cover of *F. pallescens*, increasing the number of bare soil patches that are colonized by annuals and other perennials instead of by seedlings of this species [23, 24]. Since there is today no place in this District that remained free of grazing pressure, it was hypothesized that under pristine conditions, and because this grassland steppe evolved with a few native grazers (*i.e.* the guanaco, *Lama guanicoe* Müller), *F. pallescens* plants used to grow closer one to another, providing thereby a great number of protective sites for seed germination and seedling establishment [22]. At the time of seed dispersal, there were a great number of safe sites [*sensu* 25] available for seed germination, and subsequently seedling survival and establishment. The introduction of domestic grazers (cattle, sheep, and to a lesser degree horses) may have disrupted these pristine conditions, and through defoliation and trampling, has reduced vegetation cover and modified top soil physical conditions. Since *F. pallescens* reproduces strictly from seed [22, 26, 27], its replacement in successive generations will solely depend on its capacity to cope with the environmental and disturbance pressures posed by the system (from seed germination to seedling establishment). In this chapter we explore the seed characteristics and seedling dynamics of this species, regarded as a key plant species for livestock production in the western steppes of the piedmont of the Andean cordillera. Good management schemes based on sound science are needed, then, to halt deterioration and promote the sustainable use of *F. pallescens* in this grassland steppe.

Germination and Seed Characteristics of *F. pallescens*

Festuca pallescens produces seeds annually, and after ripening their dispersal occurs gradually, north to south and from lower to higher altitudes, from November (late spring-beginning of the summer in the Southern Hemisphere) to early February (mid- to late summer) [28]. After dispersal, *F. pallescens* seed replenish its transient soil seed bank, and start germinating after the end of the dry summer season and when the first autumn rains moisten the upper soil profiles [22, 27]. *Festuca pallescens* seeds are small and light, lacking of specialized appendages to help them into the soil or to retain water for germination. These characteristics make these seeds to easily germinate when soil temperatures and moisture are favorable, but fail to be established as seedlings if any unfavorable environmental event occurs thereafter [24, 29, 30]. *Festuca pallescens* seeds are in general relatively homogeneous, with 442,477 ± 11,400 seeds per kg. In several laboratory trials (determined by the tetrazolium test), *F. pallescens* seeds showed a high viability percentages, reaching in most cases above 99%. These laboratory tests also showed that the bulk of seed germination occurs when soil temperatures are from 10 to 15°C, and upper soil moisture is above 10 %. When soil moisture is below 8%, seed germination is severely restricted [29]. These favorable conditions for germination naturally occur in the field in autumn, during a short period that occurs right after rains began and until mean upper soil temperatures fall below 5°C [29]. Non-germinated seeds remain in the soil underneath a snow mantle during the winter; germination resumes in the spring after snowmelt, and sharply decline thereafter as the dry summer season approaches. A study showed that *F. pallescens* seedlings germinated in fall have more chances of further survival during the next dry summer than those germinated in spring [22]. While seedlings germinated either in the fall or in early spring may present similar aboveground development at the beginning of the summer, excavation of their root systems showed that fall-germinated seedlings had considerable higher root biomass and had elongated it faster to deeper soil horizons than those germinated during the spring [22]. This situation gives an advantage for seedlings germinated during the fall to cope with the summer water stress and become established, and may explain why they survive better than spring germinated seeds.

Environmental and Anthropogenic Disturbances Affecting *F. pallescens* Seedling Dynamics

Among the multiple disturbances that affect *F. pallescens* seedling establishment after germination, are those categorized as "natural", or posed by the fluctuating environmental conditions (and whose consequences are beyond human influence), and others, that are directly caused by human interventions. However, neither of these factors act alone, and in general, the fate of seedlings reaching adulthood would result of how well are they able to overcome these interacting disturbances.

Grazing, with its defoliation and trampling effects could be considered as the most important human-caused disturbance that triggers other processes affecting *F. pallescens* seedling dynamics. Continuous defoliation seems to debilitate adult plants and may affect its reproductive capacity, while the combination of defoliation and trampling creates big patches around established *Festuca* bunches, reducing the survival and establishment of germinated seeds. This effect was corroborated by comparing free-range grazed areas with nearby areas excluded to grazing [16, 31]. Sheep grazing not only reduced aboveground biomass of adult plants, but also these animals tend to step on inter-space microsites (located among adult plants) rather than on, or nearby, adult *Festuca* plants. Apart from modifying top soil physical properties, trampling negatively affects seedling survival in these microsites, being this effect consistent with Balph and Malecheck [32], who found higher trampling incidence on tussock interspaces than on tussocks in similar semiarid grassland. It is worth to note that other studies showed that grazing may either promote [33-36], or reduce [31, 37, 38] seedling establishment. These differences, however, could be related to variations in the kind of grassland grazed, the grazing pressure, the season of grazing, and the stocking rate used. In *Festuca* grasslands of Patagonia, the yearlong grazing scheme generally applied, together with the usual stocking rates used, appears as a major disturbance affecting *F. pallescens* seedling dynamics, significantly reducing their survival as compared to ungrazed areas.

Another factor that affect seedling establishment is competition with adult plants. By using root exclusion tubes [similar to those described in 39], an experiment showed that competition for water with adult *Festuca* plants may be negligible right after germination and up to seedlings reach the stage of four leaves, and up to one tiller. After that stage and when seedlings become older (five leaves or more and to one tiller), intra-specific competition with adult plants severely restricts seedling survival [31]. As mentioned for other grasslands, intra-specific competition may be more detrimental for seedling survival than inter-specific competition [40, 41]. This is because seedlings and adult plants affected by competition share similar root phenology and hence water demands [31]. This competition effect was also noticeable when comparing seedlings grown in adjacent microsites (near an adult *Festuca* plant) with those grown in interspace microsites (open gaps among adult *Festuca* plants), when grazing was excluded. At early stages of seedling development (*i.e.* right after germination and until seellings develop up to four leaves), adjacent microsites showed significantly higher emergence and survival than those germinated in interspace microsites. This may indicate a protective, facilitator effect of adult plants to the young seedlings grown nearby at this early stage. However, as seedlings become older and develop a more extensive root system, the facilitation effect gradually shifts to a competition effect, and that is why, at older stages, seedling survival was significantly higher in interspace microsites [31].

In any water-limited environment, the availability of water near the soil surface appears as a key factor for seedling emergence and survival [37, 42]. Since *F. pallescens* grows in a Mediterranean type of climate, water starts to be a limiting factor (mainly in the upper soil profile) as the growing season advances. This is a time when seedlings need to become established. Higher soil water content in this profile could be found during the whole growing season near protecting adult plants than on interspace microsites, although at deeper soil profiles this difference disappears [31]. Why then seedlings grown in adjacent microsites diminish their survival rates, compared to those grown in the interspaces, as the growing season advances? As mentioned before, root competition for water is less intense at the interspaces, and allows seedlings to elongate their respective root systems toward deeper soil horizons, where more water might be found. This rapid elongation is favored by the lower root length densities in the interspaces than at the adjacent microsites, as it is commonly found in the Patagonian steppe [43, 44].

Another disturbance that may preclude *Festuca* seedling establishment is cryoturbation, or commonly called "frost heaving"[2]. In most of the Sub-Andean district and when the spring is cold and windy, the interspaces (or patches among *Festuca* bunchgrasses) can be exposed to this process. The appearance of this process and its effects on *F. pallescens* seedling survival was studied for three consecutive years [22]. Results showed that frost heaving consistently occurred every year from September to November (early- to mid-spring and after snow melts), and started to disappear during the early summer and for the rest of the year.

One particularity of this phenomenon it that it occurs with similar intensity in slopes and plains, and in grazed and ungrazed areas, although it vanishes earlier in early summer on north facing slopes (sun-side areas) than on plain areas. The effect of frost heaving on *Festuca* seedlings is that it physically pushes them out of the soil, thereby exposing their root systems to the combined effects of desiccating high winds and low water availability during the subsequent summer [18, 22]. This effect causes a high *Festuca* seedling mortality during the summer, especially for those seeds germinated during the spring.

As mentioned before, seedlings germinated earlier during the fall may increase their survival if they had the chance to elongate their root systems to deeper soil horizons. It is important to mention that frost heaving has been considered as one of the major causes of failure in seedling establishment in some North American semiarid rangelands [37], where frost heaving has accounted for up to 75% of seedling mortality during some years [45].

Fire is another disturbance that affect *Festuca* grasslands and hence their seedling dynamics. In its whole area of distribution, lightning-caused fires may occur during summer storms [46-48], while human ignitions, started in ecotone areas, is another cause of fires reaching and affecting *Festuca* grasslands. Although fire has been proven to have a little effect on adult plants (they fully recover their aboveground biomass in three to four years after a fire event), seeds of this species, which have no protecting structures, might be differentially affected by high temperatures [49]. In laboratory experiments and when seeds received a temperature treatment of 80°C for 10 min, their germination percentage was not significantly different than those of non-treated seeds [50]. But when the temperature treatment was increased to 120°C for 5 min, almost all seeds died, apparently because they could not resist the thermal shock [24]. In a field experiment, *Festuca* seeds exposed to a surface fire (which reached temperatures similar to those exposed to 80°C in the laboratory) showed similar germination percentages than those in the control unaffected by the fire. However, when these seeds germinated and developed as seedlings, their performance was very weak, showing difficulties for establishment and a very poor competition performance as compared to those germinated in fire-free areas [24]. As Hanley and Fenner [51] suggested for other grass seedlings, it is likely that *Festuca* seedlings originated from seeds exposed to high temperatures, may have suffered a deleterious effect on their embryos by the thermal shock, and that may explain why their later performance is weaker than those originated by normal seeds.

---

[2] Frost heaving is a process in which ice is formed beneath or in between upper soil particles ($\approx$ 10-20 cm) when freezing conditions occur in the atmosphere. This ice expands then in the direction of heat loss (vertically toward the surface), pushing upward the upper layers of the soil and the root system of seedlings grown in this microenvironment.

In western Patagonia, winds blow continuously throughout the year at relatively high speed, and they can have a detrimental effect on *F. pallescens* seedling survival. At Media Luna Ranch, were most of the studies about *Festuca* seedlings dynamics were carried out, measurements of wind speed at different heights were recorded during several years (see climatic diagram of Media Luna in Figure 1.c.IV).

Wind speed showed a similar trend as mean annual temperature, being significantly higher during the summer as compared to the winter. During the summer dry season, data showed that at 1.5 m aboveground, wind might blow at an average of 35.9 km h^{-1}, while at 5 to 10 cm aboveground (in between adult *Festuca* plants), it diminished to 11.5 km h^{-1} at windward sides of adult *Festuca* plants, and to 4.7 km h^{-1} at leeward sides [22]. The higher wind speed registered at windward sides caused a desiccating effect in the upper soil layers; seedlings emerged in these unprotected sides, and died earlier than those grown at leeward (protected) microsites [22].

In any grassland steppe, vegetation dynamics is greatly influenced by the interactions of multiple disturbances [52]. In the case of *Festuca* grassland, in which not only *F. pallescens,* but also other less important grass species reproduce strictly by seed, the study of seedling dynamics, and how are they affected by disturbances, is essential to understand its functioning. In this chapter we explored and provided evidence of some of the main disturbance factors affecting the recruitment of seedlings of *F. pallescens*. In general, although this grassland steppe appears to be adapted, and co-evolved with some of the described disturbances since ancestral times (*i.e.* fire, high wind speeds), a more recent disturbance (grazing) emerge as the main factor that modified its former structural and dynamic characteristics.

Changes produced by overgrazing triggered other disturbances that certainly reduced the availability of safe sites for *F. pallescens* seedling germination and establishment. Future studies should be focused at deepening the knowledge of the combined effects of several disturbances on seedling dynamics. Among them, the interaction of grazing and fire appears as one of the most relevant [24, 50, 53].

Austrocedrus chilensis (cypress)

Ecological Description and Use

In the eastern slopes of the Andes in northwestern Patagonia in Argentina, *Austrocedrus chilensis* forests occupy about 140.000 ha. *Austrocedrus chilensis* stands occur in a variety of sites in a 60 km west to east narrow strip between the 37°08'09" and 43°43'57" SL parallels [54, Figure 1]. As in the whole Andean Patagonia, this narrow strip comprises one of the most extreme West to East precipitation gradients in the world [3]. At high altitudes in humid sites in its western area of distribution, *A. chilensis* grows associated to *Nothofagus dombeyi* (Mirb.) Oerst. (locally called coihue), forming mixed forest stands.

In mesic areas at medium altitudes, *A. chilensis* could grow in either mixed or pure stands, while in xeric areas in the east toward the steppe, it grows in isolation or in groups of scattered trees (Figure 1.c.II-III). Although some trees of this species can reach up to 1,000 years old on the western slopes of the Andes in Chile [55], individuals older than 400 years are rare on the eastern Andean side in Argentina [3].

As in most of the eastern Andean slopes in Argentine Patagonia, soils upon which *A. chilensis* stands develop are derived from volcanic ashes [56]. The evolution of these soils has been mainly linked to the precipitation regime and the natural barriers that modeled ash distribution [57]. In sites with high precipitation values to the west, Andosol soils could be found, possessing high fertility and water retention [56].

Toward the east, soils are subjected to a seasonal dry period, and allophanic substances appear [56]. Soils of either area are well provided with nutrients (except phosphorous), allowing for a healthy growth and development of *A. chilensis* stands. Since about 65 years ago, however, some *A. chilensis* stands started to present a disease initially called "mal del ciprés", which causes a rapid decline and death of *A. chilensis* trees. Recent studies showed that this disease is associated to a *A. chilensis* root infection by the pathogen *Phytophtora austrocedrae* [58] and that seedlings, saplings, and adult trees of this species are all susceptible to this lethal disease [59, 60]. The low genetic diversity found in *Phytophtora austrocedrae* suggests that this pathogen may be an alien, newcomer species in Patagonia [60]. Since this disease could affect seedlings, saplings and adult trees, its effects should be considered in any future natural regeneration or artificial restoration plan.

Under natural conditions, the pattern of *A. chilensis* stands regeneration varies according to the type of habitat they develop and the associated species present in the area [61]. In pure stands grown under mesic conditions, *A. chilensis* establishment occurs in small to medium gaps among mature trees [61, 62], and is often associated to high herbaceous and shrub cover [62]. In xeric sites, instead, *A. chilensis* seedlings sporadically establish underneath shrubs canopies of the genera *Discaria, Schinus, Berberis, and Lomatia* [61, 63, 64]. The most important climatic conditions that precludes initial *A. chilensis* seedling establishment in these xeric areas is the occurrence of extreme dry periods during spring and summer [65].

As in many other temperate forests, fire is the major natural disturbance affecting *A. chilensis* forests [61, 63]. Ancient chronicles [66-69], studies about pollen and charcoal records from dated sediments [70-73], as well as analyses of tree ring and vegetation structure [47, 65, 74, 75] have allowed to reconstruct past fire regimes, and the role that humans and climatic variations have played in *A. chilensis* forest dynamics.

These studies showed that the majority of today *A. chilensis* forest stands, either mixed or pure, have been originated after natural or human-caused fires occurred mainly within the last two centuries. At the beginning of the XXth century, when European settlers colonized western Patagonia, *A. chilensis* stands started to be highly impacted by different human disturbances. *Austrocedrus chilensis* trees were cut for house construction and/or used as firewood, and many times complete stands were burned out to open areas for livestock production [76] and, to a lesser degree between the 70s and 80s, to plant exotic pines [77].

However, and considering that *A. chilensis* is a highly valuable timber species in Patagonia, several silvicultural studies [78-80], and management techniques [80, 81], have been proposed to sustainably manage *A. chilensis* stands. Besides these efforts, the actual management of *A. chilensis* forests only implies "salvage cuttings" [*sensu* 82], mainly done on stands that may have suffered the effects of fire, or that present signs of the before mentioned disease locally called "mal del ciprés" [58]. Although special logging recommendations have been proposed in areas affected by this disease to further facilitate *A. chilensis* natural regeneration [77], these recommendations have not yet been enforced. As a consequence, the combined action of logging and subsequent grazing may act in a synergistic way, greatly reducing the chances of natural reestablishment of *A. chilensis* stands.

Germination and Seed Characteristics of *A. chilensis*

Austrocedrus chilensis is a dioecius plant that reproduces almost exclusively by seed. Pollination generally occurs at the end of the spring; fruits mature during the summer [83] and the seeds, winged and very small, are dispersed during the next fall and up to mid-winter.

After the seed rain, up to 5000 seeds per m^2 could be found in the soil of pure stands. However, an important fraction of these seeds are not viable, and the number of apparently viable seeds may be reduced to nearly 3000 per m^2 [84]. Seeds remaining in the soil bank must over-winter in order to break dormancy [85] and start germinating during the next spring. Under field conditions, germination has been observed that is highly variable [84]. Laboratory tests showed that the average number of *Austrocedrus* seeds is 260,300 ± 29,400 per kg, and their germination percentages often range from 60 to 80% [86].

Since *A. chilensis* forms a transient soil seed bank [*sensu* 87] and reproduces almost exclusively by seed, regeneration is dependent upon the chance of viable seeds of finding "safe sites" [25] to successfully germinate and become established. *Austrocedrus chilensis* regeneration depends then on the combination of several limiting factors, such as the availability of viable seeds in the soil seed bank (many times constrained by the distance from the mother tree, which has low dispersal capacity [61]), its transient seed bank [84], the occurrence of favorable weather periods after the seed rain [65], and the presence of propitious soil microsites for seed germination and early seedling establishment. On unburned sites, the protection provided by understory vegetation seems to benefit *A. chilensis* seed germination and recruitment [62].

Apart from these natural factors and after germination, *A. chilensis* seedlings are affected by a diversity of natural and anthropogenic disturbances that may promote, or in some cases preclude, their survival and further establishment as saplings and adult plants.

Environmental and Anthropogenic Disturbances Affecting *A. chilensis* Seedling Dynamics

As it happens with many other species, germination and initial emergence of *A. chilensis* seedlings in the field is mainly determined by an adequate combination of soil moisture and temperature [88]. Weather conditions during the growing season after germination may also influence seedling establishment, being this effect more evident in arid or semiarid areas than in mesic environments [61, 89]. The relationship between facilitation and weather conditions during the stages of germination and early seedling establishment seems to be crucial for plant regeneration success [90, 91]. In Patagonia, studies carried out 15 years after wildfires on xeric sites, showed that *A. chilensis* natural regeneration is favored by the presence of shrubs acting as nurse plants [91]. In these sites, nurse plants seem to improve their under-canopy micro-environmental conditions (by lessening the effects of extreme temperatures and water stress), creating then safe sites for *A. chilensis* seed germination and early seedling establishment. Another study supporting these statements showed that this nurse-protecting effect may be crucial for seed germination and seedling emergence in xeric sites, but its importance diminishes in mesic sites (Figure 1.c.II-III) [92]. These authors speculate that in mesic sites, higher water availability than in xeric sites during the summer masked the potential association of plant cover with seedling emergence. The high water availability during critical periods (*i.e.* abundant rains during the growing season) may also produce similar masking effects in xeric sites. During spring and summers with above-average precipitation values, maximum emergence is achieved in areas outside shrub canopy covers. If rains are scarce and a dry period occurs during the growing season, *A. chilensis* seedling emergence increases in areas with high shrub cover [92].

Similar to what it occurs in xeric sites, *A. chilensis* seedling survival in mesic sites is benefited by the plant cover provided by shrub nurse plants if the growing season present below-average precipitation values [92]. This positive effect of plant cover on survival during dry growing seasons has also been indicated for other species grown under similar conditions [90, 93]. Besides this and during a dry summer, shrub cover may also attenuate high temperatures produced in the seedling microenvironment, counterbalancing the negative effects of the dry growing conditions and reducing mortality rates of shaded seedlings [94-98]. In general, all studies about *A. chilensis* seedling dynamics showed that mortality is very high during the first growing season after germination, being higher in xeric as compared to mesic areas, and that further *A. chilensis* establishment is strongly influenced by climatic variability [61, 65, 91, 99].

Apart from the fluctuating environmental conditions, *A. chilensis* seedling dynamic is affected by different disturbances. Among them, fire seems to play a major role. When a fire occurs in an *A. chilensis* stand, it not only affects the soil seed bank and kills seedlings and saplings, but also changes macro- and micro-environmental conditions through a drastic modification of the forest structure [100]. In *A. chilensis* forests, the natural fire regime varies according to the precipitation gradient. In general, natural fire frequency increases from humid sites in the west, towards xeric sites in the east [63] (Figure 1.b). In the past, fire frequency increased in the majority of *A. chilensis* stands after 1850, coincidentally with an expansion in the number of indigenous people that inhabited the region and the beginning of European settlement. It is important to point out that human-caused fires were set for hunting purposes by aboriginals, and from the last decades of the XIX century and up to 1930, to clear areas for grazing and/or agriculture by oncoming European settlers [76]. This frequency peaked at the end of the XIX century, and started to decline in the mid 30's of the XXth century, because indigenous population decreased, and the creation of several National Parks in the region brought about the policy of fire suppression [47]. After that and up to the present, and although a fire control policy was strictly enforced, naturally, accidentally, and/or purposely set fires still occur in *A. chilensis* forests.

In general and once a fire occurs, the duration and magnitude of post-fire changes involved in the recovery of a forest stand are directly related to the (i) environmental conditions prevailing at the time of the fire (intensity); (ii) damages on the soil environment after fire has passed (severity), and (iii) specific conditions of the site in which the fire occurred [101]. Besides these and specifically for *A. chilensis* stands, post-fire recovery will also depends on the presence of nearby surviving female *A. chilensis* trees, which would provide the seed source for stand regeneration. A study carried out in two contrasting *A. chilensis* burned areas (mesic and xeric) showed that in both sites the consumption of biomass and the reduction of vegetation cover increased soil temperatures for several years, being this effect magnified in xeric as compared to mesic areas. Soil moisture, by contrast, appeared as a non-limiting factor in either site, since burned plots had always higher water content at all soil depths than unburned ones [102]. The higher water availability found in burned sites may be due to the lack of evapo-transpiring vegetation as compared to unburned sites.

In humid sites where mixed *A. chilensis – Nothofagus dombeyi* forests prevail, fires tend to be catastrophic and infrequent, while in mesic and xeric areas limiting with the Patagonian steppe, fires are more frequent and less severe [fire interval from 15 to 32 years, 63]. After a fire event, *A. chilensis* establishment is associated to site conditions, being rapid in humid and mesic sites and more slowly in xeric areas [61]. Burned areas also showed significantly less germinating seeds than unburned areas [84], adding another factor that may reduce the availability of "safe sites" for successful seedling germination and establishment. This synergistic effect (lower seed availability and higher soil temperatures) appear to be magnified in xeric sites. In contrast, in mesic sites, this effect seems to be lessened, particularly in the first few years after the fire. These findings were corroborated by several studies [61, 65, 91], who pointed out that *A. chilensis* seedling establishment in xeric sites is greatly influenced by unfavorable environmental conditions, and mainly by high temperatures. In general for xeric and mesic sites and after a fire event, the colonization of burned areas by pioneer herbs and shrubs could be a mechanism to avoid extreme high temperatures, that otherwise could overheat and dehydrate *A. chilensis* seedlings. Because of its most favorable water balance, this protective mechanism appears to operate faster in mesic environments than in xeric ones. This is why for xeric sites, where *A. chilensis* germination, growth and establishment are more sporadic and marginal, active restoration practices are recommended to speed up successful post-fire establishment.

Apart from the set of environmental conditions and the effects of fire, *A. chilensis* seedling dynamic is also greatly affected by livestock browsing. On many stands, the development of *A. chilensis* saplings is severely inhibited by browsing [103].

These inhibitory effects are manifested in seedlings and saplings by presenting stunting and poor form rather than showing a reduction in their abundance. High cattle grazing pressure on post-fire *A. chilensis* stands could inhibit its regeneration, leading to the establishment of a degraded shrub steppe with scattered trees [104]. The mortality of *A. chilensis* trees by fire and the subsequent grazing pressure act in a synergistic way, by limiting the availability of new seeds in one case and reducing the survival of established seedlings in the other. Both disturbances significantly preclude the natural regeneration of *A. chilensis* forests.

Nothofagus pumilio (lenga)

Ecological Description and Use

Nothofagus pumilio (Poepp. et Endl.) Krasser, is the most important native tree species of the Andean forests of Patagonia, mainly because of its ecological and economical importance. In Argentina, *N. pumilio* forests cover almost the entire length of the sub-Antarctic forests on the eastern slopes of the Andean Cordillera, from the 35° 35' S latitude parallel in the province of Neuquén to the 55° S in the province of Tierra del Fuego (Figure 1). This species usually occupies the upper portion of the altitudinal limit of woody vegetation (up to 2000 m.a.s.l.) in its northern distribution area in Neuquén province, while it grows near sea level in its southern distribution area in Tierra del Fuego [105-107].

Nothofagus pumilio is a very plastic tree species adapted to grow under a great variety of soils, environmental conditions, and disturbance regimes [108]. In fact, the species could be found in water limited environments, in which average annual precipitation may reach 500 mm per year (under a Mediterranean type of climate), to others reaching 3,000 mm per year (under either iso-hygro or Mediterranean type of climates). *Nothofagus pumilio* is also able to support extreme temperatures, from mean annuals of 3.5 to 4°C in upper altitudinal areas to 7 to 9°C in milder areas at lower altitudes or also near sea level [108].

In *N. pumilio* forests in which large scale disturbances are relatively low (such as those located in the north area of distribution in Chubut, Río Negro and Neuquén provinces), trees of this species could reach 300, and exceptionally, 360 yr old [109]. Although some trees may reach older ages, at about 140 yr old most of them are started to be affected by a trunk rot disease, which in some way difficult to determine their age [110].

In the northern part of its distribution in continental Patagonia, and similarly to what occurs with *A. chilensis* or *F. pallescens*, *N. pumilio* grows under a typical Mediterranean climate, with precipitation concentrated during winter and early spring, followed by a dry and mild period during summer and early fall. Going south, this regime gradually changes to more iso-hydric conditions, being precipitation more evenly distributed along the year. In the northern part of its distribution and up to the 52° S, however, the amount of annual precipitation is greatly influenced by the barrier imposed by the Andean Cordillera. As mentioned before, western humid winds coming from the South Pacific Ocean discharge most of the precipitation on the upper parts of the Andes. They then pass to the eastern slopes as more dry air masses that rapidly lose their humidity content. This makes that upper mountain ranges near the border with Chile may receive 5000 mm of precipitation per year, while in less than 50 to 80 km toward the Patagonian steppe, precipitation sharply diminishes to *ca* 500 mm annually [106, 111, 112]. To the South of the 52 and up to the 55° S parallel in the island of Tierra del Fuego, a regular, evenly distributed pattern of annual precipitation occurs [113]. Throughout its wide distribution area, *N. pumilio* stands are clearly distinguishable from other components of the Andean forests, being composed of simple mono-specific structures with narrow ecotones [107]. However, given the different environments in which it develops, *N. pumilio* presents different structures and regenerative dynamics, mainly associated with the frequency, magnitude and severity of natural disturbances (windstorms, fires, volcanic ashes, avalanches, landslides, or the falling of senescent trees) [10, 107]. Anthropogenic disturbances, such as logging and cattle browsing, also affect *N. pumilio* forest structure and dynamics [114].

During secondary succession and as consequence of these disturbances, *N. pumilio* stands may present either even- or uneven-aged structures. These contrasting situations may represent two extremes in a range of different possible structures. At the southern end of *N. pumilio* distribution in Tierra del Fuego, *N. pumilio* dynamics is greatly influenced by tree falls usually occurring during wind storms, and this result in even-aged young structures [115]. However, the wind storms that cause large falls may also produce other small gaps in the forest canopy when over-mature *N. pumilio* trees fall. The result is then small gaps in which uneven-aged structures usually develop. In general, these small falling events are more common in mature forests located at low altitudes sites which had not previously experienced catastrophic disturbances or human interventions. In these areas, the falling out of senescent trees may promote the opening of gaps or patches of about 0.1 ha in which regeneration begins [116, 117]. These patches generally possess favorable undergrowth conditions which allow the formation of small clumps of seedlings. The result of this process is a multi-aged and multi-strata forest, even when the formation of these gaps may be an episodic phenomenon [109].

In sites with high rainfall levels (*i.e.* South of Tierra del Fuego and some south-western areas of Chubut province), *N. pumilio* regeneration is established even after major disturbances affecting up to hundreds or even thousands of hectares [10, 115]. The same occurs in forests affected by intensive forest harvesting [109, 118-120]. By contrast, in sites with water deficit during the summer, such as in the northern sector of *N. pumilio* distribution in Río Negro, Neuquén and Chubut provinces, regeneration cannot be established with low canopy coverage [121]. In these areas, seedlings establishment is strongly influenced by water availability and usually occurs in small gaps caused by falling trees [122].

Germination Characteristics of *N. pumilio* Seeds

Similar to *Festuca* and *A. chilensis*, *N. pumilio* reproduces strictly by seed. *N. pumilio* propagules are nut fruits that may contain one, and exceptionally two seeds [123]. The weight of these seeds is highly variable, ranging from 11.4 to 21.9 g the thousand seeds (between 45,600 to 87,000 seeds per kg) [86]. Seeds are wind dispersed after ripening during the month of February and up to April. Because of its morphology that does not favor its movement at the soil surface, seed dispersal is generally restricted to a maximum of 60 m from the mother tree in the direction toward which the main dominant winds blow [124]. Seed remain latent during winter and start to germinate in October. Providing enough soil water, *N. pumilio* seeds are able to germinate in a wide range of temperature conditions (from 0 to 20°C) in either light or dark conditions [123], although their viability percentages rarely surpasses the 40 %.

Environmental and Anthropogenic Disturbances Affecting *N. pumilio* Seedling Dynamics

Nothofagus pumilio possesses a transient soil seed bank, and after seed germination in spring, most seedlings remain in the soil with a minimum growth, until adequate light conditions are reached to speed up growth and development [10, 123, 125]. *Nothofagus pumilio* seedlings present very high photosynthetic plasticity, which allow them to rapidly cope with the micro-environmental changes produced after natural or anthropogenic disturbances modify stand structure. Maximum growth, however, is produced under optimal environmental conditions, which occurs when soil water content is between 40 to 60 % of field capacity, and radiation levels are between 150 and 200 μmol m^{-2} s^{-1} [126].

Among the environmental factors that determine the fate of *N. pumilio* seedlings after germination, the availability of water during the growing season is of paramount importance. In the eastern and drier areas of its distribution, *N. pumilio* seedlings are only installed on shaded (moist) microsites, or where the presence of woody debris retains soil moisture to be used during the summer drought. For that reason in these sites, seedling location within canopy gaps is a decisive factor for their recruitment and survival [122, 127-129]. In moist sites, however, and since water may not be a limiting factor, neither the forest cover nor the position in the gap seem to influence seedling survival [129], at least during their early stages of development.

Related to sapling development when regeneration is being installed, the balance of positive and negative interactions appears to change according to the site they develop, shifting in time as saplings grow older. During the first 20 years after a gap opening, growth of *N. pumilio* regeneration is mainly determined by light availability in mesic sites, achieving an average growth (in height) of 22 cm per year. In xeric sites, by contrast, water availability appears as the main determinant of *N. pumilio* growth, and only an average growth of 15 cm per year is achieved. As saplings grow older and develop a more extensive root system, differences in growth due to the precipitation levels gradually decrease. Moreover, in the period between 20 and 35 years after gap opening, saplings in both mesic and xeric sites grow faster in larger gaps having higher daily light availability and independent of water status. This means that in xeric sites, and after having developed their root systems, sapling growth gradually shifts from being water-dependent to light-dependent.

During the early stages of their development and under low radiation conditions, *N. pumilio* seedlings are able to emerge and survive at high densities (more than 100 seedlings m^{-2}) providing that soil moisture is adequate and grazing pressure negligible [130, 131]. After a number of years, however, light requirements increase and seedling density drastically diminishes, especially if these light conditions are maintained at sub-optimal levels. In this way, when seedlings (or saplings) reach about 1 m in height, and even when other environmental variables are at adequate levels, competition for light makes seedling density to greatly diminish, with mean values of 5 seedlings m^{-2}, and ranges that go from 0 to 20 seedlings m^{-2} (López Bernal, unpublished data).

In its northern area of distribution in the Province of Chubut, at least 20% of their stands have been reported as degraded [132]. The main causes of this degradation are, in decreasing order, natural (mainly fire) and anthropogenic disturbances (grazing/browsing and non-adequate silvicultural practices). The effects of these disturbances have a synergistic effect on preventing *N. pumilio* regeneration, since either its seedlings or saplings lack of adaptive mechanisms to avoid or cope with these disturbances. It has been determined that *N. pumilio* takes long time to recover in open areas affected by extreme fires, mainly due to the combined effects of a scarce (or null) amount of seed in the soil bank, and the adverse conditions for seed germination and seedling survival in the post-fire microenvironment when plant cover is very low. On the other hand and when regeneration is already installed, *N. pumilio* seedlings exposed to either grazing or browsing by either sheep, cattle or other wild introduced herbivores such as the European hare, the red deer, or the beaver in Tierra del Fuego, are severely damaged, since seedlings and saplings do not possess defense mechanisms as to resist herbivory [121, 133, 134]. The impact of these disturbances appears to be greater in the eastern than in the western area of the *N. pumilio* distribution. This is because fire frequency, grazing or browsing, and logging pressures are higher at the East than at the West of such distribution.

Nothofagus pumilio forests have been traditionally logged by using the high grading technique. This technique simply implies to select and cut the best individual trees of the stand, leaving those that do not present good lumber qualities [135]. The remaining, standing trees are then those over- mature and/or sick, that through an excessive canopy cover, diminish incoming radiation, essential to establish *N. pumilio* regeneration [122, 136, 137]. In order to overcome that problem and allow regeneration to be established, different logging techniques have been proposed. For the southern area of distribution in Tierra del Fuego, clear cutting or protection cutting techniques over large areas have been proposed [135, 138-141]. In the northern area of the *N. pumilio* distribution, where most of our data came from, small interventions have been recommended [8, 117, 142, 143]; they imply the cut of a few old, senescent or over mature trees to open small gaps. Each technique tries to mimic the natural dynamics of the species in either area, so that even aged regeneration stands will be proposed for Tierra del Fuego. At the same time, irregular gaps, with uneven age-class structures will be advised for the northern area of distribution of *N. pumilio* [131, 138]. Both techniques will provide the basis for a sustainable use of this valuable forest resouirce.

General Conclusion

Although as adult plants all three species have many differences in structural and dynamics characteristics in the environmental gradient studied, they share many similarities and also some differences linked to seeds and seedling dynamics. Related to similarities, either *F. pallescens*, *A. chilensis*, or *N. pumilio* reproduce by seed, and these seeds form transient soil seed banks that need to be periodically replenished [23, 27]. Among the differences, while *F. pallescens* produces seeds annually, *A. chilensis* or *N. pumilio* seed production is constrained to periodic cycles, in which some years of high seed production are followed by others of scarce or null production [123, 124, 144]. In relation to seed viability, *Festuca* seeds always showed high seed viability percentages (*ca.* 99%), followed by *A. chilensis* (from 60 to 80 %) [86], and then by *N. pumilio*, whose viability could be highly variable, but rarely surpasses 40% [86, 125, 145]. These numbers showed an upward trend in seed viability along this environmental gradient, being very low in humid sites where *N. pumilio* grows, relatively high in mesic environments in which *A. chilensis* lives, and very high in semiarid, steppe zones where *F. pallescens* inhabits. Perhaps the more humid environment where *N. pumilio* grows is more favorable to the development of seed diseases caused by insects, and is also probably affected by the presence of different seed predators [145]. As we move toward the steppe, the prevailing environmental conditions may be more restrictive to the development of diseases or the presence of parasites and predators. It should be taken into account, however, that for this zone and as far as we know, information about the effects of seed diseases and predation in *F. pallescens* seeds is very scarce in the scientific literature.

After germination and during the early stages of establishment, seedlings of the three species are conditioned by a set of environmental conditions that affect their further survival. Although with huge differences in the total amount of annual precipitation in this gradient, all share the same Mediterranean type of climate, being humid and cold during the winter and mild and dry during the summer growing season. Seedlings of all three species are highly vulnerable to soil water stress, especially if precipitation shows below average values during the growing season, and high wind speed is present in the establishment area. While *F. pallescens* seedlings may suffer from intraespecific competition as they grow older, young *A.*

chilensis seedlings appear to be benefited in protected microsites brought about adult shrubs during early development stages, indicating a facilitative effect of the shrub to the developing seedling. When *A. chilensis* seedlings reach a sufficient height and root biomass is adequately developed, it seems to out-compete the shrub, and the former facilitation effect changes to a competition effect, displacing the once protective shrub as succession advances. *Nothofagus pumilio* seedlings, by contrast, take many years to develop a root system big enough to supply water demands of aboveground organs.. However, and under these conditions only those seedlings and saplings able to reach higher levels of incoming luminosity will survive in the system.

Related to the disturbances affecting seedling survival and establishment in the environmental gradient studied, grazing and fire appear as the most important for precluding *Festuca* seed germination, establishment and survival of seedlings in the steppe zones, followed by other disturbances generally triggered by grazing pressure. Unplanned logging practices, browsing by introduced wild or domestic browsers, and grazers, fire, and in some cases stand replacement by introduced conifer species, are considered as the main disturbances affecting *A. chilensis* and *N. pumilio* seedling regeneration in mesic and humid areas of this gradient. In order to recuperate their ecological functions and preserve this unique ecosystem, careful management plans, including active restoration practices, should be developed based on sound science. This chapter dealt with seedlings dynamics of three of the most conspicuous species grown in the study environmental gradient, providing information that could be useful for designing the proposed recuperation plans.

References

[1]   J. Paruelo, A. Beltrán, E. Jobbágy, O.E. Sala and R. Golluscio, *Ecol. Aust.* 8, 85 (1998).

[2]   A. L. De Fina, in: M. J. Dimitri (Ed.), Colección Científica del INTA, Buenos Aires (1972).

[3]   R. Villalba, Climatic influences on forest dynamics along the forest-steppe ecotone in northern Patagonia. PhD Thesis, Dept. Geography, Univ.Colorado (1995).

[4]   T. T. Veblen, in: S. T. A. Picket and P. S. White (Eds.), Academic Press, San Diego, CA. (1985).

[5]   T. T. Veblen, T. Kitzberger and A. Lara, *J. Veg. Sci.* 3, 507 (1992).

[6]   A. Cesa and J. Paruelo, *J. Arid Environ.* 75, 1129 (2011).

[7]   T. Kitzberger, in: C. Kunst, S. Bravo and J. Panigatti (Eds.), INTA, Santiago del Estero (2003).

[8]   T. T. Veblen, C. Donoso, F. M. Schlegel and R. B. Escobar, *J. Biogeogr.* 8, 211 (1981).

[9]   M. J. Dimitri, in: M. J. Dimitri (Ed.), Colección Científica del INTA, Buenos Aires (1972).

[10] T. T. Veblen, C. Donoso Z., T. Kitzberger and A. J. Rebertus, in: T. T. Veblen, R. S. Hill, and J. Read (Eds.), Yale University Press, London (1996).

[11] A. Soriano, *Rev. Inv. Agríc.* 10, 349 (1956).

[12] G. E. Defossé, M. B. Bertiller and J. O. Ares, *J. Range Manage.* 43, 157 (1990).

[13] T. T. Veblen and D. C. Lorenz, *Ann. Assoc. Am. Geogr.* 78, 93 (1988).

[14] M. S. Sagarzazu Ansaldo and G. E. Defossé, *Study of a 2nd set of largest past fires.* I.P.F. Paradox", (2009).

[15] A. L. Cabrera, Las Regiones Fitogeográficas Argentinas. *Ed. ACME,* Buenos Aires (1976).

[16] G. E. Defossé, M. B. Bertiller and R. Robberecht, *J. Veg. Sci.* 8, 677 (1997).

[17] J. J. Morrison, La ganadería en la región de las mesetas australes del territorio de Santa Cruz. Tesis de grado, Fac. Agron. Vet., Univ. Buenos Aires (1917).

[18] A. Soriano, *Rev. Inv. Agríc.* 10, 323 (1956).

[19] J. O. Ares, A. M. Beeskow, C. M. Rostagno, M. P. Irisarri, J. Anchorena, G. E. Defossé, and C. A. Merino, in: A. Breymeyer (Ed.) Elsevier, Amsterdam (1990).

[20] M. B. Bertiller and G. E. Defossé, *J. Range Manage.* 43, 300 (1990).

[21] R. J. C. León and M.R. Aguiar, *Phytocoenología.* 13, 181 (1985).

[22] G. E. Defossé, Germination, emergence, and survival of *Festuca* spp. seedlings in a steppe of Patagonia, Argentina. PhD Thesis, College of Graduate Studies, Univ.Idaho (1995).

[23] M. B. Bertiller and F. R. Coronato, *Biodiv. Conserv.* 3, 47 (1994).

[24] J. Franzese, Impacto de la invasión de *Rumex acetosella* L. en los pastizales del noroeste de la Patagonia y su relación con el fuego. Tesis doctoral, CRUB, *Univ. Nac. Comahue* (2012).

[25] J. L. Harper, *Population Biology of Plants.* Academis Press, London (1977).

[26] A. Soriano, in: Proc. 8th Int. Grassland Congress, London, (1960).

[27] M. B. Bertiller, *J. Veget. Sci.* 3, 47 (1992).

[28] M. B. Bertiller, A. M. Beeskow and F. Coronato, *J. Arid Environ.* 21, 1 (1991).

[29] G. E. Defossé, M. B. Bertiller and R. Robberecht, *Seed Sci. Tech.* 23, 715 (1995).

[30] M. B. Bertiller, P. Zaixso, M. P. Irisarri, and E. Brevedan, *J. Arid Environ.* 32, 161 (1996).

[31] G. E. Defossé, R. Robberecht and M. B. Bertiller, *J. Range Manage.* 50, 73 (1997).

[32] D. F. Balph and J. C. Malecheck, *J. Range Manage.* 38, 226 (1985).

[33] A. G. Savory and S. D. Parsons, *Rangelands* 2, 234 (1980).

[34] A. G. Savory, *Rangelands* 5, 155 (1983).

[35] C. W. D. Gibson, T. A. Watt and K. Brown, *Biol. Conserv.* 42, 165 (1987).

[36] M. Oesterheld and O.E. Sala, *J. Veg. Sci.* 1, 353 (1990).

[37] R. N. Mack and D. A. Pyke, *J. Ecol.* 72, 731 (1984).

[38] D. O. Salihi and B. E. Norton, *J. Appl. Ecol.* 24, 145 (1987).

[39] S. G. Cook and D. Ratcliff, *J. Appl. Ecol.* 21, 971 (1984).

[40] R. I. Yeaton and M. L. Cody, *J. Ecol.* 64, 689 (1976).

[41] R. Robberecht, B. E. Mahall and P. S. Nobel, *Oecologia* 60, 21 (1983).

[42] A. M. Wilson and D. D. Briske, in: 1st Int. Rangeland Congress, Denver, CO, (1978).

[43] A. Soriano and O. E. Sala, *Isr. J. Bot..* 35, 91 (1986).

[44] A. Soriano, R. A. Golluscio and E. Satorre, *B. Torrey Bot. Club.* 114, 103 (1987).

[45] H. H. Biswell, A. M. Schultz, C. W. Hedrick and J. I. Mallory, *J. Range Manage.* 6, 172 (1953).

[46] M. C. Dentoni, G. E. Defossé, N. F. Rodríguez, M. M. Muñoz and H. Colomb, Estudio de Grandes Incendios: El caso de la Ea. San Ramón en Bariloche, Río Negro-Patagonia Argentina. P.N.d.M.d.F.-C.-. GTZ, Esquel, Chubut, (1999).

[47] T. T. Veblen, T. Kitzberger, R. Villalba and J. Donnegan, *Ecol. Monogr.* 69, 47 (1999).

[48] G. E. Defossé, M. S. Sagarzazu and M. M. Godoy, in 2ª Reunión Patagonica y 3ª Nacional sobre Ecología y Manejo del Fuego, Esquel, Argentina, (2006).

[49] S. L. González, Estrategias de regeneración post-fuego en pastizales del noroeste Patagónico: un enfoque experimental. Tesis doctoral, CRUB, Univ. Nac. Comahue (2011).

[50] M. E. Hanley and M. Fenner, *Acta Oecol.* 19, 181 (1998).

[51] R. J. Hobbs and L.F. Huenneke, *Conserv. Biol.* 6, 324 (1992).

[52] C. Gittins, L. Germandi and D. Bran, *J. Arid Environ.* 75, 986 (2011).

[53] D. Bran, A. Pérez, D. Barrios, M. Pastorino and J. Ayesa, Eco-región Valdiviana: Distribución Actual de los Bosques de "Ciprés de la Cordillera" (*Austrocedrus chilensis*) - Escala 1:250.000. Informe final, I.-A.-F.V. Silvestre, (2002).

[54] V. C. LaMarche, R. L. Holmes, P. W. Dunwiddie and L.G. Drew, Tree-ring Chronologies of the Southern Hemisphere. 1. Argentina. Chronology Series V. Lab. Tree-ring Res., Univ. Arizona, Tucson (1979).

[55] F. Colmet Dâage, M. J. Mazzarino and A. A. Lanciotti, Características de los suelos volcánicos en el S.O. del Chubut. Comunicación Técnica, I. EEA, Bariloche, (1993).

[56] P. H. Etchevehere, in: M. J. Dimitri (Ed.), Colección Científica del INTA, Buenos Aires, Argentina (1972).

[57] A. G. Greslebin, E. M. Hansen and W. Sutton, *Mycol. Res.* III, 308 (2007).

[58] A. G. Greslebin and E. M. Hansen, *Plant Pathol.* 59, 604 (2010).

[59] M. L. Vélez, P. V. Silva, O. A. Troncoso and A. G. Greslebin, *Plant Pathol.*61(5), 877 (2012).

[60] T. T. Veblen, B. R. Burns, T. Kitzberger, A. Lara and R. Villalba, in: N. J. Enright and R. S. Hill (Eds.), University Press, Melbourne (1995).

[61] M. E. Gobbi, *Bosque (Valdivia)* 28, 50 (2007).

[62] T. Kitzberger, Fire regime variation along a northern Patagonian forest-steppe gradient: stand and landscape response. PhD Thesis, Dept. Geography, Univ. Colorado (1994).

[63] T. T. Veblen, T. Kitzberger, B. R. Burns and A. J. Rebertus, in: J. J. Armesto, C. Villagrán and A. M. K. Arroyo (Eds.), Edit. Universitaria, Chile (1995).

[64] R. Villalba and T. T. Veblen, *J. Ecol.* 85, 113 (1997).

[65] F. Fonk, Viajes de Fray Francisco Menéndez a Nahuel Huapi. Niemeyer, C.F., Valparaiso, Chile (1900).

[66] B. Willis, El norte de la Patagonia. Ministerio de Obras Públicas, Buenos Aires (1914).

[67] M. Rothkugel, Los bosques patagónicos. Min. de Agr. y Gan., Ofic. de Bosques y Yervales., Buenos Aires (1916).

[68] G. C. Musters, Vida entre los patagones. Un año de excursiones por tierras no frecuentadas desde el Estrecho de Magallanes hasta el Río Negro. Ediciones Solar, Buenos Aires (1979).

[69] T. T. Veblen and V. Markgraf, *Quat. Res.* 30, 331 (1988).

[70] V. Markgraf and L. Anderson, *Rev. Inst. Geog. Sao Paulo* 15, 33 (1994).

[71] F. Schäbitz, *Quat. South Am. Antartic Pen.* 10, 17 (1994).

[72] J. G. Goldammer, P. Cwielong, N. Rodríguez and J. Goergen, *Biomass Burning and Global Change.* 2, 653 (1996).

[73] T. T. Veblen and D. C. Lorenz, *Vegetatio.* 71, 113 (1987).

[74] T. Kitzberger and T. T. Veblen, *Ecoscience.* 4, 508 (1997).

[75] L. M. Chauchard, J. O. Bava, S. Castañeda, P. Laclau, G. A. Loguercio, P. Pantaenius and V. Rusch, Manual para la Buenas Prácticas Forestales en Bosques Nativos de Norpatagonia, Min. Agric., Gan. Pesca, Buenos Aires (2012).

[76] G. Loguercio, M. F. Urretavizcaya, H. Claverie and M. Rey, *IDIA XXI (INTA).* 8, 84 (2005).

[77] I. N. Costantino, Parcelas experimentales permanentes *Libocedrus chilensis* (Don) Endl. Estudios de crecimientos y regeneración natural. Min. Agric., Gan. Pesca, Buenos Aires (1949).

[78] J. Goya, J. Ferrando, D. Boscos and P. Yapura, *Rev. Fac. Agron. UNLP.* 71, 165 (1995).

[79] G. A. Loguercio, Erhaltung der baumart "ciprés de la cordillera" *Austrocedrus chilensis* (D.Don) Florin et Boutelje, durch nachhaltige Nutzung. PhD Thesis, Universität München (1997).

[80] J. O. Bava and H. E. Gonda, in Proc. Congreso Forestal Argentina y Latinoamericano. Paraná, Argentina (1993).

[81] D. M. Smith, B. C. Larson, M. J. Kelty and P. M. S. Ashton, The practice of silviculture. Applied forest ecology. John Willey and Sons, New York (1997).

[82] J. Grosfeld, Análisis de la variabilidad morfológica y arquitectura de *Austrocedrus chilensis* (D. Don) Pic. Serm. et Bizzarri, *Fitzroya cupressoides* (Molina) I. M. Johnst., *Pilgerodendron uviferum* (D. Don) Florin y *Cupressus sempervirens* L. (Cupressaceae). Tesis Doctorado, CRUB, Univ. Nac. Comahue (2001).

[83] M. F. Urretavizcaya and G. Defossé, *Forest Ecol. Manage.* 187, 361 (2004).

[84] L. Contardi, Morfología, estructura, y calidad de semillas de *Austrocedrus chilensis* (D. Don) Flor. et Boutl. CIEFAP, Esquel (1995).

[85] M. F. Urretavizcaya and M. F. Oyharçabal, *Patagonia Forestal* 17, 15 (2011).

[86] J. L. Walck, J. M. Baskin, C. C. Baskin and S. N. Hidayati, *Seed Sci. Res.* 15, 189 (2005).

[87] A. M. Mayer and A. Poljakoff-Mayber, The germination of seeds. Pergamon Press, (1982).

[88] D. Baumeister and R. M. Callaway, *Ecology* 87, 1816 (2006).

[89] J. T. Greenlee and R. M. Callaway, *Am. Natur.* 148, 386 (1996).

[90] T. Kitzberger, D. F. Steinaker and T. T. Veblen, *Ecology* 81, 1914 (2000).

[91] M. F. Urretavizcaya, G. E. Defossé, and H. E. Gonda, *Restor. Ecol.* 20, 131 (2012).

[92] K. Tielbörger and R. Kadmon, *Ecology* 81, 1544 (2000).

[93] D. DeSteven, *Ecology* 72, 1066 (1991).

[94] D. S. Gill and P. Marks, *Ecol. Monogr.* 61, 183 (1991).

[95] G. T. Hastwell and J. M. Facelli, *J. Ecol.* 91, 941 (2003).

[96] A. R. Berkowitz, C. D. Canham and V. R. Kelly, *Ecology* 76, 1156 (1995).

[97] Y. Pueyo, C. Alados, B. García-Ávila, S. Kéfi, M. Maestro and M. Rietkerk, *Restor. Ecol.* 17, 908 (2009).

[98] R. Villalba and T. T. Veblen, *Can. J. Forest Res.* 27, 580 (1997).

[99] M. F. Urretavizcaya, Cambios ambientales y restauración eclógica post-incendios en bosques de *Austrocedrus chilensis*. Tesis Doctorado, CRUB -, Univ. Nac. Comahue (2005).

[100] C. E. Van Wagner, in: R. W. Wein and D. A. MacLean (Eds.), John Wiley and Sons, Chichester (1983).

[101] M. F. Urretavizcaya, G. E. Defossé and H. E. Gonda, *Ann. For. Sci.* 63, 63 (2006).

[102] M. A. Relva and T. T. Veblen, *Forest Ecol. Manage.* 108, 27 (1998).

[103] T. T. Veblen, T. Kitzberger, E. Raffaele and D. C. Lorenz, in: T. T. Veblen, W. L. Baker, G. Montenegro and T. W. Swetnam (Eds.), Springer-Verlag, New York (2003).

[104] L. A. Tortorelli, Maderas y bosques argentinos. Editorial ACME, Buenos Aires (1956).

[105] T. T. Veblen, D. H. Ashton, F. M. Schlegel and A. T. Veblen, *J. Biogeogr.* 4, 275 (1977).

[106] C. Donoso Z., Bosques templados de Chile y Argentina. Variación, estructura y dinámica. Editorial Universitaria, Santiago de Chile (1995).

[107] J. E. Schlatter, *Bosque* 15, 3 (1994).

[108] A. J. Rebertus and T. T. Veblen, *J. Veg. Sci.* 4, 641 (1993).

[109] P. P. Cwielong and M. Rajchenberg, *Eur. J. Forest Pathol.* 25, 47 (1995).

[110] E. G. Jobbágy, J. M. Paruelo and R. J. C. León, *Ecol. Aust.* 5, 47 (1995).

[111] V. R. Barros, V. H. Cordon, C. L. Moyano, R. J. Méndez, J. C. Forquera and O. Pizzio, Cartas de precipitación de la zona oeste de las provincias de Río Negro y Neuquén. Fac. Cs. Agr., Univ. Nac. Comahue, Cinco Saltos (1983).

[112] J. J. Burgos, in O. Boelcke, D. M. Moore and F. A. Roig (Eds.), CONICET (Argentina), Royal Society (Gran Bretaña) e Instituto de la Patagonia (Chile) (1985).

[113] P. López Bernal, G. Defossé, C. P. Quinteros and J. Bava, in: J. M. García and J. J. Diez Casero (Eds.), InTech, Croatia (2012).

[114] A. J. Rebertus, T. Kitzberger, T. T. Veblen and L. M. Roovers, *Ecology* 78, 678 (1997).

[115] T. T. Veblen and C. Donoso, *Bosque* 8, 133 (1987).

[116] J. O. Bava, Aportes ecológicos y silviculturales a la transformación de bosques vírgenes de lenga en bosques manejados en el sector argentino de Tierra del Fuego. CIEFAP, Esquel (1999).

[117] G. Gea-Izquierdo, G. M. Pastur, J. M. Cellini and M. V. Lencinas, *Forest Ecol. Manage.* 201, 335 (2004).

[118] E. Mutarelli and E. Orfila, *Rev. Forest. Arg.* 15, 109 (1971).

[119] J. M. Rosenfeld, R. M. Navarro Cerrillo and J. R. Guzman Alvarez, *J. Environ. Manage.* 78, 44 (2006).

[120] J. O. Bava and C. J. Puig, in: Actas del Seminario de Manejo forestal de la lenga y aspectos ecológicos relacionados. CIEFAP, Esquel (1992).

[121] V. Rusch, in: Actas Seminario de Manejo forestal de la lenga y aspectos ecológicos relacionados. CIEFAP, Esquel (1992).

[122] J. G. Cuevas and M. T. K. Arroyo, *Rev. Chil. Hist. Nat.* 72, 73 (1999).

[123] J. M. Cellini, Estructura y regeneración bajo distintas propuestas de manejo de bosques de *Nothofagus pumilio* (Poepp. et. Endl.) Krasser en Tierra del Fuego, Argentina. Tesis doctoral, Fac. Cs. Nat. Museo, Univ. Nac. La Plata (2011).

[124] V. Albarracín, Banco de semillas en bosques quemados de *Nothofagus pumilio* en bosques de Chubut y Santa Cruz. Tesis de grado, Fac. Cs. Nat., Univ. Nac. Patagonia (2011).

[125] G. Martinez Pastur, M. V. Lencinas, P. L. Peri and M. Arena, *Forest Ecol. Manage.* 243, 274 (2007).

[126] D.C. Rechene, in: CIEFAP, Esquel (1995).

[127] K. Heinemann, T. Kitzberger and T. T. Veblen, *Can. J. Forest Res.* 30, 25 (2000).

[128] K. Heinemann and T. Kitzberger, *J. Biogeogr.* 33, 1357 (2006).

[129] G. Martinez Pastur, P. L. Peri, M. C. Fernández and G. Staffieri, *Bosque.* 20(2), 39 (1999).

[130] P. M. López Bernal, J. O. Bava and S. H. Antequera, *Bosque* 24, 13 (2003).

[131] J. O. Bava, J. D. Lencinas and A. Haag, Determinación de la materia prima disponible para proyectos de inversión forestal en la provincia del Chubut. C.F.I. (2006).

[132] L. A. Cavieres and A. Fajardo, *Forest Ecol. Manag.* 204, 237 (2005).

[133] C. Quinteros, P. López Bernal, M. Gobbi and J. Bava, *Agroforest. Syst.* 84, 261 (2012).

[134] H. Schmidt and J. Caldentey, Apuntes del tercer curso de silvicultura de los bosques de lenga. CONAF-CORMA-Universidad de Chile, Valdivia (Chile) (1994).

[135] C.J. Puig, Efecto de distintos grados de cobertura, el pastoreo y la liebre sobre la densidad, distribución y calidad de la regeneración natural de la lenga (*Nothofagus pumilio* (Poepp. et Endl.) Krasser) en la provincia del Chubut. Tesis de grado, Dept. Ing. Forestal, Univ. Nac. Patagonia (1993).

[136] G. A. Loguercio, in: CIEFAP, Esquel (1995).

[137] S. H. Antequera, Efecto del tipo de corta sobre el rendimiento en madera rolliza y la regeneración natural de un bosque de lenga (Nothofagus pumilio) de calidad media en la provincia del Chubut, Argentina. Tesis de Maestría, Univ. Götingen (2002).

[138] J. Arce, P. L. Peri and G. Martinez Pastur, in: Actas 1er Congreso Latinoamericano IUFRO, Valdivia, Chile (1998).

[139] G. Martínez Pastur, M. V. Lencinas, J. M. Cellini, P. L. Peri and R. Soler Esteban, *Forest Ecol. Manage.* 258, 436 (2009).

[140] M. Davel and J. O. Bava, Estudio comparativo de distintos tratamientos silvícolas en bosques de lenga (*Nothofagus pumilio* (Poep. et Endl.) Krasser). INTA, (1999).

[141] T. T. Veblen, F. M. Schlegel and B. Escobar R., *J. Ecol.* 68, 1 (1980).

[142] V. Rusch, Estudio sobre la regeneración de la lenga en la Cuenca del Río Manso Superior, Río Negro. C.N.C.T., Buenos Aires (1987).

In: From Seed Germination to Young Plants
Editor: Carlos Alberto Busso

ISBN: 978-1-62618-653-8
© 2013 Nova Science Publishers, Inc.

Chapter 12

Water Stress and Temperature Effects on Germination and Early Seedling Growth of *Digitaria eriantha*

Roberto Brevedan[1,2], *Carlos Busso*[1,2,*], *María Fioretti*[1], *Mirta Toribio*[1], *Sandra Baioni*[1], *Yanina Torres*[1,2], *Osvaldo Fernández*[1,2], *Hugo Giorgetti*[3], *Diego Bentivegna*[2], *José Entío*[4], *Leticia Ithurrart*[1,2], *Oscar Montenegro*[3], *María Mujica*[4], *Gustavo Rodríguez*[3] *and Guillermo Tucat*[2]

[1]Dept. Agronomía, Universidad Nacional del Sur, Bahía Blanca, Argentina
[2]CERZOS (CONICET), Bahía Blanca, Argentina
[3]Chacra Experimental de Patagones, Ministerio de Asuntos Agrarios, Carmen de Patagones, Argentina
[4]Facultad de Ciencias Agrarias, Universidad Nacional de La Plata, La Plata, Argentina

Abstract

We assessed seed germination and seedling survival of the grass *Digitaria eriantha* cv. Irene to determine its potential for re-vegetation in the arid parts of Argentina. We evaluated the effects of (1) water stress and temperature on the germination, and (2) water stress on the early seedling growth under controlled conditions. During the first 24 hrs the germination at constant temperature exceeded by 50 % (one-way ANOVA: $F3,20=67.40$, $p<0.001$) the germination at the 30—10°C alternating temperatures, although the total germination was about 80% in all temperature treatments. Germination percentages and coefficients of velocity were lower (two-way ANOVA main effect of water potential: Germination: 15 h, 18 h, 2006: $F6,35=4.69$, $p<0.0013$, 2007: $F6,35=38.99$, $p<0.0001$; 24 h, 2006: $F6,35=10.66$, $p<0.0001$, 2007: $F=44.99$, $p<0.0001$; 36 h, 2006: $F6,35=6.05$, $p<0.0002$, 2007: $F6,35=23.54$, $p<0.0001$; 60 h, $F6,70=2.46$, $p<0.0325$; Coefficient of velocity: $F6,70=30.33$, $p<0.0001$) as water potentials decreased.

* E-mail: cebusso@criba.edu.ar.

Early seedling growth was lower (three-way-ANOVA main effect of water potential: $F_{5,120}=19.14$, $p<0.001$) at lower water potentials. The high sensitivity of the two studied processes in *D. eriantha* suggests that it is quite susceptible to be considered for re-vegetation purposes in the rangelands of arid Argentina.

Introduction

The rangelands of central Argentina have few warm-season, native perennial grasses that are palatable to domestic livestock [1]. In the rangelands of the southern Buenos AiresProvince [e.g., the Chacra Experimental de Patagones, Patagones, Argentina (40°39'49.7"S, 62°53'6.4"W; 40 m.a.s.l.)], within the phytogeographical province of the Monte [2], *Pappophorum vaginatum* is the primary native, warm-season, C_4 palatable perennial grass for livestock [3]. Thus, there is a need to increase perennial grass species in this region [4] for additional forage for the needs of the increasing cattle production.

Digitaria eriantha ssp. *eriantha* is a C_4, palatable perennial grass that was introduced toArgentina from Africa in 1991 [5], and has been introduced to the USA [6], Australia [7], China and Europe [8], and South Africa [9]. This species is characterized byhigh drought resistance, high warm-season forage production and adaptability to different, especially sandy, soils [10-11].The cultivar Irene is better adapted than the spp. *eriantha* to the soil and climatic conditions of the semiarid regions in Argentina [12].

Periods of water stress are common in the rangelands of north-western Patagonia [13]. They can occur from early spring to late summer (e.g., years 2008 and 2009 compared to the long-term mean in Figure 1). Therefore, *D. eriantha*, with its high drought resistance, is a potential species for introduction to these rangelands. Lacking, however, are studies on the water potential and temperature effects on the seed germination and early seedling growth [14].Plants are particularly susceptible to water stress during seed germination, seedling emergence and early growth [14, 15].

In arid and semiarid environments, where favourable conditions of water availability and temperature regimes may be temporally limited [16-17], time to germination is also important for successful plant establishment. Time to germination requires high water potentials [18] and temperature determines the optimal and minimum water potentials for germination in various species [19]. High temperatures can shorten the time to germination, and are especially important when favourable moisture conditions occur [20].

There is a paucity of information on germination ecology of *D. eriantha*. In the arid and semiarid areas of Argentina, the optimum humidity and temperature for seed germination varies within and among years [21]. With increasing water stress, total germination and time to germination both decline [14]. Failure of seed germination occurs when the specific humidity and temperature demands are not met [22]. For example, oxygen diffusion to the embryo may reduce germination under excess soil moisture, because the space between the lemma and palea might be filled with water [23]. Excessive soil moisture lowered the germination of *Oryzopsis holciformis* [24]. As for temperature, germination is often higher in alternating temperatures. For example, *Elymus cinereus* cv. Magnar exhibited enhanced germination when the temperatures alternated between 15 and 25°C [25]. Similarly, *Achnatherumrobustum* had optimal germination at 20°C for 8 h, and then at 15°C during the remaining 16 h [26]. Also*Leymus cinereus* exhibited higher germination percentages at alternating temperatures of 15/25°C [27].

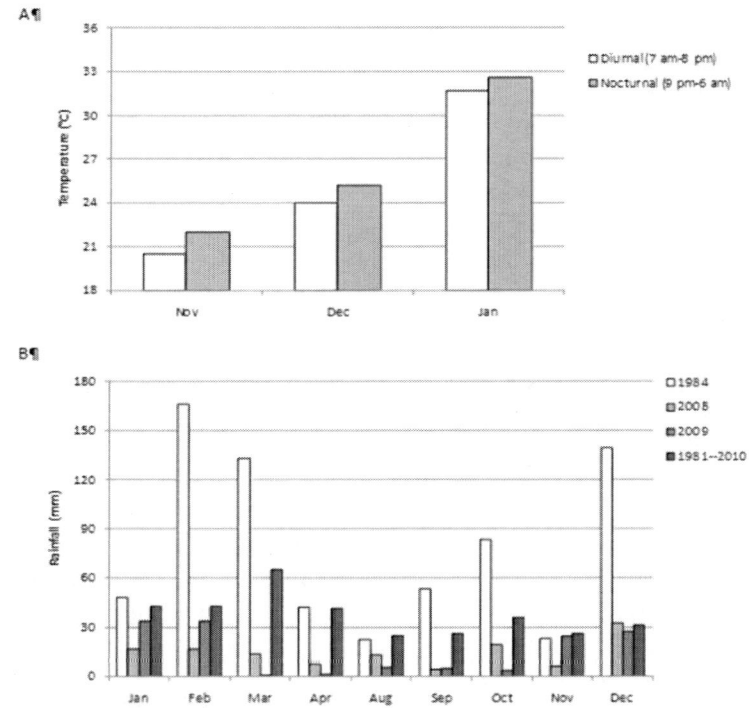

Figure 1.A)Mean monthly diurnal and nocturnal temperatures (°C) during November, December and January in 2006-2010. B) Monthly precipitation (mm) at various years in the Chacra Experimentalde Patagones, Argentina (40°39'49.7"S, 62°53'6.4"W; 40 m.a.s.l.).

Our objectives were to (i) identify germination responses of *D. eriantha* to various temperatures and water potentials, and (ii) determine the effect of water stress on its initial seedling growth. The specific hypothesis tested were (a) germination percentages and coefficients of velocity are higher at greater than at lower water potentials, and at high compared to alternating temperatures, and (b) early seedling growth is higher at greater than at lower water potentials. This information will be useful for the people developing strategies for re-establishing and managing the degraded rangelands of north-western Patagonia, Argentina.

Materials and Methods

Storage Time and Conditions

Seeds of *D. eriantha* cv. Irene were harvested in 2006 and 2007. The total weight of 1000 seeds was 0.339 g in 2006 and 0.329 g in 2007. The seeds were kept in a storage room at 20°C. Seed viability was tested using the 2,3,5-triphenyl tetrazolium chloride test (TTC) [28]. Five sets of 50 seeds each were immersed in water in Petri dishes during 24 hours. The seeds were then incubated in 20 ml-Petri dishes containing a 0.1% solution of TTC at 35°C under darkness, and the embryos were examined to establish their viability.

Temperature Effect

Germination was evaluated at different temperatures using seeds from the 2007 harvest. Four replicates of a 100 seeds each on each temperature treatment were placed on a filter-paper sheet and exposed to a 13.9-36.9°C temperature gradient. The mean monthly temperature and monthly precipitation data at the *Chacra Experimental de Patagones*, within the Phytogeographical Province of Monte, are presented in Figure1. A hollow, empty rectangular plate of stainless steel (0.95 m length, 0.20 m wide, 0.50 m height) was used to create the thermal gradient (Figure 2).

This was achieved submerging one end of the plate in cool water (4°C) and the other end in warm water (40°C). Thereby, since the legs of the apparatus were placed in separate, constant temperature water baths, the temperature gradient was established by heat transfer through the plate. A refrigerator engine was connected to the cold end of the apparatus to allow maintenance of a constant temperature in the water bath (i.e., 4°C). Ethylene glycol, a cooling liquid, was also added into the water bath at the cold end. The warmest extreme of the plate was maintained at that temperature using an electrical resistance that imposed an opposition to the flow of electric current. Thermocouples and an infrared thermometer were used to measure various areas of different isothermal conditions each along the rectangular plate (Figure 2).

These areas were perpendicular to the plate length. The rubber included in the lid of the thermal plate made it hermetic, so that water losses via evaporation were avoided (Figure 2). An absorbent paper sheath that remained saturated with water was placed on a stainless steel surface along the length of the thermal plate; this surface was located near the lid of the thermal plate (Figure 2). The paper sheath was watered twice a day to keep it saturated. Thermal registration was obtained using 10 thermocouples connected to a Delta T multi-channel data-logger (Figure 2). It means that 10 temperature readings were taken at the same time along the thermal plate with the data logger within a range of 13.9°C to 36.9°C.

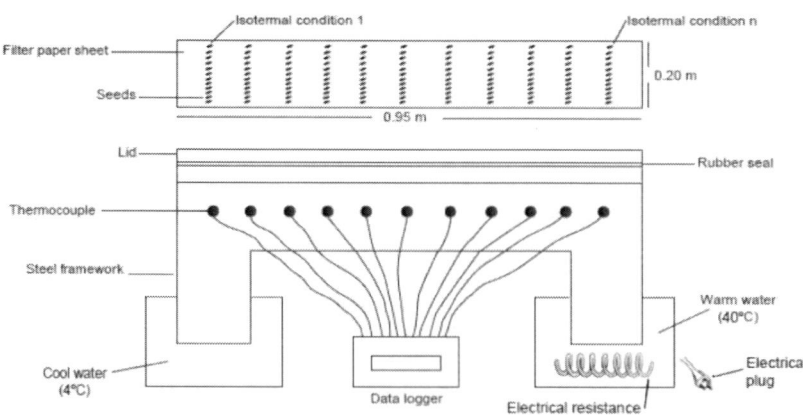

Figure 2. Diagram of the stainless steel equipment use to create a thermal gradient for the germination studies under laboratory conditions.

Both cumulative germination and the rate of germination were recorded by counting and removing the germinated seeds every 12 hours for six days. A seed was considered germinated once its radicle reached a length of at least 2 mm [29].In the next experiment, the seeds were placed on filter-paper sheets in plastic boxes (53 mm length, 59 mm wide, 15 mm height). Distilled water was added, and the boxes were placed in a germination cabinetthat was built at the Regional Center of Biochemical and Biological Research in Bahía Blanca, Argentina. Measures of such cabinet were 45 cm length, 45 cm wide and 135 cm height; it had temperature control from 4 to 40°C and provided 25 to 45 µmol m^{-2} sec^{-1} photosynthetically active radiation. Boxes received the following temperatures: (1) constant 30°C, (2) constant 35°C, (3) alternated: 30°C, 14h light; 10°C, 10h darkness, and (4) alternated: 35°C, 14h light; 10°C, 10h darkness. Each temperature treatment was in a different compartment within the cabinet, and there were six replicates (boxes) of 50 seeds each per treatment. Observations were carried out every 24 hours during four days.

Water Potential Effect

The studies of water-potential effects on germination were carried out using seeds from the 2006 and 2007 harvests. The seeds were germinated in plastic boxes similar to the ones used in the temperature experiment. The experimental solutions were prepared by adding polyethylene glycol 6000 (PEG 6000) to distilled water [30]. PEG does not reduce seed germination [29]. The osmotic potential was determined with a Wescor 5500 osmometer, after calibration with standard KCl solutions. The water potential treatments were 0, −0.4, −0.6, −0.8, −1.0, −1.2, −1.5 and −2.0 MPaexposed to 35±1°C. Six replicates of 50 seeds each were used for each water potential. Seeds were on paper-sheets saturated with the corresponding water potential solution. Each box was sealed with Parafilm to prevent water evaporation. Counting of the germinated seeds was conducted using a Forma Scientific 3770 germination chamber, where the germination study was carried out.Both cumulative germination and the rate of germination were recorded by counting and removing the germinated seeds every 12 hours for six days. A seed was considered germinated once its radicle reached a length of at least 2 mm [29]. The coefficient of velocity (CV) was calculated as:

$$CV = 100 \ [\textstyle\sum Ni / \sum NiTi] \tag{1}$$

where N is the number of seeds which germinated on day i, and T is the number of days from sowing [31]. One advantage of using the coefficient of velocity as a measure of the rate of germination is its capacity to take into account the germination speed; one limitation is that it does not take into consideration the distribution of germination events within a time frame [31].

The root and shoot lengths of three randomly selected seedlings were measured per box on each water potential treatment. An average of these measurements was considered to be one replication.

Statistical Analysis

We used a completely randomized experimental design with six replications. The temperature data were analyzed using one-way ANOVA. The percentage germination data were transformed to arc-sin of the square root to comply with the normality and homoscedasticity assumptions of variance.

Within any given study time from imbibition, two-way ANOVA (year × water potential) were used to analyze the germination and coefficients of the velocity data.

When the interaction was significant, each year or water potential was analyzed separately. The growth of young seedlings was analyzed using three-way ANOVA (year × plant part × water potential). The untransformed values are presented in Figures. When F-test results were significant (p<0.05), the treatments were compared using LSD. The data were analyzed using the statistical software INFOSTAT version 2009 [32].

Results

Storage Time

The seeds harvested in 2006 and 2007 exhibited 33% and 93% viability, respectively.

Germination Percentage

Temperature

After one day of the study initiation, the lowest (one-way ANOVA: $F_{3,20}$=67.40, $p<0.001$) germination percentage was reached at the 30-10°C alternating temperature (Figure 3).

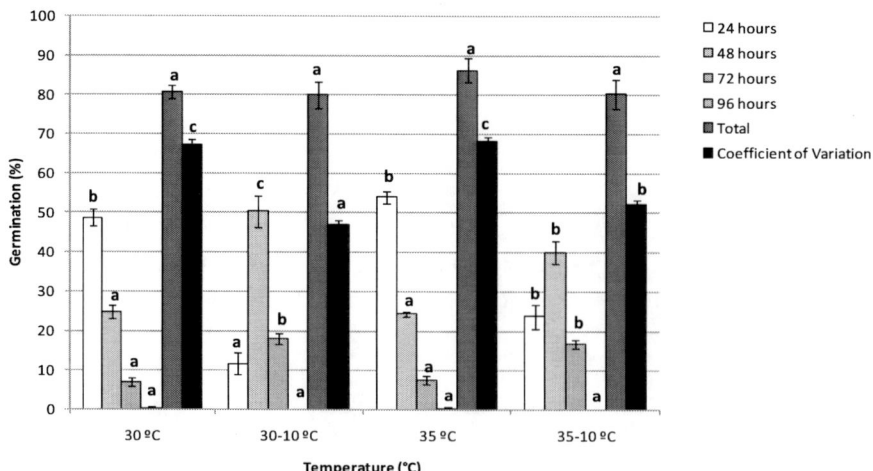

Figure 3. Germination percentage and coefficient of velocity of *Digitaria eriantha*cv. Irene as a function of different temperature treatments. Means with the same letter above bars are not significantly different according to the LSD test (p≤0.05). Values are the means ±1 SE of n=6.

At the same time, the percentage germination was about 50% at 30 and 35°C, although this percentage did not differ compared to the alternating temperature of 35-10°C. The germination percentages were greater (one-way ANOVA: 48 h: $F_{3,20}=25.39$, $p<0.0001$; 72 h: $F_{3,20}=23.32$, $p<0.0001$) at alternating temperatures compared to constant temperatures after two and three days (Figure 3).

Despite these differences among temperature treatments, the total germination was similar (approx. 80%; one-way ANOVA: $F_{3,20}=0.67$, $p=0.5823$) across treatments after four days (Figure 3). The coefficients of velocity were highest (one-way ANOVA: $F_{3,20}=87.70$, $p<0.001$) at constant temperatures and lowest (one-way ANOVA: $F_{3,20}=87.70$, $p<0.001$) at 30--10°C (Figure 3).

Water Potential

Changes in germination percentages with time: Cumulative germination was greater (two-way-ANOVA main effect of year: 0 MPa: $F1,10=293.33$, $p<0.0001$; 0.2 MPa: $F1,10=43.59$, $p<0.0001$; 0.4 MPa: $F1,10=63.70$, $p<0.0001$; 0.6 MPa: $F1,10=57.37$, $p<0.0001$; 0.8 MPa: $F1,10=4.64$, $p=0.0566$; 1 MPa: $F1,10=11.89$, $p=0.0062$; 1.2 MPa: $F1,10=0.39$, $p=0.5447$) in 2007 than in 2006, except at -0.8 and -1.2 MPa (Figure 4).

In 2006, the germination percentages were greater (two-way-ANOVA main effect of year: $F6,35=4.69$, $p=0.0013$) at 0 and -0.2 MPa than at -0.8 MPa and greater water potentials (Figure 4).

At 24 and 36 h after initiation in 2006, the germination percentages were similar [two-way-ANOVA main effect of water potential (24 h: $F6,35=10.66$, $p<0.0001$; 36 h: $F6,35=6.05$, $p=0.0002$)] between 0 and -0.8 MPa with these germination percentages greater than those at -1.0 and -1.2 MPa (Figure 4). The germination percentages did not differ (two-way-ANOVA main effect of water potential: $F6,35=1.22$, $p=0.3199$) after 72 h across the different water potentials.

The germination percentages among the water potential treatments were more marked in 2007 than in 2006 (Figure 4). Until 2400 hrs from the study initiation, the germination percentages did not differ between 0 and -0.2 MPa ($p>0.05$), but they did decrease as the water potentials decreased (Figure 4). Twelve hours later, the germination percentages were greater (two-way-ANOVA main effect of water potential: $F_{6,35}=23.54$, $p<0.0001$) at 0 than at -0.2 MPa. The water potentials of 0 and -0.6 MPa were greater (two-way-ANOVA main effect of water potential: $F_{6,35}=23.54$, $p<0.0001$) than those at or below than -0.8 MPa.

Cumulative germination: Cumulative germination percentages were similar ($p>0.05$) among the treatments from 0 to -0.8 MPa in 2006 (Figure 4). However, the percentages decreased (two-way-ANOVA main effect of water potential: $F6,35=7.75$, $p<0.0001$) at lower water potentials (Figure 4). Cumulative germination in 2007 was greater (two-way-ANOVA main effect of water potential: $F6,35=24.14$, $p<0.0001$) at 0 than -0.2 MPa, but the values were similar ($p>0.05$) between -0.2 and -0.6 MPa (Figure 4).

At water potentials lower than -0.6 MPa, the germination percentages decreased (two-way-ANOVA main effect of water potential: $F6,35=24.14$, $p<0.0001$) as the water potentials also decreased (Figure 4).

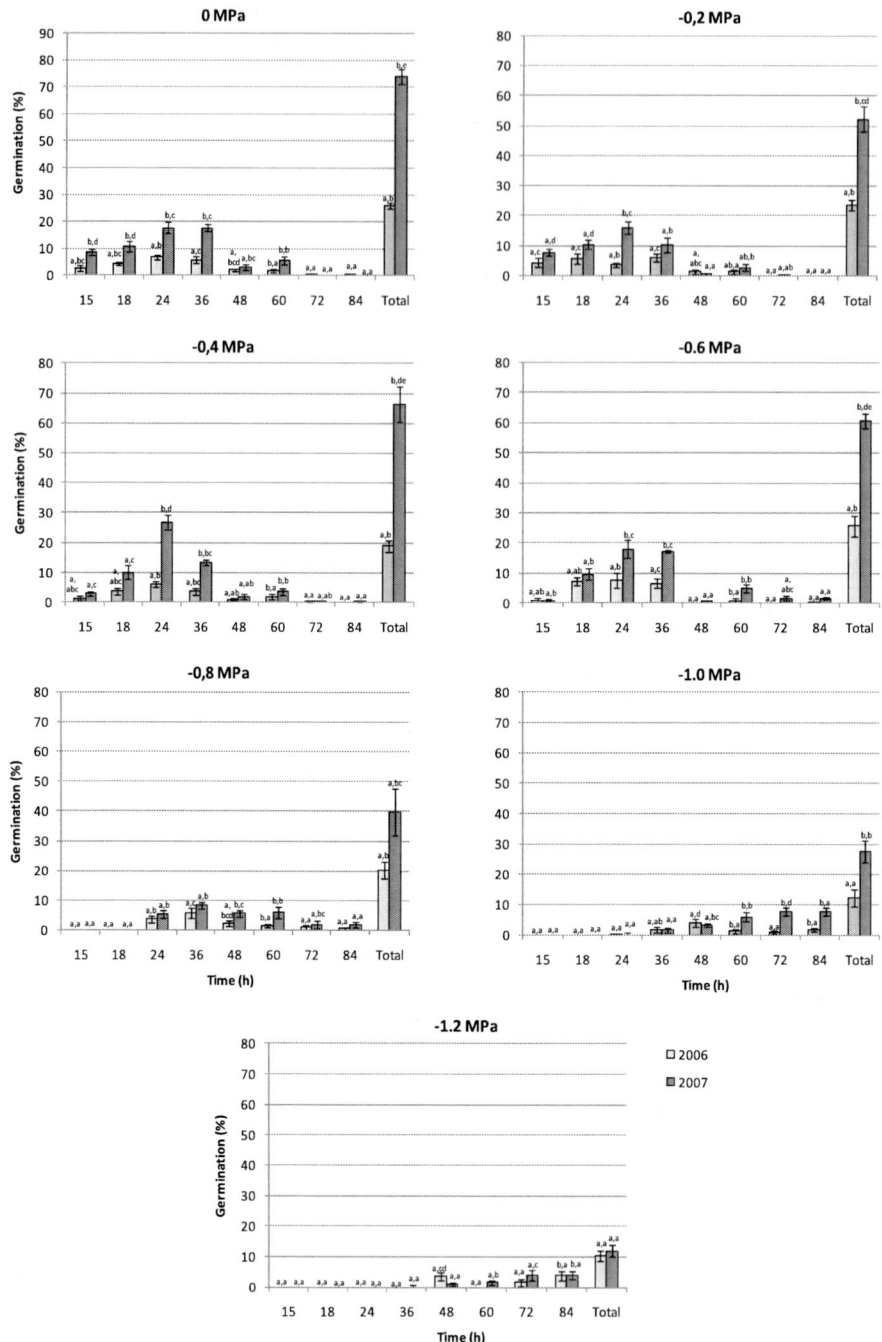

Figure 4. Percentage of germination of *Digitaria eriantha* seeds in 2006 and 2007 (on an hourly basis from study initiation, and Total, cumulative germination) as a function of water potential. Each value is the mean±1 SE of n=6. Different letters to the left of the comma indicate significant differences between years, and those to the right of the comma indicate significant differences among water potential treatments. Differences between means were tested using the LSD test (p≤0.05).

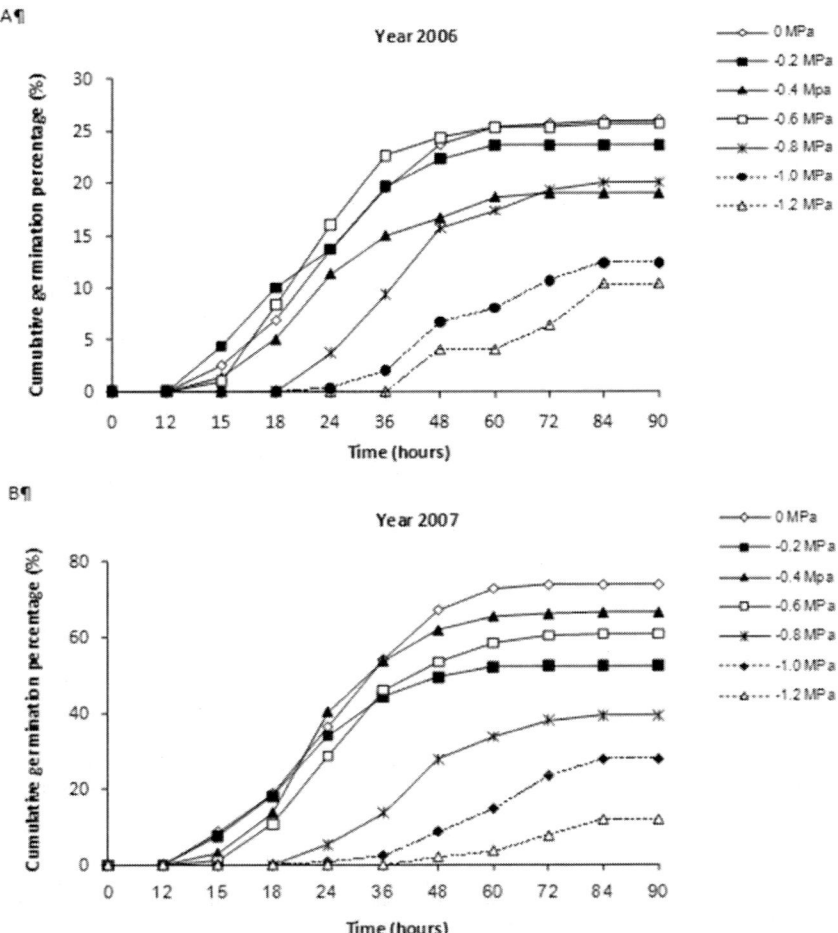

Figure 5.Cumulative germination percentages in *Digitariaeriantha* versus time from study initiation at various water potentials in 2006 and 2007. Maximum cumulative germination is shown under the column 'Total' in Table 3 for 2006 and 2007. Each value is the mean of n=6. Note the different scales in the Y axis for 2006 and 2007.

Patterns in changes of germination percentages: Similar patterns of changes in the germination percentages with time occurred in both study years (Figure 5). The germination percentages followed a similar pattern with time from initiation of the study at water potentials from 0 to -0.6 MPa. At lower water potentials (from -0.8 to -1.2 MPa), germination was delayed (Figure 5).

Coefficients of Velocity of Germination

No differences (two-way-ANOVA main effect of year: $F_{1,70}$=1.95, p=0.1673) were found between the study years in the coefficients of velocity across the water potential treatments (Figure 6). In both study years, the coefficients of velocity were greater (two-way-ANOVA main effect of water potential: $F_{6,70}$=30.33, $p<0.0001$) from 0 to -0.6 MPa than at lower water potentials. The lowest (two-way-ANOVA main effect of water potential: $F_{6,70}$=30.33, $p<0.0001$) coefficients of velocity were found at -1.0 and -1.2 MPa (Figure 6).

Early Seedling Growth

Early seedling growth was similar between years (three-way-ANOVA main effect of year: $F_{1,120}=0.71$, $p=0.401$; 2006: 6.33 mm, n=72; 2007: 5.40 mm, n=72), and plant parts (three-way-ANOVA main effect of plant part: $F_{1,120}=0.41$, $p=0.522$; shoots: 5.51 mm, n=72; roots: 6.22 mm, n=72) (Figure 7). It was greater (three-way-ANOVA main effect of water potential: $F_{5,120}=19.14$, $p<0.001$) at 0 than at -0.6 MPa and lower water potentials (Figure 7).

Discussion

Storage Time

Maintenance of seed quality in storage from the time of production until the seed is planted is imperative to assure its planting success and value. There was a marked decrease in seed viability with storage time in *D. eriantha*. This might have been partially the result of the storage conditions,and it can be diminished by reducing the storage time.

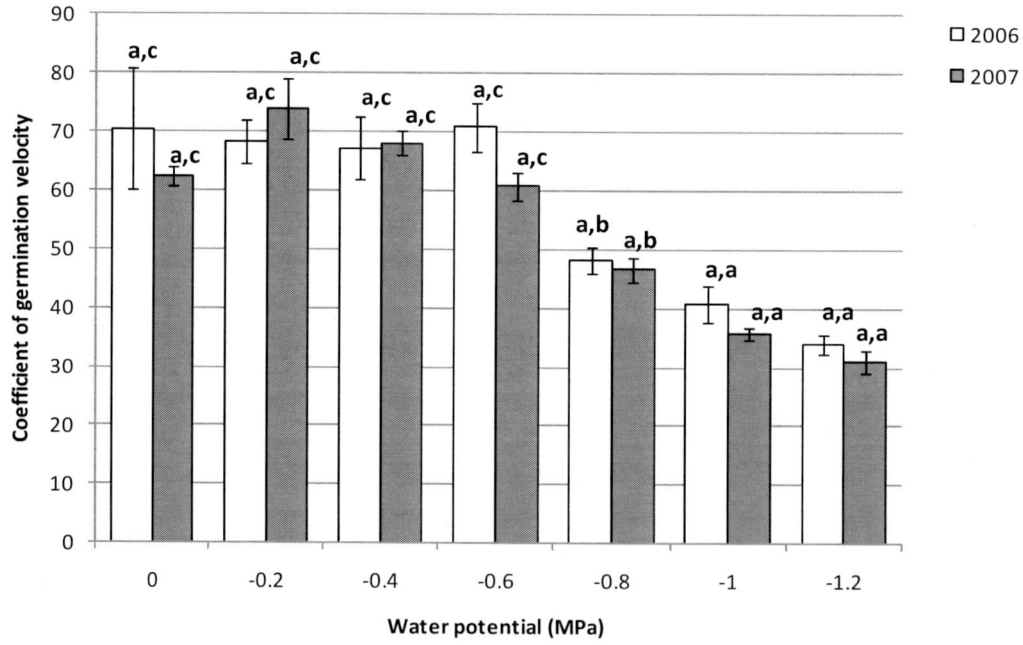

Figure 6. Coefficients of velocity of germination in the various water potential treatments in 2006 and 2007. Each value is the mean ± 1SE of n=6. Letters to the left of the comma indicate significant differences between years, and those to the right of the comma indicate significant differences among water potential treatments.Differences between means were tested using the LSD test (p≤0.05).

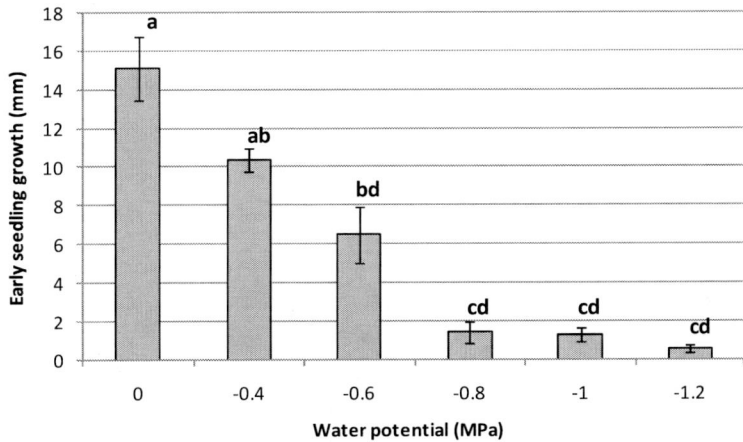

Figure 7. Early growth of *Digitaria eriantha* seedlings (shoot + root) at various water potentials. Each value is the mean±1 SE of n=24. Different letters indicate significant differences (p≤0.05) among water potentials according to the LSD test.

For example, the grass seed industry in Oregon ships seeds within a few months after harvesting [33]. Other examples to reduce storage period are: i) harvesting wheat seed in the Highlands of Bolivia in April and planting in May in the lowlands, and ii) twice yearly production of rice in Colombia [33]. In situations where seeds need to be stored, harvesting efforts should capture the seeds with low moisture [34] and subsequently storage these seeds in cool and dry warehouses to lower the risks in storage [34].

The storage of the *D. eriantha* seeds at 20°C likely caused the loss of seed viability two years post-harvest. Losses of viability because of storage conditions likely contributed to the lower germination percentages in 2006 than in 2007.

Temperature Effects

Germination percentages and coefficients of velocity were favored by constant compared to alternating temperatures, but only during the first 2400 hours. Both variables were similar under constant or alternating temperatures at the end of the study.

These results imply that seeding of *D. eriantha* could occur from late spring to early summer (i.e., January: Figure 1) when the soil temperatures are 30°C. Precipitation impactsthe seed germination and seedling growth, however (i.e., see the dry 2008 and 2009, and the long-term precipitation data in Figure 1).

Constant and alternating temperatures differed with respect to their influences on germination. More than 50% germination occurred at constant rather than at alternating temperatures during the first 2400 hrs. The coefficients of velocity data showed that more seeds germinated at constant than at alternating temperatures, at least early in the germination phase. These results differ from most other studies on the germination of perennial grass species, as in those germination was stimulated more by alternating than by constant temperatures [35].

Our results have site-specific inferences as high germination can be desirable at constant temperatures after seeding due light rainfall, which is common (61% of rainfall events are

<5mm) [36]. Although constant temperatures were advantageous for the first three days of germination, the similar total germination after four days suggests that temperature is not a major impediment to germination of *D. eriantha* as the seeds can be expected to germinate over a wide thermal gradient.

Water Potential Effects

In contrast to temperature, moisture availability imposed substantial limitations on the seed germination of *D. eriantha*, which has similar germination requirements as many mesophytic crops [37-39]. The germination percentage and speed of germination decreased as the water potential lowered. Decreased coefficients of velocity at lower water potentials indicate longer germination times [31]; in fact, the seeds started to germinate later at lower than at higher water potentials (Figure 1). With high soil moisture, the seeds of *D. eriantha*can germinate over a wide range of temperatures; however, water stress overrides temperature influences, and narrows the conditions where germination will occur. This response presumably reflects an adaptive strategy, because *D. eriantha* is generally restricted to habitats with moister conditions than those in the Phytogeographical Province of Monte [40]. This strategy protects against germination under conditions of transient or low soil moisture, resulting in germination mostly occurring in periods with protracted conditions of high soil moisture. Relatively warm soil temperatures and water stress are usually simultaneous events in the *Chacra Experimental de Patagones*, within the Phytogeographical Province of Monte. As such, dormancy induced by this combination may act to preserve a portion of the seedbank for germination at a later date. Induction of dormancy under these conditions may also serve to block germination that would otherwise predispose the seedlings to temperature and moisture conditions that may not be conducive for their growth and survival.

Seedling Growth and Recommendation for Planting Time

Our data collectively demonstrate that *D. eriantha* seeds can germinate over a broad range of temperatures, but severe restrictions are imposed by reduced moisture availability.

Seeding of this species in the rangelands of central Argentina (e.g., the *Chacra Experimental de Patagones*) will most likely fail under water stress (e.g., *see* the dry 2008 and 2009 inFigure 1). Therefore, the recommendation is to plant from late spring to early summer, when the seedbed temperatures are increasing and the soil moisture is still adequate (e.g., *see* year 1984 in Figure 1). The timing of planting will be an essential management practice regarding the use of this species for re-vegetation purposes in the rangelands of central, arid Argentina.

Acknowledgment

We thank Dr.Justin Derner, Research Leader High Plains Grasslands Research Station, Cheyenne, WY, who completely edited a late version of the manuscript.

References

[1]   C. A. Busso, H. D. Giorgetti, O. A. Montenegro and G. D. Rodríguez, *Phyton, Int. J. Exp. Bot.* 53, 9 (2004).

[2]   A. L. Cabrera, In: E. F. Ferreira Sobral (Ed.). Enciclopedia Argentina de Agricultura y Jardinería. ACME, Buenos Aires (1976).

[3]   H. D. Giorgetti, O. A. Montenegro, G. D. Rodríguez, C. A. Busso, T. Montani, M. A. Burgos, A. C. Flemmer, M. B. Toribio and S. S. Horvitz, *J. AridEnviron.* 36, 623 (1997).

[4]   D. L. Anderson, *Ecol. Arg.* 4, 9 (1980).

[5]   G. Di Giambatista, M. Garbero, M. Ruiz, M. Giulietti and H. Pedranzani, *Pastos Forr.* 33, 1 (2010).

[6]   M. A. Sanderson, P. Voigt and R. M. Jones, *J. Range Manage.* 52, 145 (1999).

[7]   J. B. Hacker, G. P. M. Wilson and L. Ramírez, *Euphytica*68, 193 (1993).

[8]   E. Maclin, *Genus and Species of Crabgrass.* Available: http://www.ehow.com/info_ 8407286_genus-species-crabgrass.html (2011).

[9]   P. C. V. Du Toit, *J. Range Manage.* 53, 529 (2000).

[10] C. S. Dannhauser, Tydskrif van die Weidingsvereniging van SuidelikeAfrika5, 193 (1988).

[11] C. S. Dannhauser, *J. Grassl. Soc. South. Afr.* 8, 120 (1991).

[12] P. Rimieri, Creación de cultivares mejorados e identificables de *Poa ligularis* y *Digitaria eriantha.* Inf. Técn. Proy., Área Prod. Animal. INTA EEA, San Luis (1997).

[13] Y. A. Torres, C. A. Busso, O. A. Montenegro, L. Ithurrart, H. Giorgetti, G. Rodríguez, D. Bentivegna, R. Brevedan, O. Fernández, M. M. Mujica, S. Baioni, J. Entío, M. Fioretti and G. Tucat, *Appl. Soil Ecol.* 49, 208 (2011).

[14] R. W. Brown, In: D. J. Bedunah and R. E. Sosebee (Eds.), Wildland plants: Physiological Ecology and Developmental Morphology, pp. 635--710. Soc. Range Manage., Denver (1995).

[15] M. Almansouri, J. M. Kinet and S. Lutts, *Plant Soil* 231, 243 (2001).

[16] D. A. Johnson, In: N. C. Turner and P. J. Kramer (Eds.), Adaptations of Plants to Water and Temperature Stress, pp. 419-433, John Wiley, New York (1980).

[17] D. W. Owens and C. A. Call, *J. Range Manage.* 38, 336 (1985).

[18] J. A. MacMahon and D. J. Schimpf, In: D.D. Evans and J. L. Thames (Eds.), Water in Desert Ecosystems. Hutchinson and Ross Inc., Dowden. *USIBP Synthesis Series* 2, 114 (1981).

[19] M. L. Sharma, *Agron. J.* 68, 390 (1976).

[20] W. T. McDonough and R. O. Harniss, *J. Range Manage.* 27, 204 (1974).

[21] J. T. Romo and L. E. Eddleman, *J. Range Manage.* 41, 491 (1988).

[22] P. H. Raven, R. F. Evert and S. E. Eichhorn, The biology of plants, Worth Publishers Inc., New York (1986).

[23] R. R. Black and J. A. Young, *J. Range Manage.* 45, 205 (1992).

[24] S. Dasberg and K. Mendel, *J. Exp. Bot.* 22, 992 (1971).

[25] R. A. Evans and J. A. Young, *J. Range Manage.* 36, 395 (1983).

[26] J. A. Young, C. D. Clements and T. A. Jones, *J. Range Manage.* 56, 247 (2003).

[27] S. E. Meyer, J. Beckstead, P. S. Allen and H. Pullman, *Int. J. Plant Sci.* 156, 206 (1995).

[28] G. O. Throneberry and F. G. Smith, *Plant Physiol.* 30, 337 (1955).

[29] W. E. Emmerich and S. P. Hardegree, *Agron. J.* 82, 1103 (1990).

[30] B. Michel and M. Kaufmann, *Plant Physiol.* 51, 914 (1973).

[31] S. J. Scott, R. A. Jones and W. A. Williams, *Crop Sci.* 24, 1192 (1984).

[32] J.A. Di Rienzo, F. Casanoves, M.G. Balzarini, L. Gonzales, M. Tablada and C.W. Robledo, INFOSTAT versión 2009. Grupo INFOSTAT, FCA, Univ. Nac. Córdoba (2009).

[33] S. Elias, A. Garay, B. Young and T. Chastain. Seed Viability in Storage: A brief review of management principles with emphasis on grass seeds stored in Oregon,Oregon St. Univ. Seed Lab., Technical Brochures-Maintaining, Available: seedlab.oregonstate.edu /book/export/html/123

[34] L. Copeland and M. McDonald, Principles of seed science and technology, Chapman and Hall, New York (1995).

[35] R. A. Distel, D. V. Peláez and O. A. Fernández, *Austr. Rangel. J.* 14, 49 (1992).

[36] A. Páez, C. A. Busso, O. A. Montenegro, G. D. Rodríguez and H. D. Giorgetti, *Phyton, Int. J. Exp. Bot.* 74, 1 (2005).

[37] J. Levitt, Responses of plants to environmental stresses, Academic Press, New York (1980).

[38] J. S. Choinsky and J. M. Tuohy, *Ann. Bot.* 68, 227 (1991).

[39] G. Bonvisutto and C. A. Busso, *Phyton, Int. J. Exp. Bot.* 76, 119 (2007).

[40] E. Cano, Pastizales naturales de La Pampa. Descripción de las especies más importantes.Conv. AACREA-Pcia. de La Pampa, Buenos Aires (1988).

Humid, Subtropical Ecosystems

In: From Seed Germination to Young Plants
Editor: Carlos Alberto Busso

ISBN: 978-1-62618-653-8
© 2013 Nova Science Publishers, Inc.

Chapter 13

Effect of Salinity Stress on Germination of Woody Species in the Santa Fe Forest Wedge (Humid Chaco Region), Argentina

Nélida Carnevale[1,2], Claudia Alzugaray[1] and Rodrigo Freire[1]*

[1]Facultad de Ciencias Agrarias. Universidad Nacional de Rosario;
[2]Consejo de Investigaciones de la Universidad Nacional de Rosario,
Zavalla, Santa Fe, Argentina

Abstract

The Santa Fe Forest Wedge is dominated by woody plant communities. This region has been exploited intensively for wood extraction during the last decades specially *Schinopsis balansae* ("quebracho"). As a result of that mismanagement, the forest area has been higlhy reduced, and additional negative effects have occurred such as top-soil salinization in large areas. The species present in Forest Wedge (humid Chaco) develop in soils with halo hydromorphic characteristics. However, the effects that salt could have on their germination and later plant life stages have not been studied in this region. It has been established that the adaptation of plants to salinity during the seed germination and early stages of seedling development can be crucial for their establishment. Furthermore, salinity can affect the distribution of some species even in the last stages of their development. Therefore, the distribution of individuals in the forest could be determined by traits of each plant species. The study on the conditions for germination of woody species in the humid Chaco started on the assumption that any species is tolerant to varying salt concentrations, particularly sodium chloride. Treatments consisted in different solution concentrations of NaCl and polyethylene glycol at the same molal concentrations. We studied 7 tree and 2 shrub species. For the tolerance levels evaluated, all species were glycophytes or slightly halophytes. Some of them showed a possible adaptation to salinity increases. Changes in land use as deforestation followed by

---

[*] E-mail address: nelincita@gmail.com.

agriculture (e.g., found in the Chaco-Pampeana Plains) might (i) be hardly predictable, and (ii) produce changes in the overall forest dynamics.

Introduction

The Forest Wedge is located in the eastern part of the *Chaco santafesino* (the fraction of the Chaco region which belongs to the territory of the Province of Santa Fe), between 28° and 30° SL. It extends eastward up to the Paraná River, and to the west as far as the "Golondrinas" stream and stream bed, which are part of a system of interconnected shallow lakes. The total study area is 10.000 km^2.

The climate is seasonally subtropical. The mean annual temperature ranges from 18° C in the south to 26° C in the north, with maximum temperatures of 48°C in the summer, and frosts in the winter [1]. Rainfall occurs during the summer, and there is a marked dry season in winter of varying duration which lasts for about seven months. Mean annual precipitation is 1.100 mm with wide interannual variations, (e.g., from 1.300 to 678 mm [2]).

The Chaco-Pampean Plain is neither geologically nor morphologically homogeneous. Also, it is not uniformly flat. Instead, it is a structured, dynamically active system. The hydrographic landscape is considered part of the Chaco region [3], and it is covered by numerous stream bed systems running from northeast to southeast which connect to each other in periods of heavy rainfall. The runoff patterns of these streams are fairly inefficient, which results in large flooding areas [3]. The terrain is flat to slightly undulated, and the depressions are drainageways which become flooded in the summer.

Soils are associated with the geomorphology of the landscape, with slight topographic variations determining changes in their physical and chemical properties as well as in the frequency of floodings; they make up a complex mosaic of different types, often halo-hydromorphic in character [4]. This means that their properties can vary significantly within a few meters. The parent material is constituted by deposits alluvial-lacustrine; these soils are formed by fine sediments such as fine sand, lime and clays, and they frequently present impermeable layers, which result in the formation of swamps [5]. At the higher topographic elevations there are Typic Argiudoll soils, in the middle there are Aquic Argiudolls, and in the lower, Alfisols [6].

The general landscape of the Forest Wedge is flat, with a terrain predominantly quasi-normal, with not very defined gradients which hinders the normal runoff of water [4]. The soils of the extensive, gently undulating or flat hillocks can be classified as Typic Natraqualfs. In these soils grow the study species.

The Forest Wedge is located at the east of the Chaco Phytogeographical Province [7] and is characterized by the presence of large tree populations alternating with marshes and grass openings. The landscape is formed by variations in plant populations along humidity and salinity gradients associated with topography [8]. This region is clearly dominated by woody plants, which range from low shrubs to very tall trees. Even though most of the region is covered by forest, there are differences in plant cover due to the water constraints, which become stronger to the west.

In the lowest part of the gradient, there are forests of *Prosopis nigra* (*P. nigra* var. *ragonesei*) called "algarrobales" growing near marshes or streams; at higher altitudes along the gradient there are populations of *Schinopsis balansae* ("quebrachales"), and at the highest elevation, the Bosque chaqueño (Chaqueño Forest) or Transitional Southern-Brazilian Forest, or high forest [8-11].

In addition to the woody plant communities that characterize the Forest Wedge, there is a variety of other woody communities known as "azonal"; these communities include not only populations of *Prosopis* sp. or *Copernicia alba* ("palmares") but also populations of *Stetsonia coryne* ("cardonales") or *Prosopis ruscifolia* ("vinalares") [12]. There are also savannas or extensive grass openings with a tall herbaceous layer made up of *Sorghastrum agrostoides*, *Leptochloa chloridiformis*, *Elyonurus muticus*, *Paspalum intermedium* and *Spartina argentinensis*, among other species, and marshes. These are senescent, flooding areas covered by hygrophillous plants and sometimes flanked by *Copernicia alba* palm trees.

Native forests have been exposed to intensive exploitation since 1906 for tannin extraction, mainly from "quebracho colorado chaqueño" trees (*Schinopsis balansae*). The most important tannin-extraction companies settled at the north of Santa Fe, since this area showed the highest *Schinopsis balansae* density. When its extraction was no longer profitable, fields were given to settlers who have continued exploiting and degrading the resource, by cutting wood for making posts, firewood, and charcoal.

Also, parts of the forests have been cleared for agriculture and cattle raising [2]. The forested area in the Forest Wedge has decreased sharply between 1976 and 2008, from 790.000 to 385857 ha; it means a decrease of 51.2% within a period of three decades. In the period 1976-2008, the average deforestation rate was -2,30,-2,32, one of the highest in the world [13] (Table 1).

Table 1. Changes in area covered by natural forest in the ForestWedge (Santa Fe)

Period	Total deforestation (has)	Annual deforestation (has)
1976 – 1986	143.006	14.301
1986 – 2000	127.011	9.072
2000 – 2005	57.118	11.426
2005 - 2008	58.722	19.574
Total: 1976 – 2008	385.857	12.058

The decrease in tree cover and the process of forest fragmentation have brought about the appearance of bare soil sectors with an increased content of salt in the topsoil. The deforested area covered with salt increased more than 37% in 29 years [13].

Soil salinization, which limits crop yields and has adverse effects on plant germination and vigour, is an increasingly significant problem in the world [14], and it is expected to increase within the next decades [15]. This is a highly relevant problem in Argentina, since this country ranks third in soil area affected by salinity, following Russia and Australia [16].

In the areas surrounding the urban centers of the Forest Wedge (such as the Reconquista and Vera cities, in the Province of Santa Fe), the forest has been replaced by cropfields, especially since 2001, when the increasing demand for agricultural land brought about an expansion of the agricultural frontier.

Previous studies have shown that agriculture increases deep drainage, which in turn causes a gradual rise of the water table, with a subsequent strong mobilization of dissolved salts, which affect soil fertility when they reach the topsoil. This process has been reported for the Espinal and Chaco Seco (Dry Chaco) regions [17]; it can also affect the hydrologic regime and the salinization of water and soils, as has happened in the dry forests of Australia, Africa and North America [17].

Numerous researchers have pointed out that the adaptation of plants to salinity at the germination and early seedling stages is crucial for their establishment [18-19]. Salinity can affect the distribution of certain plant species even in their last developmental morphology stages [20]. The seedling stage is the most vulnerable period in a plant's life cycle. Also, since germination determines the time and site where a seedling starts to develop [21-22], the temperature and water requirements for seeds of a given species will vary according to the time when germination begins. Consequently, an adequate germination response of halophytes to environmental conditions will determine their distribution in saline soils. Thus, the distribution of individuals in the forest might be determined by the soil properties and the characteristics of each plant species.

Soil Tolerance Mechanisms

The word *halophyte* literally means "salt plant", but it is used to refer to those plants which can grow in the presence of high concentrations of sodium salts. Those which cannot grow under such conditions are called *glycophytes* ("sweet" plants). Halophytes are further subdivided into the extreme euhalophytes and the moderate oligohalophytes [23]. Salinity-induced stress is the most severe abiotic stress [23].

The effects of salinity can be classified as osmotic, nutritional, and toxic. The first two are secondary salt-induced stresses, while the third is a primary salt injury [24]. The most noticeable effect is a decrease in growth caused by numerous biochemical mechanisms; they result from (i) the toxicity of specific ions, (ii) an increase in osmotic pressure, or (iii) an increase in alkalinity. These mechanisms can limit water availability or affect the physiology of cells or affect metabolic processes [25]. Most of the salt-induced stresses in nature are caused by sodium salts, especially sodium chloride.

Different plant families have different salinity tolerance limits. For example, the salinity tolerance limit is low in legumes (beans, peas), medium in cereal grasses (rye. wheat. oats. barley), and high in some forage or "technical" plants, such as Sudan grass, alfalfa, sunflower, and sugar beet [23]. This limit can be indicated by interrupted growth or death of tissues [26], necrosis or marginal leaf burn [27] followed by loss of turgor, falling of leaves and ultimate plant death [28].

To our knowledge, no studies have been carried out on salinity tolerance in the germination of native tree species of the Santa Fe Forest Wedge. This aspect has been studied in some "algarrobo" species, but that are common in other areas, such as *Prosopis chilensis* and *Prosopis flexuosa* [29] and *Schinopsis quebracho-colorado* [30].

In the secondary forest called "quebrachal", most seeds that fall from the trees remain on the topsoil, where they can be affected by the water regime or by the presence of animals and microorganisms, which can affect germination in different ways.

The distribution of adult individuals of the dominant species in the forest has been attributed to hydrological and saline variables. However, it has not been investigated how these factors affect the first stages of plant development, and the permanent establishment of tree and shrub species.

The study on the conditions for germination of woody species in the Forest Wedge, located in the humid Chaco, was started on the assumption that such species are tolerant to varying salt concentrations in the soil, particularly sodium chloride.

Seeds were subjected to vigour and germination tests using different salt (NaCl) and polyethylene glycol (PEG) concentrations in order to determine the type of damage either primary (toxic) or secondary (osmotic or nutritional), under the treatments.

Materials and Methods

In general, the same procedures were used for all the study species.

Treatments consisted in solutions of NaCl and polyethylene glycol (PEG) (8000) at the same molal concentration, which was obtained by levelling off their osmotic pressure measured in bars. These solutions were used to determine whether there were different responses to toxicity-induced or osmotic damages.

The concentrations used for measuring the effect of each solution on the germination of the different species were determined on the basis of the usual concentrations found on the soil types where they grow [6].

Germination and Vigor Tests

Germination tests were carried out following the guidelines in the Tree and Shrub Seed Handbook [31] using 4 replicates of 50 seeds each. Seeds were disinfected with a 2% sodium hypochlorite solution for two minutes and sown in sand-filled trays at field capacity, which were placed in translucent polyethylene bags. Seeds of Leguminosae were scarified for 10-20 minutes using concentrated sulfuric acid. The studies were conducted in germination chambers with a 12-hr photoperiod at 25°C. Seedlings were assessed every 7 days for 28 days, following the specifications in the seedling assessment manual [32-33], and the germination rates were expressed as a percentage of normal seedlings.

Vigor was determined using the Index of Germination Velocity (IGV) and Mean Germination Time (MGT). IGV was determined using the same procedure as in the germination tests, with 4 replicates of 50 seeds each in which physiological germination was recorded every day for 30 days or more. Seeds were considered germinated when the primary roots reached a length of 2 mm. IGV was calculated using the methodology cited in [34]. IGV was used to calculate MGT.

$$IGV = \frac{\sum C_i}{\sum C_i * T_i} * 100 \quad MGT = \frac{\sum C_i * T_i}{\sum C_i}$$

Ci= Number of seeds germinated per day

Ti= Number of days from the start of the study to Ci seed germination

Treatments for *Acacia praecox, Achatocarpus praecox, Caesalpinia paraguarienses, Lycium cuneatum, Maytenus vitis-idaea, Prosopis alba* and *Schinus fasciculatus,* consisted of solutions of NaCl and PEG at 0.2, 0.25 and 0.3 M concentrations. Two additional treatments, 0.4 and 0.6 M concentrations were used for *Prosopis nigra* and *Schinopsis balansae.*

Schinopsis balansae and *Maytenus vitis-idaea* were subjected to the tetrazolium test to evaluate salt-induced damage. Seeds were removed from the fruits and hydrated for 24 hours on paper towels soaked with distilled water and 0.3M saline solution; the seed coat was cut to facilitate entry of the reagent. Thereafter, they were treated with a 1% solution of 2. 3. 5-triphenyl tetrazolium chloride. Seeds were incubated in an oven at 30°C for 24 hours in the TTC solution, following the methodology of Craviotto and Mudrovitsch et al. [35-36]. A digital seed analyzer was used to capture images of the stained seeds, and the resulting patterns were described.

One way analysis of variance was applied to determine whether IGV and MGT differed significantly between treatments. Then, a multiple range test was used to determine differences among treatments (Statgraphics 5.0 for Windows).

The study species in this work were: *Acacia praecox. Achatocarpus praecox, Caesalpinia paraguarienses, Lycium cuneatum, Maytenus vitis-idaea, Prosopis alba, Prosopis nigra, Schinopsis balansae* and *Schinus fasciculatus.* All of them are native woody species in the study region.

General Characteristics of the Study Species

Acacia praecox Griseb. - Family: Fabaceae - Common name: "garabato"

Small tree or shrub, spinose, although in very rare occasions it does not bear spines, 3 to 8 metres high, trunk up to 20 cm diameter; alternate, compound bipinnate leaves. Inflorescences are arranged in capitula, with creamy white, strongly scented flowers 1.5 cm in diameter. The fruit is a dehiscent pod with a papery, membranous texture, light brown in colour, 5 to 14 cm long and 1.5 to 2.5 cm wide. Seeds are light brown, smooth and ellipsoidal in shape.

Uses

The wood is very hard and heavy, and it is used for manufacturing farm tools and as a fuel [37]. It has potential value as an ornamental due to its appearance and flowering habit. Medicinally, it is used as a healing agent, antisiphilitic, disinfectant, astringent, anti-inflammatory, inhalant, and for kidney diseases.

Pods from several individuals were collected in the first half of January 2012. Seeds were removed from the pods and only those which did not present signs of insect damage were selected.

Achatocarpus praecox Griseb. - Family: Achatocarpaceae - Common names: "tala negro", "palo tinta", "palo de tinta", "rumi-caspi", "ivirá-hü", "virazú"

The common name of this species comes from the colour of its leaves as they dry.

It is a small dioecious tree growing 3 to 7 m tall, the main stem of which is sometimes replaced by a group of smaller stems. Leaves are simple, alternate or arranged in fascicles from 3 to 5, attached to brachyblasts, generally elliptically lanceolate in shape. Infloresecences are borne in racemes. The small unisexual flowers can grow up to 3 cm long. The fruit, an ovoid berry 5 mm in diameter, becomes translucent at maturity so that the black, lenticular seeds, about 3 mm in diameter, can be seen from the outside through the pericarp. It is one of the typical species of the region.

Uses

It is used for manufacturing tool handles and as a fuel. Its wood is soft and light, dark yellow in color. The native people of the Chaco region ascribe medicinal properties to this plant. A decoction of its leaves is used as a dye, and it was formerly used by indigenous people for tattooing and for dying fabric. It is a honey plant, and native people maintain that honey made from its flowers is one of the best. It bears potential as an ornamental due to the beauty of its branches at the fruiting stage.

Caesalpinia paraguariensis (D. Parodi) Burkart - Family: Fabaceae - Common names: "guayacán", "guayacán-hü", "ivirá-berá"

It is a stout deciduous tree without spines, growing 10 to 15 m height (occasionally 20 m), with a diameter of up to 1m. Its bark is characterized by a greenish brown color, which peels in irregular strips exposing the new grayish green bark. The leaves are bipinnate, and the yellow to orange flowers are clustered in groups of 8 to 15 in simple racemes. The fruit is a thick woody indehiscent pod containing 2 to 8 seeds, orbicular or elliptical in shape, 3 to 4 cm (sometimes up to 6 cm) long, 2.8 cm wide and 7 to 12 cm thick, with a bright dark colour.

Uses: Its hard wood is used for railway ties, posts, and knife handles as well as for making musical instruments as a substitute for ebony. Its wood and fruits have tanning properties. This species is excellent as a fuel. Its leaves are eaten by cattle.

Lycium cuneatum Dammer - Family: Solanaceae

Shrub or small tree 1 to 4 m height, with a greyish or brown branching stem with branches bent downwards, which sometimes get to the forest floor. It changes from thick pubescent to glabrescent, with simple hairs that branch several times, becoming more glabrous at maturity; it bears spines 2 to 23 mm long, but occasionally may be spineless. The leaves are flat, membranous, hairy on both sides, ovate, elliptic, quasi-orbicular or obovate in shape, with an obtuse or acute apex, growing 4-40 mm × 3-32 mm, with 3-15 mm long petioles. Tetramerous flowers are attached to brachyblasts in groups of 3 to 20; the corolla is white or creamy-yellow, sometimes with greenish (or more rarely purplish) spots at the base of the lobes. The fruit is a dark purple, fairly black, globoid, kidney-shaped berry measuring 3-4 mm × 3.5-4 mm. There are 3 to 6 reniform 2-2.5 mm x 1.5-2 mm seeds in each locule [38].

Fruits (berries) were collected in March 2010. They were washed, and seeds were dried at room temperature in the lab.

Uses

Used as a plaster or for washing and making frictions for skin affections.

Maytenus vitis-idaea Griseb. - Family: Celastraceae - Common names: "tala salado", "sal de indio", "ibirá-yuquí", "colquiyuyo", "carne gorda", "chaplán", "palta"

Spineless, crooked shrub or small tree, 2 to 5 m height, with a tree trunk diameter up to 20 cm, and wrinkled dark-gray bark. The fleshy persistent leaves are deep green or glaucous, simple, alternate, obovate or circular in shape, growing up to 9 cm long. The small yellowish-green flowers are monoecious, and are arranged in short racemes. The fruit, a dehiscent ellipsoid, three-valved capsule measures 15 mm × 7 mm. It typically bears three bright purple, quasi-ellipsoid arillate seeds, measuring 9 × 4 mm.

Uses

Among the many applications of this species, its fleshy leaves are used as a fodder, and it has medicinal and insecticidal uses [39]. Used as an astringent and also for ophthalmic affections, it also has anti-carcinogenic properties [37], even though with a low concentration of the active principle.

Fruits were collected from trees located in a site 15 km away from Vera city (Province of Santa Fe) in July 2002. Seeds were washed to remove the aril.

Prosopis alba Griseb. Pl. Lor. - Family: Fabaceae - Common names: "algarrobo blanco", "taco", "ibopé-pará"

Named "algarrobo" by the European conquerors due to its likeness with the European carob tree (*Ceratonia siliqua*) [40]. It is one of the native species with the largest geographical distribution, being found both in the Dry and the Humid Chaco.

It is a stout tree with a rounded, parasol-shaped crown, 18 m height or more. The trunk, which can grow up to 70 cm in diameter, bears a small number of axillary spines 1 to 3 cm long. Deciduos foliage made up of compound leaves. The yellowish-green hermaphrodite flowers are arranged in spike-like racemes which cluster in groups of up to 8 on a brachyblast. The leathery, woody indehiscent fruits are falcate or curved in shape and yellow in color. The seeds are orthodox and long-lived.

Uses: Its high-quality, hard wood is used for flooring, wine casks, tools, doors and windows. It makes excellent firewood, and its flowers are frequented by bees and yield good honey. The sweet pods are eaten by people and used as a fodder for cattle. Flour made from the pods is used to make cakes and a typical beverage, called "aloja" or "patay". A decoction of the pods is used as a medicine, for dissolving bladder stones. It is highly appreciated for its shade and as an ornamental.

Collecting procedure: Mature fruits were collected from trees by the end of May 2004. Seeds were removed by hand from the pods and scarified with acid before subjecting them to vigour tests.

Prosopis nigra Griseb. - Family: Fabaceae - Common names: "algarrobo negro", "algarrobo chico", "ibopé-pará", "yana tacu"

Tree with a few spines or altogether spineless, of variable size, 7 to 10 m high, more or less parachute-shaped crown, and a short, many-branched trunk. Compound, bipinnate leaves arranged fascicles on alternate brachyblasts. Inflorescences are arranged in spike-like cylindrical racemes. The flowers are yellowish, hermaphroditic, 5 to 6 cm long, and resemble those of other species of the *Prosopis* genera. The fruits are flat, quasi-woody, coriaceous pods, commonly with purple mottles, straight or slightly curved in outline; they are shorter than those of "algarrobo blanco", measuring 7 to 18 cm long, 1 cm wide and 7 - 8 mm thin. Each pod contains 10-20 glossy brown, flat, ovoid seeds.

Uses

Its hard, heavy wood has similar characteristics and uses as wood from "algarrobo blanco". It is used for its shade and wood, and as a fuel, food and drink. A sweet, highly nutritious flour called "patay" is made by grinding the dry pods. This flour is used to make a honey-like jelly, as well as two beverages, "añapa" and "aloja", the latter alcoholic. Medicinally, it is used for treating malaria, rheumatic affections, and as an antipyretic and diaphoretic.

Collecting Procedure

Mature fruits were collected before they fell from the trees in May 2002 and seeds were removed from the pods.

Schinopsis balansae Engl. - Family: Anacardiaceae - Common names: "quebracho", "quebracho colorado santafesino", "ialán", "urunday-pitá"

Stout polygamous, dioecious tree, which can reach a height of up to 24 m and a diameter of 1.40 m with a straight trunk and no lateral branches. Its crown is similar in shape to a truncated inverted cone. The greyish-brown bark presents deep cracks. Leaves are simple, alternate, and have a leathery texture. Inflorescences are arranged in terminal pyramidal panicles. The flowers are small and yellowish or reddish in color. The fruits are reddish woody samaras, which turn brown at maturity. The seminiferous portion is flat and ovoid, and the wing is oblong, obtuse, about 20 mm in length and 9 mm in breadth.

Uses

Hard, heavy wood; reddish-brown heartwood very developed. Being resistant to decay, it is used for outdoor structures, and for making posts, beams, railway sleepers, and carvings. Its high tannin content encouraged its exploitation during the nineteenth and twentieth centuries.

Collecting Procedure

Seeds were collected from the trees at maturity time, by the end of March 2003.

Schinus fasciculata Griseb. - Family: Anacardiaceae - Common name: "molle pispita"

Bushy, spinose species growing 1 to 6 m height. The thick, persistent leaves are polymorphic (different leaf shapes occur on the same individual). Small, whitish flowers are arranged in racemes. The fruits, strongly scented violet or purple drupes, have a diameter of 5 to 6 mm and contain a single orthodox seed.

Uses

Used locally as a fuel, as an ornamental or for fencing; the essential oils extracted from its leaves are used in perfumery, and its sweet, pungent fruits are eaten as a substitute for pepper. Though not widely used as a medicinal plant, its leaves contain important active antifungal agents [41]. Also used as a laxative, as a balsam and for the treatment of catarrh [42].

Collecting Procedure

Fruits were washed and dried at room temperature in the lab.

Results

Acacia praecox

Germination percentage was 59% in the control vs. 32% in 0.3M NaCl (Table 2). However, neither Mean Germination Time (MGT) nor the Index of Germination Velocity (IGV) could be calculated because seeds of this species germinate within 12 hours approximately. It meant that the divisor stayed at 0 and the formula remained indeterminate.

Significant differences ($p > 0.01$) were found between the control and all treatments, except treatment 1 (NaCl 0.2M). Differences were also significant between 0.2 M NaCl and all the other treatments; and between the following solutions: 0.25 PEG and 0.25 NaCl; 0.25 PEG and 0.3 NaCl; 0.25 PEG and 0.3 PEG. However, a higher germination percentage (35%) was obtained with the last concentration (0.3M) than with the other two PEG treatments with lower molal concentration (22% and 12% germination under 0.20M and 0.25M PEG, respectively) (Table 2).

Achatocarpus praecox

Germination percentage changed from 78% in control vs. 5% in 0.3 M NaCl (Table 2).

Significant differences in IGV were found between (i) the control and all NaCl concentrations; (ii) 0.2 and 0.3 M NaCl; (iii) the control and 0.25 M PEG; (iv) 0.25 M NaCl and 0.2 M PEG; (v) 0.3 M NaCl and 0.2 M PEG; (vi) 0.3 M NaCl and 0.25 M PEG; and (vii) the 0.2 and 0.25 M PEG treatments. No seeds germinated at the highest PEG concentration (0.3 M), so no values were recorded for IGV or MGT. Average IGV and MGT values for the control were 8.57% and 12 days, respectively.

Significant differences in MGT were found between (i) the control and the 0.25 M NaCl treatments; (ii) 0.25 M NaCl and 0.2 M PEG; (iii) 0.3 M NaCl and 0.25 M PEG; and (iv) 0.2 and 0.25 M PEG.

MGT value was zero under the 0.3 M treatment (Fig 1) (Table 2).

Table 2. Germination, IGV, and MGT germination, of woody species of Forest Wedge (Santa Fe province)

Species		Control	NaCl					PEG		
			0.2 M	0.25 M	0.3 M	0.4 M	0.6 M	0.2 M	0.25 M	0.3 M
Acacia praecox	G %	59.00	55.00	34.00	32.00	-	-	22.00	12.00	35.00
	IGV	-	-	-	-	-	-	-	-	-
	MGT	-	-	-	-	-	-	-	-	-
Achatocarpus praecox	G	78.00	59.00	18.00	5.00	-	-	56.00	27.00	0.00
	IGV	8.90	5.80	4.21	4.03	-	-	6.70	4.25	0.00
	MGT	11.80	17.07	25.14	24.75	-	-	15.09	23.67	0.00
Caesalpinia paraguariensis	G %	39.00	31.00	21.00	13.00	-	-	39.00	20.00	0.00
	IGV	18.41	6.55	5.87	8.11	-	-	11.74	6.20	0.00
	MGT	5.63	11.13	12.41	12.33	-	-	9.08	12.04	0.00
Lycium cuneatum	G %	16.50	2.00	0.50	0.00	-	-	-	-	-
	IGV	11.19	2.36	0.76	33.00	-	-	-	-	-
	MGT	-	-	-	-	-	-	-	-	-
Maytenus vitis idaea	G %	83.00	33.00	22.00	4.00	-	-	70.00	22.00	12.00
	IGV	7.98	4.77	4.17	5.08	-	-	6.83	5.60	3.84
	MGT	12.50	21.08	20.00	21.25	-	-	14.70	18.66	15.26
Prosopis alba	G %	94.50	-	-	98.50	-	-	-	-	34.50
	IGV	46.65	-	-	25.17	-	-	-	-	16.45
	MGT	2.14	-	-	3.96	-	-	-	-	6.62

Table 2. (Continued)

		Control	NaCl					PEG		
			0.2 M	0.25 M	0.3 M	0.4 M	0.6 M	0.2 M	0.25 M	0.3 M
Prosopis nigra	G %	51.00	9.00	-	3.00	1.00	0.00	-	-	-
	IGV	4.20	5.70	-	9.20	1.05	0.00	-	-	-
	MGT	24.50	21.00	-	7.00	6.00	0.00	-	-	-
Schinopsis balansae	G %	45.00	26.50	-	9.00	-	-	-	-	-
	IGV	10.00	6.35	-	4.75	0.00	0.00	-	-	-
	MGT	10.45	16.75	-	22.75	0.00	0.00	-	-	-
Schinus fasciculata	G %	46.00	19.00	5.00	1.00	-	-	8.00	0.00	6.00
	IGV	21.32	5.75	3.75	2.90	-	-	3.50	0.00	4.30
	MGT	4.49	19.27	28.9	34.00	-	-	29.10	0.00	25.13

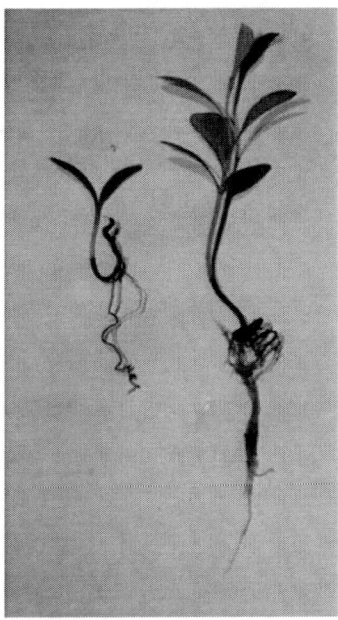

Figure 1. *Achatocarpus praecox* seedlings. Left: abnormal, caused by salinity; right: normal.

Caesalpinia paraguariensis

The percentage of germination changed from 39 % in control vs. 13 % in 0.3M NaCl and 0% in 0.3M PEG (Table 2). Significant differences in both IGV and MGT were found between the control and all treatments. IGV decreased by more than half under the NaCl treatments (5.87%) compared to the control (18.41%). The negative effect of PEG increased with increasing concentrations. There was a twofold increase in MGT values under the NaCl treatments (12 days) and the two lower PEG concentrations. No seeds germinated under 0.3 M PEG. Our results show that this species is more sensitive to osmotic stress than to toxic damage (Figure 2).

Figure 2. *Caesalpinia paraguariensis*: Normal seedlings, control solution.

Licyum Cuneatum

Average germination percentage was 16.5% in the control vs. 0.5% under 0.25M NaCl.

Significant differences in IGV were found between the control and the treatments, but not between treatments. No significant differences were found for MGT. No germination was recorded in two replications of the treatment with the lowest concentration (0.2M), and in three replications of the 0.25M treatment; under 0.3M there was no germination in any of the replicates (Table 2).

Maytenus Vitis-Idea

The germination percentage varied significantly ($p< 0.05$) between the control (83%) and the treatments. (Table 2) No significant differences in IGV were found under any of the treatments, but MGT values differed significantly between the NaCl and PEG treatments. The treatments affected not only the germination percentage but also seed MGT (Figure 3, 4 and 5).

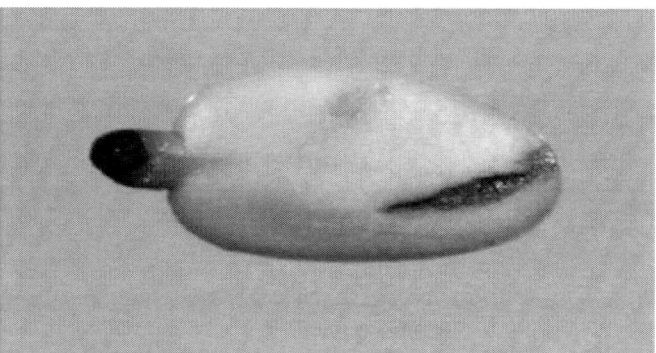

Figure 3. *Maytenus vitis-idaea.* Primary root damaged by salinity.

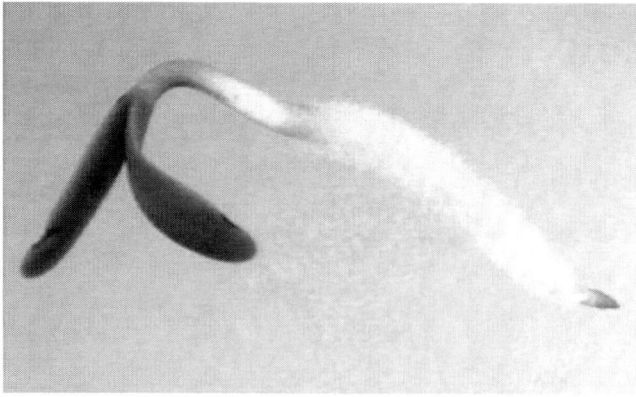

Figure 4. Normal seedling of *Maytenus vitis-idaea.*

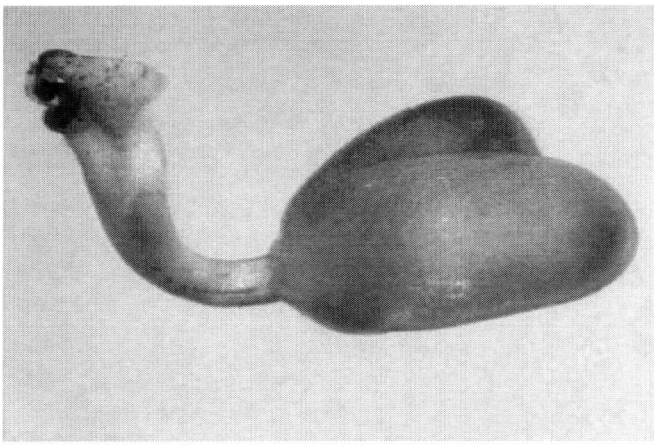

Figure 5. Abnormal seedling of *Maytenus vitis-idaea* with damage by salinity.

Prosopis alba

Average germination percentage was 94.5% in the control vs. 34.5% in the 0.3M PEG treatment (Table 2).

Significant differences (p < 0.05; 95% confidence level) were found between the control and the NaCl and PEG treatments. At the same confidence level, no significant differences in MGT were found between the control and the NaCl treatments, whereas results differed significantly between the PEG treatments and the control, and between the PEG and NaCl treatments

Prosopis nigra

The germination percentage was 51% in control, vs. 9% in 0.25M NACl and 0% in 0.3M PEG. In addition to determining germination percentage, Index of Germination Velocity (IGV) and Mean Germination Time (MGT) for this species, the percentage of abnormal seedlings from the total number of germinated seeds was also calculated immediately after seedling emergence. Significant differences (p<0.05) were found between all treatments in germination percentage and MGT but not IGV values. The highest germination percentage was obtained in the control, and no seeds germinated under the 0.4 and 0.6 M concentrations. Almost 4% (i.e., 3.9%) of the seedlings were abnormal (not having roots and presenting abnormalities in the cotyledons). Seedlings with a thickened hypocotyl and thinning in the upper part were found under the 0.3 M treatment, in contrast to the normal seedlings in the control (Figure 6).

Figure 6. *Prosopis nigra.* Left and middle: abnormal seedlings caused by salinity. Right: normal seedling (control).

Schinopsis Balansae

The germination percentage varied from 45 % in the control to 26.5 % in the 0.2M, 9 % in the 0.3M, and 0% in the 0.4 and 0.6 M NaCl treatments (Table 2).

IGV and MGT decreased considerably with increasing concentrations. Significant differences in germination, IGV and MGT were found between treatments. With higher NaCl concentrations, plant roots were thinner, darker, dull in color, and had a lower density of root hairs or no hairs at all. The percentage of abnormal plants was 2.22% in the control, 4% under the 0.2 M treatment, and 22.2% with the 0.3 M treatment (Figure 7, 8 and 9).

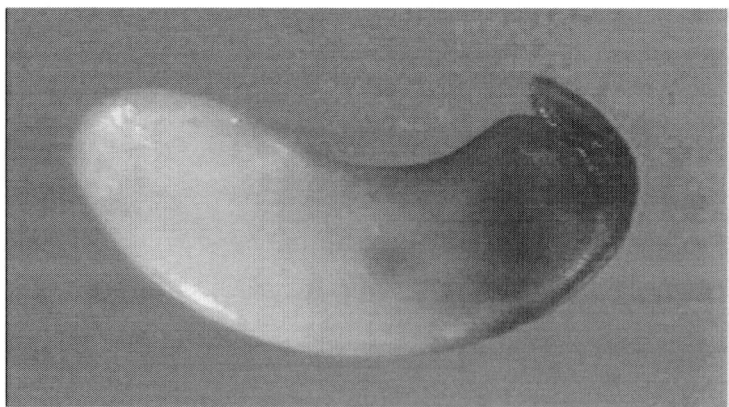

Figure 7. *Schinopsis balansae* seeds; colored area: damage caused by salinity (TTC test).

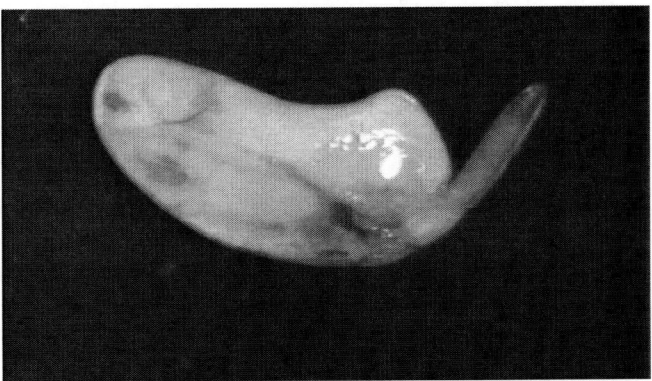

Figure 8. *Schinopsis balansae* seeds; colored area: damage caused by salinity (TTC test).

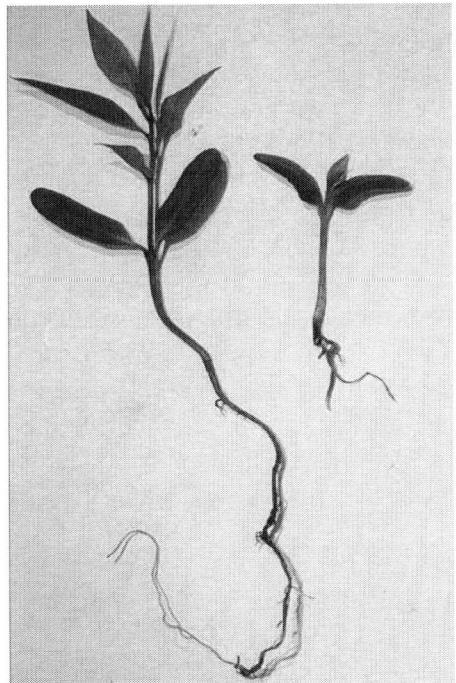

Figure 9. Left: normal seedling of *Schinopsis balansae*. Right: abnormal seedling caused by salinity.

Schinus fasciculatus

Germination percentage was 46% in the control vs. 1% in the 0.3M NaCl treatment (Table 2). Significant differences in IGV were found between the control and all treatments, as well as between the 0.2 and 0.3 M concentrations. However, no differences were found between the NaCl and PEG treatments at those concentrations. No differences in MGT were found between the control and 0.3M NaCl, but MGT values differed between the control and 0.2M NaCl; and between 0.3M NaCl and 0.2M NaCl. The same differences were recorded for the PEG treatments. These data suggest that salt concentration has a gradual effect on the germination velocity of *S. fasciculus*, though not on its mean germination time. Both NaCl and PEG treatments had negative effects on germination, except for the PEG treatment with the highest concentration, where the percentage of germinated seeds was higher than under the 0.25M treatment.

Discussion

The studies on the effect of salt (NaCl) and PEG on seed germination of the nine species in the Forest Wedge analyzed in this chapter show that there were severe limitations to germination, IGV and MGT [25]. Even though these species occur in saline soils (Natraqualfs. Ochraqualfs), most of them can be classified as glycophytes (or non-halophytes) [23], since they do not resist salinity as typical halophytes do. There is usually a salt concentration threshold above which glycophytes begin to show signs of inhibited growth, foliage discoloration and weight loss. An unusually high Na content along with a high total salt concentration cause enzyme inactivation and inhibition of protein synthesis [23].

These effects were observed in most of the study species which did not germinate at NaCl concentrations higher than 0.25 or 0.30 M and presented either primary damage by toxicity, as *Achatocarpus praecox* (Figure 1) and *Caesalpinia paraguariensis*, or secondary damage by osmotic stress as in *Maytenus vitis- idaea*. In this last specie, a similar percentage of germination as in the control was observed under a concentration of 0.2 M PEG. The results on *Lycium cuneatum* show that this species is highly sensitive during germination, even to a slight increase in salt content, and will be affected by low concentrations of salts in the soil.

In some of them, however, as in the case of *Acacia praecox* and *Prosopis alba* a higher germination percentage was obtained with the higher (0.3 M) than with the lower concentrations (0.2 and 0.25 M). This result is probable due to an increase in antioxidant enzymes to counteract the negative effects of oxidative stress, as has been found for other woody species [30].

In some species, such as *Cassia montana* the osmotic adjustment that occurs in the plant as a result of the NaCl concentration increase in the soil solution, leads to an increased growth [43].

The main mechanism underlying salinity tolerance can be due to a plant's ability to restrict the access of potentially damaging ions to metabolically active sites, both at the organ and cell levels [44].

Temperature can also act synergically with salt, inhibiting germination or seedling growth, as has been reported for some species from other genera [45-47]. It can also be due to other physiological mechanisms such as in *Schinopsis quebracho-colorado* [30], which showed different levels of salinity tolerance at different growth stages: it was more tolerant at germination than at the seedling stage. The researchers attributed this positive effect to the concentration of the compatible solute proline in the roots which might balance Cl⁻ and Na⁺ accumulation in saline soils, *Schinopsis quebracho-colorado* is a glycophyte [29] which can withstand concentrations of up to 0.2M. According to our results, *Schinopsis balansae*, the most emblematic species of the Humid Chaco, might have a similar tolerance mechanism. An increase in salinity had negative effects not only on the germination but also on the seedlings of this species.

Furthermore, one of the characteristics of this type of forests is that many of their species are xerophytes. Since these plants allocate a large portion of their energy to defense purposes, they usually have high contents of phenols, tannins and lignin in their green foliage [48]. Furthermore, the stress induced by high NaCl concentrations increases the activity of antioxidant enzymes and leads to the synthesis of polyphenols in the leaves to counteract the negative effects of oxidative stress, as in the case of the "vinal" *Prosopis ruscifolia* [30]; the same process might occur in *S. balansae* as suggested by the high phenolic content in its leaves, as well as in other species of this region [49].

In *P. alba* with a higher content of NaCl (0.3M), the percentage germination was even higher than in the control; but there was absolutely no germinated seed with 0.4 and 0.6M. Thereafter, it cannot be considered a halophyte as quoted by other authors [50] for *P. chilensis,* which showed higher germination and seedling development at high concentrations of NaCl (0.8M). For this reason we consider that *P. alba* would be a glycophyte or may be a slight halophyte.

Adaptive mechanisms of plants to salt tolerance are extremely complex, and these processes and mechanisms of plant response to salinity are multigenic [25].

Changes in land use in a flat terrain as the one found in the Chaco-Pampean Plains (e.g., deforestation followed by agriculture) might bring about hardly quantifiable changes in the overall forest dynamics, by altering the original conditions (light intensity, increased soil temperature, atmospheric humidity) [13], as well as changes in the flow and transport of underground water and salts [17].

The distribution of species in the humid Chaco Region, and even in the Great Chaco, is determined largely by their different mechanisms of adaptation to salinity and to environmental changes (temperature, soil condition) [13, 17]; these adaptations are also of crucial importance for the conservation of this region, which has also been subjected to an increased anthropogenic pressure.

Acknowledgments

We are grateful to Ignacio Barberis for the manuscript revision.

References

[1]   J. J. Burgos, *Bol. SAB* 11(Supl.), 37 (1970).

[2]   W. Grafe, M. Brassiolo, M. Simón, A. Fumagalli and R.Renolfi, *Explotación Eficaz y Protección de Recursos en la Región de la Cuila Boscosa, Depto. Vera, Pcia. Santa Fe.* Área Sist. Silvopast., Fac. Cs. Forestales. Univ. Nac. Stgo. Estero (1991).

[3]   J. Gollán and D. A. Lachaga, *Aguas de la provincia de Santa Fe*, Inst. Experim. Inv. Fomento Agríc. Gan. Public. Técn N°12. Santa Fe (1939).

[4]   L. M. Espino, M. A. Seveso and M. A. Sabatier, *Mapa de Suelos de la provincia de Santa Fe.* Tomo II. MAG Santa Fe-INTA EERA Rafaela (1983).

[5] E. Popolizio, Y. P. Serra and G. O. Hortt, *Grandes Unidades taxonómicas de Chaco.* Ctro. Geociencias Aplic., Serie C- Investigación Resistencia (1980).

[6] M. Mussetti and D. Alconchel, *Mapa de suelos del centro operativo Dr. Tito Livio Coppa*, Las Gamas (Dpto. Vera), Dir. Gral. Ext. Invest. Agrop. Santa Fe (1986).

[7] A. L. Cabrera, *Territorios Fitogeográficos de la República Argentina. Enciclopedia Argentina de Agricultura y Jardinería,* Vol. 2, Ed. Acme, Buenos Aires (1976).

[8] J. P. Lewis and E. F. Pire, *Serie Fitogeográfica N°18*, INTA, Buenos Aires (1981).

[9] J. Morillo and J. Adámoli. *Serie Fitogeográfica N°13*, INTA, Buenos Aires (1974).

[10] D. Prado, *Tesis doctoral*, Univ. Saint Andrews, Escocia (1991).

[11] J. P. Lewis, E. Pire and J. Vesprini, *Candollea* 49, 159 (1994).

[12] D. Prado, *Candollea* 48, 615 (1993).

[13] N. J. Carnevale, C. Alzugaray and N. Di Leo (*ex aequo*), In: *El Chaco sin bosques,* pp. 203- 228, *INTA-UNESCO-GEPAMA* (2009).

[14] R. Munns and M. Tester, *Ann. Rev. Plant Biol.* 59, 651 (2008).

[15] FAO, *Land and plant nutrition management service,* Available: www.fao.org/ag/agl/agll/spush/ (2008).

[16] R. Lavado, In: E. Taleisnik, K. Grunberg and G. Santa María (Eds.), *La salinización de suelos en la Argentina: su impacto en la producción agropecuaria,* pp. 11-16, EDUCC (Edit. Univ. Cat. Córdoba), Córdoba (2008).

[17] E. G. Jobbágy, M. D. Nosetto, C. Santoni and G. Baldi, *Ecol. Austr.* 18, 305 (2008).

[18] I. A. Ungar, *Bot. Rev.* 44, 233 (1978).

[19] C. Alzugaray and N. J. Carnevale. *Libro de Semillas de Especies Leñosas Autóctonas. Chaco Húmedo: Cuña Boscosa Santafesina.* Min. Aguas, Serv. Públ. Medio Amb., Secret. Medio Amb., Santa Fe (2009).

[20] K. Tobe, K. Li and K. Omasa, *Austr. Bot.* 2, 163 (2000).

[21] Y. Gutterman, *Seed Germination in Desert Plants,* Springer-Verlag, Berlin (1993).

[22] J. Kigel and G. Galili, *Seed Development and Germination*, Marcel Decker, New York (1995).

[23] J. Levitt, *Responses of Plants to Environmental Stresses*, Academic Press, New York (1972).

[24] L. Bernstein, *Plan Anal. Fert. Prob.* 4. 25 (1964).

[25] V. Mudgal, N. Madaan and A. Mudgal, *Int. J. Bot.* 6, 136 (2010).

[26] B. P. Strogonov, *Physiological Basis of Salt Tolerance of Plants (as affected by various types of salinity).* Acad. Sci. USRR. Davey and Co., New York (1964)

[27] C. F. Ehlig and L. Bernstein, *Proc. Amer. Soc. Hort. Sci.* 72, 198 (1958).

[28] E. M. Kovalskaia, *Fisiol. Rast.* 5, 437 (1958).

[29] P. E Villagra, A. Vilela, C. Giordano and J. A. Alvarez, In: R. G. Ranawat (Ed.), *Desert Plants. Biology and Biotechnology*, pp. 322-354, Springer-Verlag, Berlin (2010).

[30] D. A. Meloni, M. R. Gulotta and M. A. Oliva Cano, *Quebracho* 15, 27 (2008).

[31] International Seed Testing Association, ISTA News Bulletin N° 126, (2003)

[32] Association Official Seed Analysis (AOSA), *Seedling Evaluation Handbook*, Lincoln, NE (1992).

[33] G. D. Marino, M. V. Mas and M. J. Orlandoni. *Bol. SAB* 43(1-2), 67 (2008).

[34] J. B. Silva and J. Nakagawa, *Informativo ABRATES* 5(1), (1995)

[35] R. M. Craviotto, *Prueba topográfica por tetrazolio. Patrones para la especie soja. Laboratorio de semillas,* INTA Centro Regional Santa Fe. EEA Oliveros (1995).

[36] P. R Mudrovitsch de Bittencourt and D. R Vieira, In: F. C. Krzyzanowski, R. Dalton Vieira and J. Frama Neto (Eds.), *Vigor de Sementes: Conceitos y Testes,* pp. 1-28, ABRATES, Londina Pr. (1999).

[37] J. M. Jozami and J. Muñoz, *Árboles y Arbustos Indígenas de la Provincia de Entre Ríos*, (1982).

[38] L. M. Bernardello, *Bol. Acad. Nac. Cs. Córdoba* 57, 243 (1986).

[39] N. Alvarenga, C. A. Velásquez and N. C. Alvarenga, *Rev. Ccia. Tecnol.* UNA 1(3), 51 (2001).

[40] M. J. Dimitri, *Celulosa Argentina.* Ed. Buenos Aires (1973).

[41] N. J. Carnevale, M. Raimondi, M. Di Liberto, S. Álvarez and S. Zacchino, In: *II Congresso de Fitoterápico do Mercosul-VI Reunião da Sociedade Latinoamericana de Fitoquímica.* Diciembre 2008. (Minas Gerais. Brasil.) 43-44 pp.

[42] E. V. Carrizo, M. O. Palacio and L. D. Roic *(ex aequo)*, Univ. Nac. Stgo. del Estero. *Dominguezia* 18(1), 26, Available: www.dominguezia.org.ar/volumen /articulos/1813.pdf (2002).

[43] A. D. Patel and A.N. Pandey, *J. Arid Environ.* 70, 174 (2007).

[44] S. Chen, J. Li, S. Wang, A. Huttermann and A. Altman, *Trees- Struct. Func.* 15, 186 (2001).

[45] A. M. Aiazzi, P. Carpane, J. Di Rienzo and J. A. Argüello, *Seed Sci. Technol.* 30, 329 (2002).

[46] L. Catalán, M. Balzarini, E. Taleisnik, R. Sereno and U. Karlin, *For. Ecol. Manage.* 63, 347 (1994).

[47] P. E. Villagra, *J. Arid Environ.* 37, 261 (1997).

[48] J. D. Horner, R. G. Gates and J. R Gostz, *Oecologia* 72, 515 (1987).

[49] A. E Ragonese and V.A. Milano, *Enciclopedia Argentina de Agricultura y Jardinería.* Tomo II. Ed. Acme. Buenos Aires (1984).

[50] C. Cazebonne, A.Vega, D. A. Varela and L. A. Cardemil. *Rev. Chil. Hist. Nat.*72(1), 83 (1999).

Mediterranean High Mountain Ecosystems

In: From Seed Germination to Young Plants
Editor: Carlos Alberto Busso

ISBN: 978-1-62618-653-8
© 2013 Nova Science Publishers, Inc.

Chapter 14

Effects of pH, Light and Exogenous Plant Growth Regulators on Seed Germination and Early Seedling Growth of Native Shrubs from High Mediterranean Mountains

*Francisco Serrano-Bernardo[*], José de la Torre-Betts,
María Beltrán-Hermoso, Kelly Garcete and José Rosúa-Campos*
Department of Civil Engineering, University of Granada,
Campus Fuentenueva, Granada, Spain

Abstract

The Sierra Nevada (S Spain) constitutes a unique mountain system among Mediterranean high mountains for its richness in endemic species. The ski station on the mountain has strongly altered the landscape, making necessary measures for vegetation recovery. Several experiments were made under controlled conditions (pH, temperature, light and soil) using seeds from some native shrubs plants of Sierra Nevada and Betic Sierras. In this research, we tested the combined effects of light and plant hormones on germination of these species using a specific pH value, which is characteristic in the soils of Sierra Nevada. The seeds were pretreated with different concentrations of known hormonal growth regulators and different conditions of light and temperature. Germination, rooting, and seedling growth was monitored in all plants. The exogenous application of hormonal growth regulators combined with light affected germination in all the study species. The application of the study growth hormones improved growth of the shoot, and/or root, confirming its suitability for restoring plant cover. It is expected that, when transferred to the field, these treatments might be an effective aid for plant recruitment in recovery programs.

[*] E-mail address: fserber@ugr.es.

Introduction

Plant Cover Restoration and Landscape Integration in Alpine Systems

Many natural environments have been devastated in recent decades as a result of development, urbanization and industrialization [1]. The problem of ecosystem damage is international, and probably most countries in the world concern about it. In Western Europe, several million hectares of land need urgent attention. In Eastern Europe and Russia, devastation is very high, and in many regions it causes serious public health problems and constraints economic development. In many parts of the Third World, the situation is even more serious. Clearing of tropical forests, for example, has been very high in the last 25 years, and it has produced a dramatic increase in soil loss for crop culture [2]. A few years ago, it was estimated that 43% of the land area had reduced its ability to provide benefits to man because of the direct impacts on land use [3]. If this data is added to the current rates of world population growth, the only solution to prevent further environmental degradation lies on the ability to sustainably increase production of renewable natural resources. In this way, world demand could be satisfied without depleting those resources. The implementation of policies, plans and programs of ecological ecosystems restoration is one of the most important issues to reverse damages to the environment.

The Mediterranean region has one of the greatest plant diversities in the world [4], including the Mediterranean high mountain areas [5]. The Sierra Nevada (S Spain) is a single mountain range between the high mountains of the Mediterranean for its richness in endemic species. Also, this is the southernmost massif of Europe, and its proximity to Africa gives it certain climatic characteristics, that combined with the altitude (among other factors), makes it the only Mediterranean mountain in Spain, which has five bioclimatic belts [6], reflecting the great biodiversity of the mountain. The Sierra Nevada national park, which has more than 2000 vascular plants (species and subspecies), representing almost 30% of the flora of mainland Spain in 0.4% of the surface, is 7% of the flora of the Mediterranean region, with 0.01% of the area. From that large number of vascular plants, more than 80 are endemic to the Sierra Nevada, especially in the area of the peaks, where the percentage of endemic species increases to 30%, or unique ecological niches, such as rocky areas or gravel, where the percentage rises to 80% [7]. Located in the Sierra Nevada bordering of the National Park, but within the Natural Park, is the ski station "Sol y Nieve", the southernmost of Europe. The ski station is divided into 119 ski pistes, with a total length greater than 100 km skiing, and occupying an approximate area of 400 hectares, representing 0.4% of the Natural Park. It has 32 lifts with capacity for about 47000 people / hour and a half ski slope of 1200 m., from 2200 m. to almost 3400 m. [8]. The conditioning and opening of the ski runs has gravely altered the ecosystems of this massif, provoking a heavy loss of biodiversity and a major increase in erosion. To counteract these effects, over the last 15 years, different plant-restoration experiments have been conducted.

Macyk [9] have described the alpine ecosystems as one of the most difficult to restore due to unique physical determinants of the environment, such as low temperatures, high levels of solar radiation, persistent winds, irregular topography and steep slopes. These extreme conditions strongly influence the physical environment of plants, their physiological responses [10], and succession processes in the Alps. The result is that only a few species have adapted to survive (and reproduce) in these ecosystems or to colonize and remain there [11]. Alpine species richness decreases with altitude [12], and only a few plants can complete their life cycle in one year. Therefore, the sequence is more complex in more temperate areas of the mountain [11], increasing the difficulty of obtaining seeds or other plant propagules [13]. In these areas, the recovery of vegetation cover without human assistance is impossible for at least 40 years, and only if the ski station is closed [14,15]. Revegetation attempts often fail because, generally, do not provide lasting results at high altitudes [16]. The colonization of the ski slopes by seeding with native species is extremely slow and uncontrolled since, in the best cases, small groups may be irregularly arranged on the slopes. At the worst, these crops will have no success because they are "virtually wiped out" each year by the continuous passage of machinery and skiers, also causing increased erosion on the ground [15].

Generally, for conditioning the ski slopes, the vegetation is first subtracted, and thereafter the first few centimeters of soil, depositing them in places with coarse topography, and the areas are subsequently modeled with waste materials from other areas. This technique produces a serious deterioration in ecosystem functioning, particularly with respect to vegetation dynamics and physicochemical soil properties [17]. Several studies have shown that due to the construction of ski slopes, there have been changes in plant biomass, in the composition and colonization of species, and in the amount of available nutrients in the soil [18]. The direction and degree of qualitative changes in the soil depends on climate, soil conditions and its use thereof [19]. Thus, in mountain ecosystems exposed to extreme weather conditions, bare soil is susceptible to erosion. To reduce this erosion quickly, sometimes is required a restoration of a more favorable soil structure, biological functions of the same native species, and re-vegetation with native or exotic plants [20]. The main objective of re-vegetation is to minimize degradation of resources, and promote the reestablishment of a functional soil-plant system in the long term [17]. Soil properties resulting from amendments made to its structure can also influence the restoration of endogenous vegetation. Spontaneous re-vegetation of degraded pastures begins with the invasion of pioneer or opportunistic species [21], plants that are tolerant to extreme temperatures, soil disturbance and nutrient depletion [22]. These plants prepare the ground for the subsequent succession of green grasses that are more sensitive to the altered conditions. Thus, stabilizing of soil structure, increases of the organic matter and nutrients, and restoration of the soil moisture and biological function [23], can be achieved through the beneficial effects of the plant roots [24], in the context of ecological restoration and conservation. This highlights the importance of studying the interactions of plants with the abiotic environment, including the effects of soil structure or the level of nutrients in the colonization of plants. However, other authors suggest that it is also important to consider the co-evolution between plants and its biotic, especially microbial activities [17]. These theories involve a number of impacts in the context of mechanical preparation of ski slope, and entail serious yield losses and changes in vegetation.

The strategy for plant cover restoration is generally based on imitating nature according to the guidelines for ecological succession. Classically, it follows a fixed sequence of communities that inevitably occur in a region to a stable community, called "climax". A balance appears in this succession with the regional climate [25]. This succession dynamics includes changes in the structure and functioning of communities according to the scheme proposed by Bradshaw [26], following a single, linear pattern. However, this sequence often deviates from the monoclimax linear progression [27], so that multiple succession models are best suited to environments with frequent disturbances [28]. It provides support to restoration alternatives from the specific characteristics of the system and objectives. In degraded environments, restoration strategies should be based on the development of treatments that overcome the limitations that any type of disturbance [29].

Thus, in areas where alpine skiing is one of the biggest changes in the natural dynamics of succession, vegetation plays an important role in erosion control, stabilization, and ecological and landscape integration, especially in the ski slopes. Thus, it is necessary to improve, as far as possible, the morphology and substrate, using the most appropriate implementation techniques, and perform maintenance until vegetation is self-implanted [30].

If the plant cover restoration is successful, a stable and diverse natural environment will be created that will evolve and adapt to possible changes of use. The basic objectives that can be achieved with this restoration include [31]:

a)   to help soil stabilization, reducing the risks of landslides and mudslides,
b)   to stabilize and protect the soil, preventing soil erosion and improving soil properties, especially in the first few inches of it, where soil-vegetation interactions are strong;
c)   to reduce and control the effects of surface water erosion, accentuated by the absence of vegetation, preventing runoff, gully formation, etc., and in turn generating "lines" of natural drainage;
d)   to allow the conservation of water surface and underground resources;
e)   to integrate disturbed surfaces on the surrounding environment to reduce visual impact; and
f)   to provide work of an ornamental value and landscape character.

Study Area

Sierra Nevada has 16 peaks over 3000 m altitude, derived from the glacial and periglacial morphology of the Pleistocene epoch. The highest peaks in the Iberian Peninsula: Mulhacén, 3481 m; Veleta 3396 m or La Alcazaba 3364 m, are clear examples of these forms of relief, and converted to solid in one of the peninsular ranges of greater singularity. The proximity to the Mediterranean Sea and the magnitude of the peaks, generated a system of steep slopes with values greater than 20°, especially in areas of medium mountains that dominate nearly 60% of the area. Sixteen percent of the medium mountains has slopes of more than 30° [32]. According to petrological, structural and geophysical aspects [33], the evolution of Sierra Nevada began during the Paleozoic. From this age, various geological processes followed, especially the alpine orogeny and Quaternary glaciations that caused the deformation and metamorphism of these materials. Sierra Nevada's climate is typically Mediterranean High Mountain, with a marked summer drought [34]. The height of the massive development of the unique Mediterranean, mountainous peninsula, with five bioclimatic zones (thermo-, meso-. supra-, oro- and crioro-mediterranean), sets primarily in terms of a thermicity index [6]. The low latitude of Sierra Nevada (37°N) gives it unique characteristics of sunlight, making that the high mountain weather events occur here at altitudes much higher than anywhere else in Europe [35]. With regard to ombroclimate, Sierra Nevada also has a wide variety, from areas belonging from the ombrotype semiarid (200-350 mm year^{-1}) to ombrotype wet (1000-1600 mm/year) in the high peaks. The thermocline is presented as a typical mediterranean climate of high mountains, with a low mean annual temperature (3°C), half in winter -3°C, and half in summer 15°C in areas of peaks, and moderate values as we descend in height.

Sierra Nevada presents a wide range of soils. The main associations are given on carbonate rocks, which have resulted in soils rich in bases, or siliceous, decarbonated rocks, giving rise to acid soils. Lithosols, Regosols, Cambisols and Ranker, are the most often soil types, above 2,000 m altitude. Soils are generally acidic (soils Dystric) with a degree of saturation below 50%. Due to the presence of the ski station have occurred, especially in the last decade, some changes in the substrates, caused by intense human activity modeling and designing the ski slopes.

The main plant communities are in Sierra Nevada, without regard to the formations introduced by man. The holm oak (*Quercus ilex* subsp. *ballota*) and *Quercus pyrenaica* are in the lower bioclimatic zones (termo- meso- and supra-mediterranean). In the oro-mediterranean zone appear cushion-like scrubs such as *Juniperus sabina* L., *Juniperus communis* L. and *G. versicolor* Boiss., with patches of grassland (*Festuca indigesta* Boiss., *Festuca scariosa* Lag. Ascherson & Graebner, *Festuca elegans* Boiss., *Dactylis glomerata* L.), scrub (*Thymus mastichina* L., *Thymus serpylloides* L., *Ulex parviflorus* Pourret, *Halimium viscosum* (Willk.) P. Silva, *Cistus laurifolius* L.), and native pastures (*Festuca clementei* Boiss., *Agrostis nevadensis* Boiss.) at the highest altitudes. It is in these areas where the species adopt a peculiar form of cushion to protect from the low winter temperatures and icy winds. In the many waterways that run through the solid hygrophilous series there are riparian forests, willow, alder, ash and brambles. In the high mountain pastures, there are numerous endemic species of Sierra Nevada such as *Artemisia granatensis* Boiss., *Viola crassiuscula* Bory, *Chaenorrhinum glareosum* (Boiss.) Willk. or *Leontodon boryi* Boiss. [7]. The most common vegetation of the slopes is grassland dominated by *Festuca indigesta* Boiss.

Seed Germination and Latency

Germination activity begins when the seed initiates water intake or imbibitions, and ends with the elongation of the embryo axis [36]. The visible sign that germination has been completed is usually the rupture of the structures surrounding the embryo by the seminal root, so that visible germination is marked by root emergence [37]. The germination process can be affected by environmental factors such as temperature, light or pH [38], or plant growth regulators (PGRs).

Temperature

Seeds germinate in a range from 0 to 45°C. The approximate optimal temperature for not finding dormant seeds in most plants is 25 to 30°C. However, these intervals will depend largely on the morphological development of plants and the geographic origin of the species [39].

pH

pH effects have been studied on germination of several species. The seeds of many of these species, germinate in a wide range of pH values [40], but others only reach high germination percentages at specific pHs [41]. The International Association of Seed Experimentation [42] reported that the recommended pH for germination tests in water-sand systems, should be between 6.0 and 7.5. Acidophilic species (characteristic of acid soils), germinate well at pH 6.5 [43]. Turner et al. [44] conducted a study with *Paulonia* seeds subjected to various pHs. They found no-germination when pH was between 1.5 and 3.5. However, germination percentages were between 79%-98% at pHs from 4.0 to 7.0, respectively. Germination percentages of seeds transferred from pH 3.5 to 3.0 or 6.5, were 59 or 74, respectively. However, when seeds were transferred from pH 1.5 to 6.5 or 2.5, they did not germinate, indicating that these seeds probably die because of the acid pHs. In calcareous species, a pH of 4.0 significantly reduced germination [45].

Light

Light plays an important role in the germination of seeds. Many species define the optimum environment for growth and development depending on the radiation received [46]. In seed germination, this response is regulated mainly by the action of the photoreceptor Pr/Pfr, which was first observed in the germination of lettuce seeds [47]. The absorption of red light (wavelength range 650-680 nm.) stimulates germination, but this can be reversed if far-red light (between 700-730 nm λ) is received. This is explained by the fact that absorption of red light converts the phytochrome Pr, which is inactive and inhibits germination, to the Pfr form, which is active and promotes germination. However, far red light absorption causes the opposite effect; it converts Pfr to Pr [48]. There are seeds that germinate at similar rates in both light and darkness [49], others germinate more successfully under light conditions [50] and even others germinate in darkness [51].

Plant Growth Regulators

The effects of an exogenous application of PGRs in the seed germination process should be distinguished of the influence of the endogenous levels of these PGRs in that process [39]. Because of this, it is necessary to know the endogenous levels of any PGR before applying it exogenously; however, it is not always possible to correlate the physiological state of a seed with its hormonal content [39]. PGRs can have a dual effect, because while a rapid response would involve some modification at the of membrane level, a delay in this response could be due to an effect on gene expression. However, these effects depend on the endogenous level of these PGRs, as well as the metabolic activity and sensitivity of the tissue in question [52].

Physiologically, ethylene is involved in a number of plant processes as vegetative growth and floral development, fruit senescence and embryogenesis [53], as well as responses to mechanical injuries, pathogen invasion, abiotic stresses and auxin application [54]. While many studies highlight the importance of ethylene during seed germination, its precise role is unclear due to contradictory data. In many species, concentrations of ethylene between 0.1 to 200 ml/l are sufficient to stimulate germination [55] [64].

Cytokinins (Cks) are also important growth regulators that control cell division and growth in the presence of auxins, and delay leaf, stem, and flower senescence [56]. In addition, Cks also help diminish the inhibitory activity of abscisic acid (ABA), and promote seed germination [57]. Numerous kinetin derivatives, such as BA or Benzylaminopurine, where the furfuryl group is replaced by other groups, also stimulate germination [58].

Gibberellins, as cytokinins, can promote and control seed germination of many species [59]. Their exogenous application increases seed germination, and is able to break seed dormancy, especially in seeds sensible to low temperatures or light [36]. However, this dormancy might occur after the inhibition action of abscisic acid fails.

Auxins are compounds chemically related to 3-indolacetic-acid (AIA) which seems to be the most important auxin in many plants. The action of AIA on seed germination is not well known, however it is recognized that auxins induce Ethrel production, and some of the AIA effects on seed germination are involved with this Ethrel activation [60].

Laboratory experiments have been conducted usingn different germination pretreatments for some endemic species of Sierra Nevada, like the exogenous application of known PGRs that encourage germination in numerous species [62-63], e.g. (i) Ethrel -E-: 2-chloroethylfosfonic acid, a compound that at pH>4-5, releases ethylene after penetrating plant tissues; (ii) N^6-Bencyladenine –BA-: one of the most well-known cytokinins; (iii) Gibberellic-3-Acid -GA$_3$-: a compound based on the ring structure *ent-gibberellane* [61], and (iv) Inabarplant IV -IP IV- a commercial mixture of two phytoregulators: the auxins **Indole-3-butyric** Acid (IBA) at 0.4%, and 1-Naphthaleneacetic Acid (NAA) at 0.4%, plus Captan fungicide (15%).

In this research, some of the most representative plant species of the ski station were included for study. Some of the environmental requirements that these species need to optimize their germination and growth, and the possibility of improving them by the application of various plant growth regulators were evaluated under laboratory conditions. The major idea was the application of these results to the ski station to contribute in the (i) recovering of degraded areas, and (ii) restoring the vegetation cover. Three experiments were performed with this purpose, with the idea of improving germination and development of four species native to the Mediterranean high mountain, for further uses in re-vegetation works in the ski station of Sierra Nevada. The first experiment evaluated the pH influence on seed germination. The second experiment studied the effects of light and exogenous growth regulators on that process. The third experiment investigated the PGRs effects on rooting and growth potential.

Materials and Methods

Seeds were used from four endemic species of the Betic Sierras and the Sierra Nevada: *Genista versicolor* Boiss., *Hormathophylla spinosa* (L.) P. Küpfer, *Reseda complicata* Bory, and *Thymus serpylloides* Bory, collected in different areas near the ski station, between 2000 and 3000 m altitude.

Seed Preparation for Germination Studies

Different lots of *G. versicolor, H. spinosa, R. complicata* and *T. serpylloides* seeds, preserved for at least 3 months at 4°C in darkness, were sorted for size and similar external characteristics. Seeds with malformations or anomalies were discarded. *Genista versicolor* seeds, having a tough seed coat typical of legumes, were previously soaked in sulphuric acid (100%) for 20 min [64]. After this scarification, the seeds were vigorously washed with distilled water and rinsed with sterilized water. The *H. spinosa, R. complicata* and *T. serpylloides* seeds were surface sterilized by immersion in sodium hypochlorite (1%) for 5 min, and afterwards washed as in the former case. A seed was considered germinated after root emergence.

Substrate Preparation

In the third experiment, the substrate for germination, rooting and growth was vermiculite, a natural mineral with aluminium-magnesium silicate (SiO_2: 38-46%, Al_2O_3: 10-17%; MgO: 12-35%, among others). This is the most common form having small laminar grains of brownish mica with water between layers. This mineral can retain water far beyond its weight while maintaining a loose, aerated substrate texture [65].

Measurement of Soil pH

The four study plant species were collected mainly in the acid soil of the siliceous materials at the Nevado- Filábride area of the Sierra Nevada. We took several samples from certain places within different ski slope areas (between 2590- 2960 m.a.s.l.). To confirm this fact, some of these samples had diverse alterations because of human-related activities.

Soil samples were collected following this protocol at each sampling location: Firstly, we selected a surface area of 0.5 x 0.5 m. Then, the first, outer 5 cm. of soil were scraped, including stones, weeds or anything else what could be found within this area. Finally, we turned over everything and took our sample (i.e., a few kg of soil). For pH analysis, two identical samples were selected from each soil sample. For each group, a solution of 40 ml distilled and sterilized water and 40 ml of 1N potassium chloride (KCl) was made, and distributed into two different sterilized glass beakers. Thirty grams of each sample were placed into each of two glass beakers (n=2). We stirred the solutions for 5 minutes, and left them standing for 60 min. Subsequently, the solutions were stirred again for 2 minutes and the pH was measured.

Experimental Procedures

First Experiment

Following the protocol of seed germination previously described seeds of the study species were incubated in distilled and sterilized water and exposed to pH 5.2, 6.2, or 7.

Working temperatures were 15°C for *G. versicolor* and *T. serpylloides,* and 15°C and 25°C for *H. spinosa* and *R. complicata*. The whole experiment was performed under dark conditions.

Second Experiment

Lots of 50 seeds were placed on filter paper in sterilized Petri dishes (10 cm diameter), and different pretreatments were applied with hormonal growth regulators (as shown in Table 1). Temperature conditions were 15°C and/or 25°C, all under a light/darkness photoperiod of 12/12 h. A subsample of seeds were tested for viability with 2, 3, 5-triphenyl tetrazolium chloride [66], and a positive staining reaction was obtained in 80–90% of the seeds. These treatments were chosen according to the results from prior studies made on other species in the environment of the Sierra Nevada ski station [67], and in the laboratory [64].

Table 1. Pre-treatments, plant growth regulators (PGR) concentration, immersion time and temperature applied to the study species (Second experiment)

Species	Pre-Treatment	PGR Concent-ration (ppm)		Immersion time (hours)		Temperature (ºC)	
Genista versicolor	CONTROL						
	ET	10	100	12		15	
	BA	1		12		15	
	BA	10		24		15	
Hormathophylla spinosa	CONTROL					15	25
	BA	10	100	12		15	
	GA$_3$	10		12	24	15	25
Reseda complicata	CONTROL					15	25
	IP IV	10		12	24	25	
	ET	100		12	24	15	
	ET	10		12		25	
	ET	100		24		25	
	BA	10		12	24	15	
	BA	1		24		25	
	BA	10		12		25	
	GA$_3$	100		12	24	25	
Thymus serpylloides	CONTROL					15	
	IP IV	1	10	24		15	

E= Ethrel; BA= N^6-Bencyladenine; GA$_3$= Gibberellic-3-Acid; IP IV= Inabar plant IV.

Third Experiment

Forest flats were used, each with 25 wells of about 300 cm^3 well^{-1} (5 x 5 x 12 cm). Each well was filled with vermiculite to within 1 cm from the top, and watered with 100 ml of distilled water buffered at pH 6.2. Afterwards, were prepared different seed lots with the best results obtained of each PGRs, from previous germination tests carried out in Petri's dishes [64]. Subsequently, we deposited 5 seeds well^{-1} (25 plants m^{-2}), covered with vermiculite, that were lightly watered.

The experimental temperatures, based on prior tests on each species [64] were 15°C and/or 25°C too (Table 2), all under a light/darkness photoperiod of 12/12 h. Every 7 days, nutrients were applied in a nutrient solution at a rate of 50 ml per well [68], supplemented with 5mM KNO$_3$. The pH of the solution was adjusted to 6.2. After 28 days, the emerged plants were counted, carefully uprooted, washed to eliminate substrate remains, and dried in an oven at 80°C for 24 h. Dry weights were recorded for roots (R) and shoots (S), and the length of each plant part was measured.

Table 2. Pre-treatments, plant growth regulators (PGR) concentration, immersion time and temperature applied to the study species (Third experiment)

Species	Pre-Treatment	PGR Concent-ration (ppm)		Immersion time (hours)		Temperature (°C)	
Genista versicolor	CONTROL						
	ET	10	100	12		15	
	BA	1		12			
	BA	10		24			
Hormathophylla spinosa	CONTROL					15	25
	BA	10	100	12		15	
	GA$_3$	10		12	24	15	25
Reseda complicata	CONTROL					15	25
	IP IV	10		12	24	25	
	ET	100		12	24	15	
	ET	10		12		25	
	ET	100		24			
	BA	1		24		25	
	BA	10		12			
	GA$_3$	100		12	24	25	
Thymus serpylloides	CONTROL					15	
	IP IV	1	10	24			

E= Ethrel; BA= N^6-Bencyladenine; GA$_3$= Gibberellic-3-Acid; IP IV= Inabar plant IV.

The results were analysed using one-way ANOVA and the Duncan's multiple range test for each species, comparing the differences between the control and the pre-treatments; $P<0.05$ was considered significant.

Results and Discussion

Mediterranean, high-mountain ecosystems are especially sensitive to various biotic and abiotic stresses (e.g., overgrazing, agriculture, increase in tourist pressure), factors that act synergetically [69], and are also vulnerable to natural climatic variations [70]. As a result, the interaction between light and temperature, for example, can influence the germination both time and capacity of seeds of many species. Thus, some seeds require light at a certain temperature but not to another [71]. The characteristics of light that affect germination include photon length, quality, and irradiance reaching the seed [72]. For many species in different habitats, the responses to light are measured by the phytochromatic perception [73-74].

It is generally recognized that high-mountain species do not need a specific temperature to promote germination [10]. However, some studies reported that most of them usually prefer high temperatures [75]. This appears to occur in the case of Mediterranean high-mountain species in the oro-mediterranean level [5]. Nevertheless, this does not apply to the four species of the present study, which were located at this bioclimatic level. This might be due to the extraordinary combination of natural factors that mix and interact in the Mediterranean mountain type of ecosystem. In this sense, the Sierra Nevada is a zone where solar radiation is especially intense in the dry months, creating a great thermal contrast during the day between the atmosphere, which is always cold, and an overheated soil. In the study area, soils are rocky and present dark colours; because of this, they reflect little incident radiation [76]. The fruiting period of the study species spans from the end of July (*H. spinosa*) to the end of September (*R. complicata*). This implies, as occurs in other Mediterranean high-mountain species [77], that light sensitivity can be a major factor inducing possible secondary dormancy [38]. This is because of the few weeks that the seeds have to germinate after snowmelt, and the harsh climatic conditions of the Mediterranean high mountain [5].

First Experiment: Effects of pH on Seed Germination

In this research, all study species were located in the ski station of Sierra Nevada, dominated by siliceous materials with acidic soils. Our results showed that some of them could have suffered a slight increase in pH above 7, as a result of soil movements and conditioning works at the slopes, causing deposition of basic aggregate substrates.

Faced with the possibility that pH of the medium could have significant differences on seed germination, the effects of three different pH values were observed. There were significant effects of pH on seed germination. However, there was in general a better performance at pH 6.2. This led us to set the following experiments at this pH.

Second Experiment: Effects of Light and Exogenous PGRs on Seed Germination

The interactive effects of light and exogenous PGR's were studied on the germination of various native Mediterranean high mountain species under laboratory conditions. The study PGR's included Ethrel (E), N^6-Bencyladenine (BA), Gibberellic-3-Acid (GA_3) and IPIV.

In *G. versicolor* germination percentages and the speed of germination were improved considerably with 10 ppm Ethrel during 12 hours or N^6-Bencyladenine during 24 hours (Figure 1; Table 3).

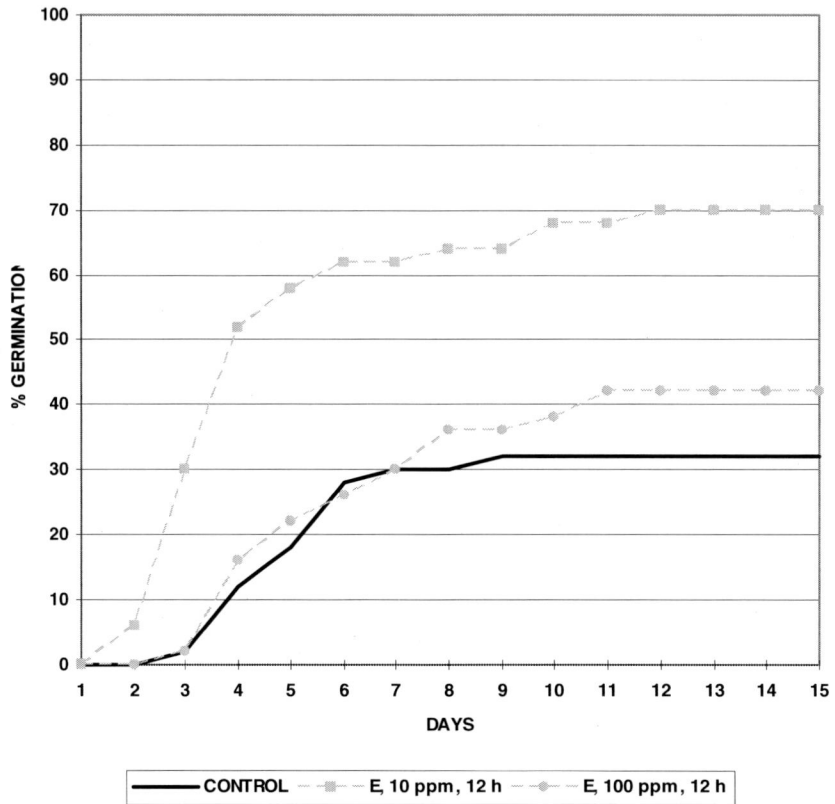

Figure 1. Effect of Ethrel application on the seed germination of *Genista versicolor* at 15°C after 12h light period.

Percentage germination was significantly greater (p<0.05) under both pre-treatments with Ethrel than in the control in *R. complicata* (Figure 2). However exogenous application of Ethrel at 25°C, did not improve germination in this species (Figure 3; Table 4), which confirms the increased susceptibility to light of the germination process in *R. complicate*. This was already tested by Baskin et al. [78] on seeds of *Schoenoplectus hallii*.

Table 3. Germination percentage and T_{50} of *Genista versicolor* at 15ºC and under light conditions

Treatments	Germination (%)	T_{50} (Days)
CONTROL	32.0±5.4 c	4.6
E, 10 ppm,12 h	70.0±2.6 a	3.3
E, 100 ppm, 12 h	42.0±7.5 bc	4.8
BA, 1 ppm, 12 h	50.0±2.6 b	4.8
BA, 10 ppm, 24 h	70.0±7.3 a	4.7

Each value is the mean ± 1 SE (n=6). Means followed by same letter are not statistically different (p=0.05). E= Ethrel; BA= N[6]-Bencyladenine.

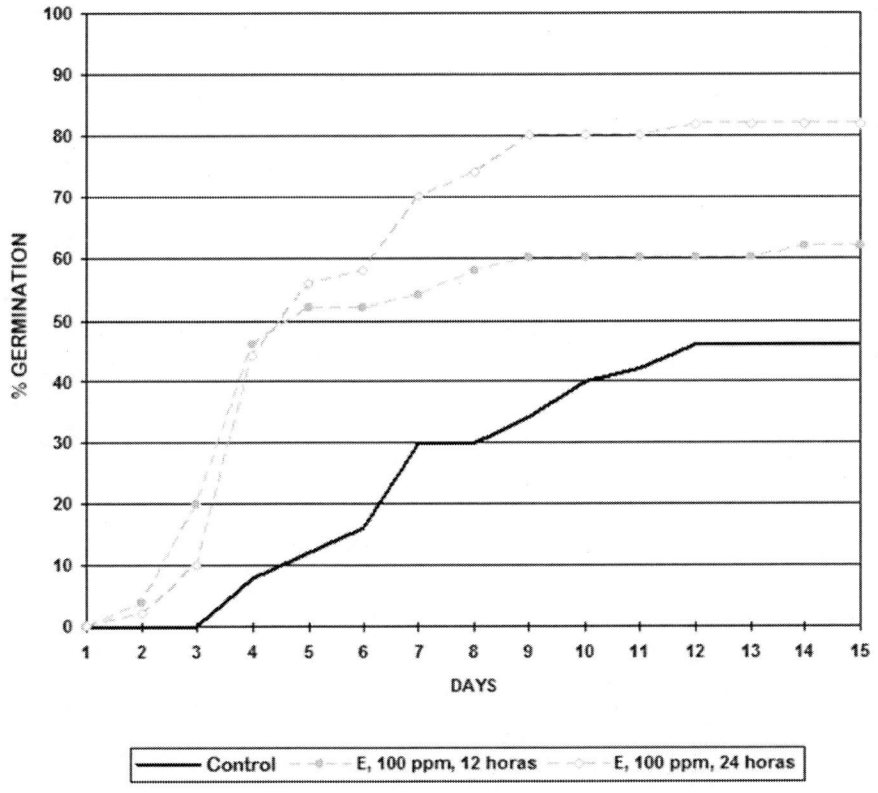

Figure 2. Effect of Ethrel application on the seed germination of *Reseda complicata* at 15°C after 12h light period.

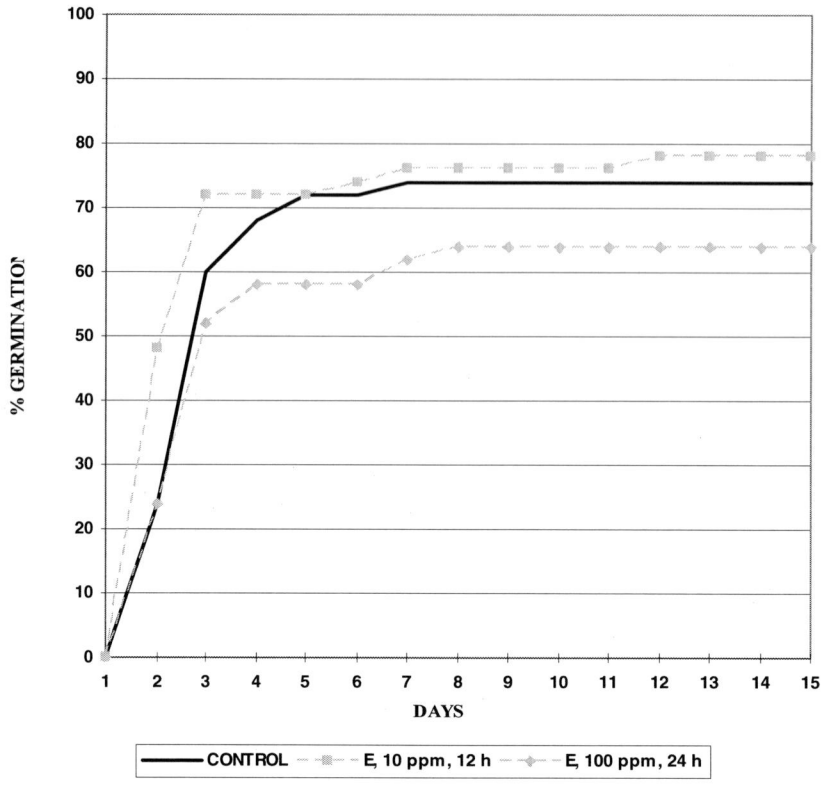

Figure 3. Effect of Ethrel application on the seed germination of *Reseda complicata* at 25°C after 12h light period.

Table 4. Germination percentage and T_{50} of *Reseda complicata* at 25°C and under light conditions

Treatments	Germination (%)	T_{50} (Days)
CONTROL	74.0±3.3 ab	2.4
IP IV, 10 ppm, 12 h	52.0±3.1 d	3.0
IP IV, 10 ppm, 24 h	56.0±4.2 cd	2.3
E, 10 ppm,12 h	78.0±5.4 a	1.8
E, 100 ppm, 12 h	64.0±4.2 bcd	2.3
BA, 1 ppm, 24 h	66.0±3.3 ab	2.5
BA, 10 ppm, 12 h	72.0±4.0 abc	2.6
GA$_3$, 100 ppm, 12 h	76.0±2.0 ab	2.3
GA$_3$, 100 ppm, 24 h	58.0±6.0 cd	2.6

Each value is the mean ± 1 SE (n=6). Means followed by same letter are not statistically different (p=0.05). E= Ethrel; BA= N[6]-Bencyladenine; GA$_3$= Gibberellic-3-Acid; IP IV= Inabar plant IV.

However, application of GA$_3$ to *H. spinosa* and *R. complicata* seeds did not improve germination of these species in the presence of light under any of the study temperatures respect to control values (Figure 4-6). Light effects have been suppressed by treating the seeds with inhibitors of the synthesis of GAs (a.e. paclobutrazol) [79], and light requirements have been replaced by the application of exogenous GAs [80-81]. However, Hillhorst and Karssen [80] speculated that the mediation of GA in the response of seeds to light is given because the application of exogenous GAs "imitates" the effect of red light in promoting germination.

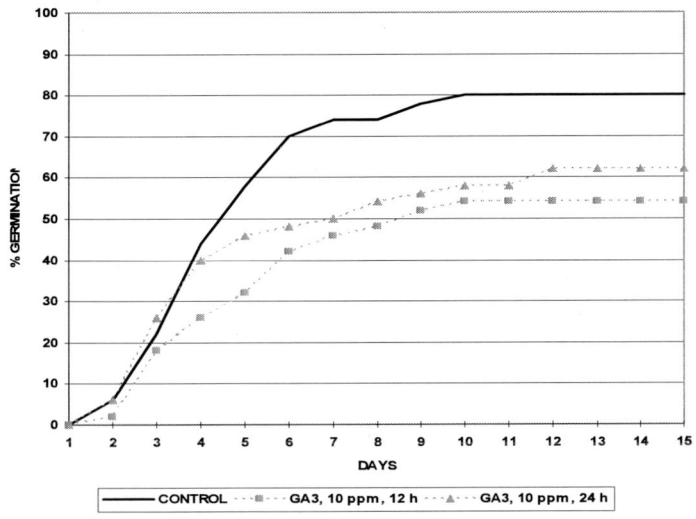

Figure 4. Effect of Gibberellic Acid application on the seed germination of *Hormathophylla spinosa* at 15°C after either 12 or 24 h light period.

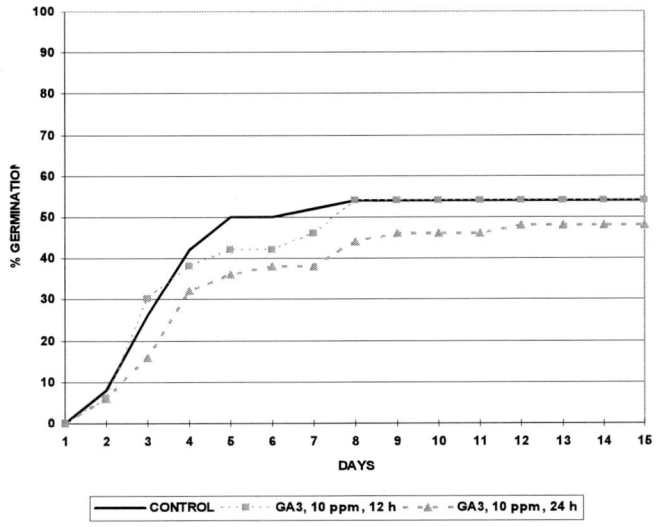

Figure 5. Effect of Gibberellic Acid application on the seed germination of *Hormathophylla spinosa* at 25°C after either 12 or 24 h light period.

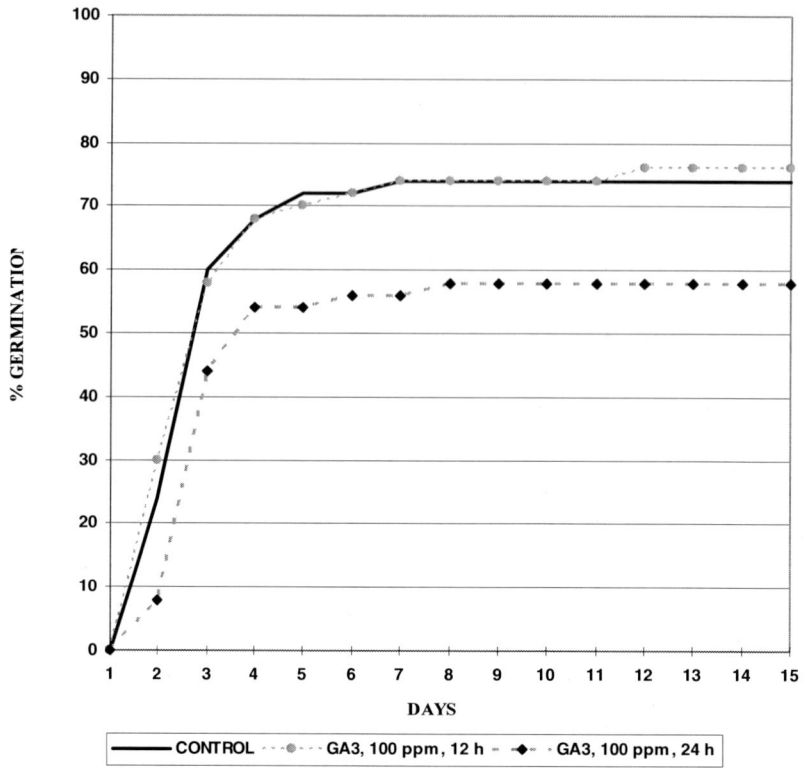

Figure 6. Effect of Gibberellic Acid application on the seed germination of *Reseda complicata* at 25°C after either 12 or 24 h light period.

Cytokinins (CK's) and light increased seed germination on *G. versicolor*, at the two study concentrations and soaking times in comparison to the control (Figure 7; Table 1). At the same time, *H. spinosa* and *R. complicata* showed similar or lower germination values than untreated seeds at 15 and 25°C (Figure 8-10; Tables 4-6).

Within the light absorption spectrum of plants, red light is the only capable of breaking dormancy of seeds treated with CK's [78]. The synergistic effects of CK´s with the light or GA's may allow germination under conditions of high temperatures [82]. However, some authors claim that high temperatures reduce the endogenous levels of CK's [83].

At least in the case of seeds of *G. versicolor*, the release of dormancy induced by light appears to be associated with the presence of CK's; these would be found when determining the levels of endogenous CK's for breaking dormancy [84]. Thus, levels of extractable CK's increased in lettuce seeds 24 hours after exposure to light [85], and in *Rumex* seeds 10 minutes after exposure to red light treatment [86]. In contrast to this, Steadman et al. [87], found that after application of CK's, the release of dormancy in seeds of *Lolium* was faster when germination tests were carried out in darkness.

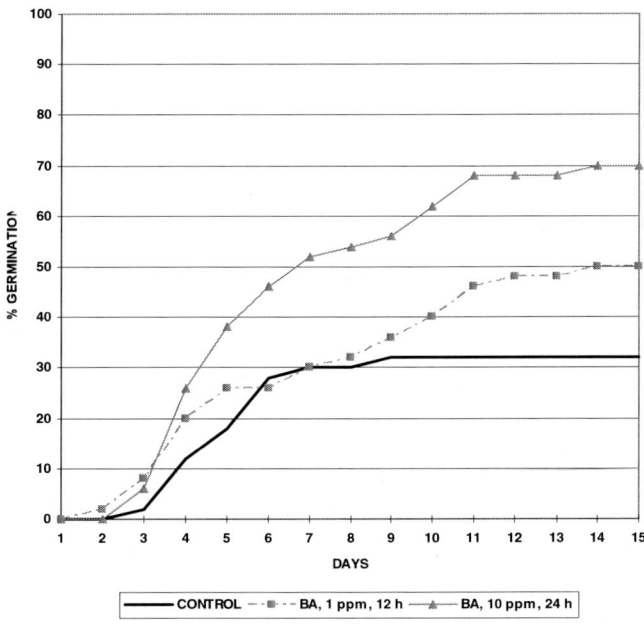

Figure 7. Effect of BA application on the seed germination of *Genista versicolor* at 15°C after either 12 or 24 h light period.

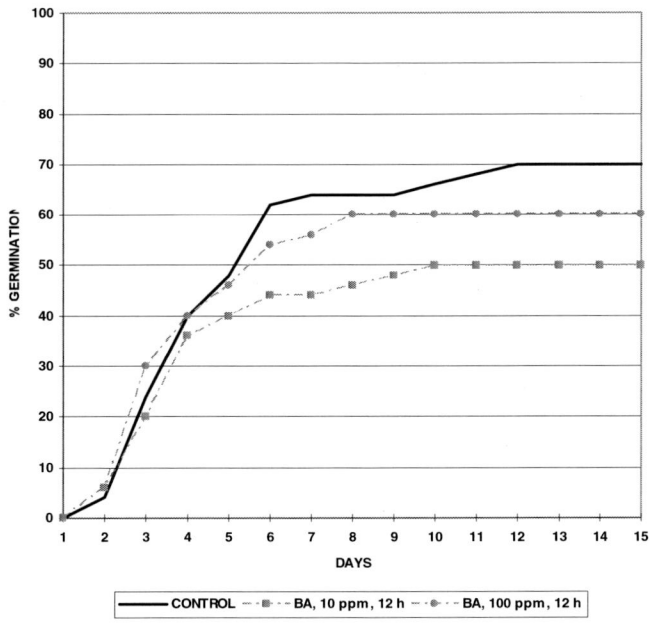

Figure 8. Effect of BA application on the seed germination of *Hormathophylla spinosa* at 15°C after 12h light period.

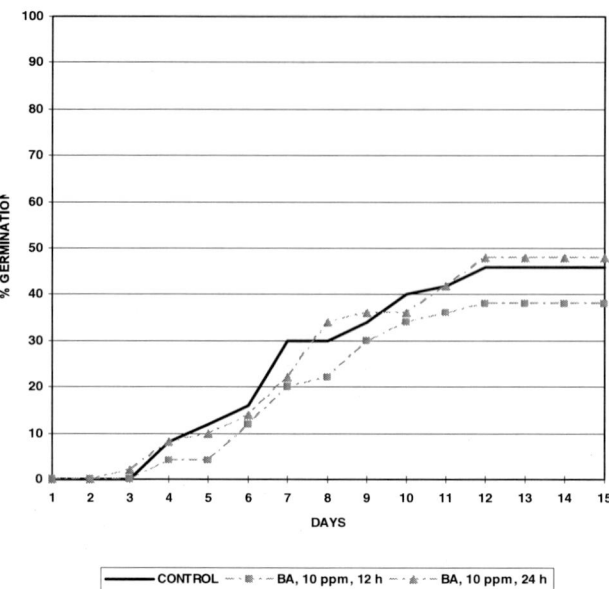

Figure 9. Effect of BA application on the seed germination of *Reseda complicata* at 15°C after either 12 or 24 h light period.

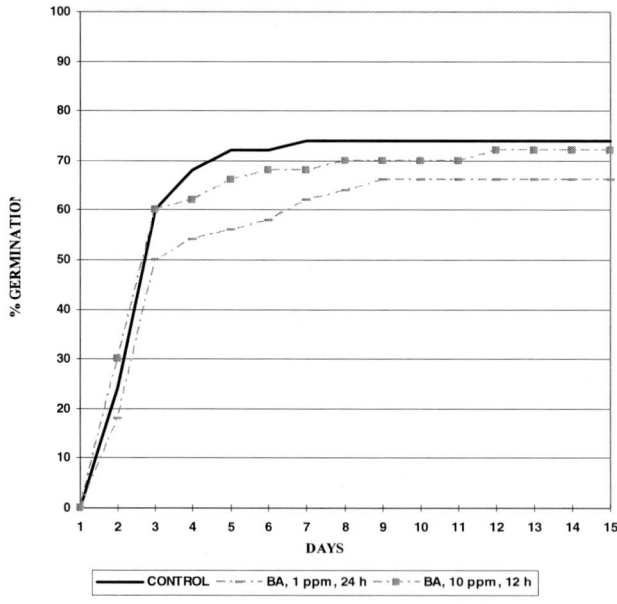

Figure 10. Effect of BA application on the seed germination of *Reseda complicata* at 25°C after either 12 or 24 h light period.

Table 5. Germination percentage and T_{50} of *Hormathophylla spinosa* at 15°C and under light conditions

Treatments	Germination (%)	T_{50} (Days)
CONTROL	70.0±5.8 cd	3.6
BA, 10 ppm, 12 h	50.0±5.8 ab	3.3
BA, 100 ppm, 12 h	60.0±2.6 a	3.0
GA$_3$, 10 ppm, 12 h	54.0±4.2 abc	4.3
GA$_3$, 10 ppm, 24 h	62.0±7.0 abc	3.3

Each value is the mean ± 1 SE (n=6). Means followed by same letter are not statistically different (p=0.05). BA= N^6-Bencyladenine; GA$_3$= Gibberellic-3-Acid.

Table 6. Germination percentage and T_{50} of *Reseda complicata* at 15°C and under light conditions

Treatments	Germination (%)	T_{50} (Days)
CONTROL	46.0±3.3 bc	6.5
E, 100 ppm, 12 h	62.0±8.7 b	3.4
E, 100 ppm, 24 h	82.0±4.8 a	3.9
BA, 10 ppm, 12 h	38.0±4.8 c	6.9
BA, 10 ppm, 24 h	48.0±5.4 bc	7.2

Each value is the mean ± 1 SE (n=6). Means followed by same letter are not statistically different (p=0.05). E= Ethrel; BA= N^6-Bencyladenine.

The interaction of light and IP IV did not increase germination of *T. serpylloides* and *R. complicata* compared to the untreated seeds of these species (Figure 11; Table 4, 7).

These results suggest that light can modulate the production or transport of auxin, and therefore that these concentrations are quantitatively related to the development or growth rates of different tissues that are part of the seed [88].

Table 7. Germination percentage and T_{50} of *Thymus serpylloides* at 15°C and under light conditions

Treatments	Germination (%)	T_{50} (Days)
CONTROL	68.0±6.5 a	2.3
IP IV, 1 ppm, 24 h	72.0±4.0 a	1.7
IP IV, 10 ppm, 24 h	80.0±6.8 a	1.8

Each value is the mean ± 1 SE (n=6). Means followed by same letter are not statistically different (p=0.05). IP IV= Inabar plant IV.

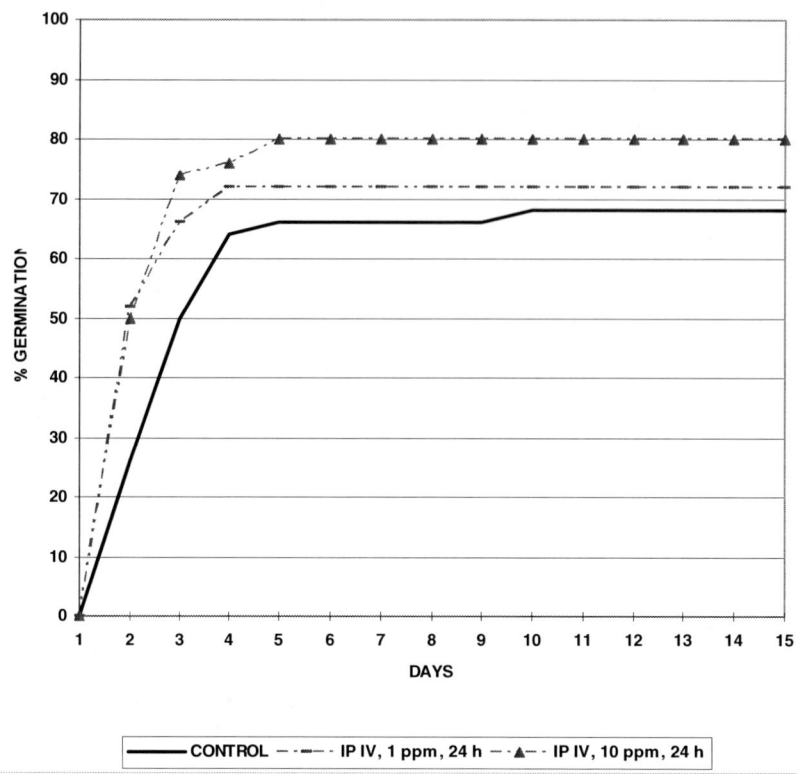

Figure 11. Effect of IP IV application on the seed germination of *Thymus serpylloides* at 15°C under a 24h period.

Third Experiment: Effect of Exogenous PGRs on Rooting and Growth Capacity

Germination data for *G. versicolor* seeds after 28 days from imbibition in the different pre-treatments with growth regulators (Figure 12) showed that most seeds exceeded the control values, especially those pre-treated with E_{10-12}. In *R. complicate*, however, seeds pre-treated with that growth regulator (e.g., E) showed lower germination percentages than the untreated control (Figure 13). The data are relevant because the final percentage of germination reached by the control was low when it was compared with the seed pre-treatment at E_{10-12} (Figure 12).

The analysis of growth for *G. versicolor* showed that BA treatments gave similar stem lengths than the control (Figure 14). However, root growth was shorter by the E application.

Shoot and root lengths were similar among treatments on *R. complicata* at 15°C (Figure 15). These results were similar to those obtained at 25°C (Figure 16). The only exception was the application of GA_3 on root growth at 25°C, where root length was increased (Figure 16).

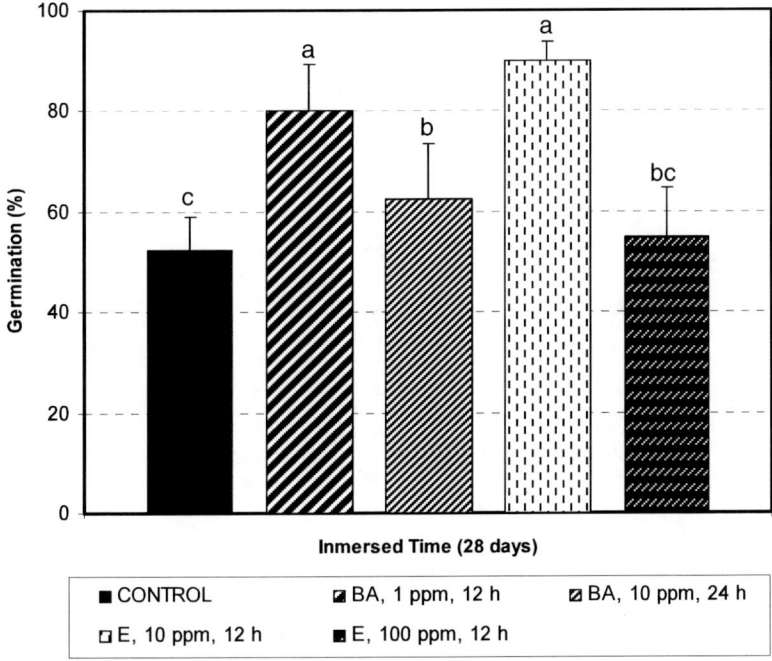

Figure 12. Effect of different treatments on the seed germination of *Genista versicolor* at 15°C sown in vermiculite. Seeds received light during either 12 or 24h

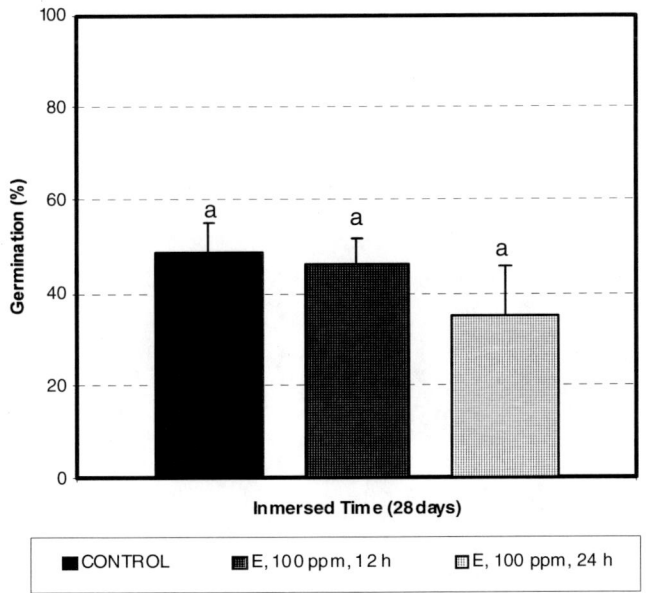

Figure 13. Effect of different treatments on the seed germination of *Reseda complicata* at 15°C sown in vermiculite. Seeds received light during either 12 or 24h.

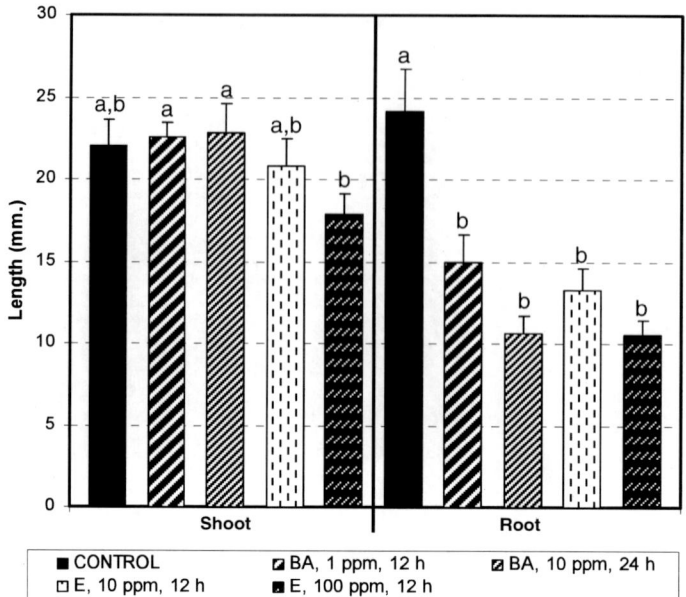

Figure 14. Root and shoot length on *Genista versicolor* after seed germination at 15°C in vermiculite. Seeds received light during either 12 or 24h. Vertical bars indicate ±1SE of the mean.

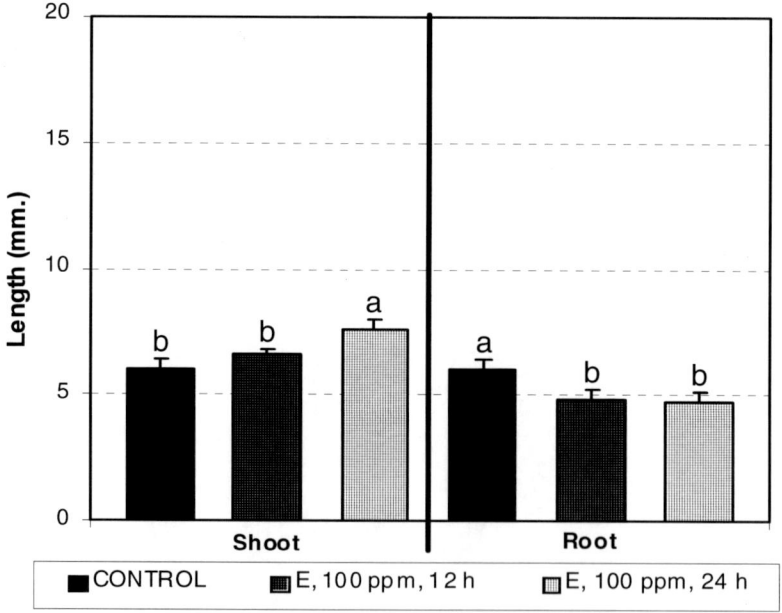

Figure 15. Root and shoot length on *Reseda complicata* after seed germination at 15°C in vermiculite. Seeds received light during either 12 or 24h. Vertical bars indicate ±1SE of the mean.

One of the most relevant growth parameters of plants is S:R ratio [89]. Differences in S:R can explain the different survival ability of various plant species [90]. However, there is no universal ideal S:R for restored vegetation, but rather each species has its own optimal range, as stated by Villar-Salvador [89].

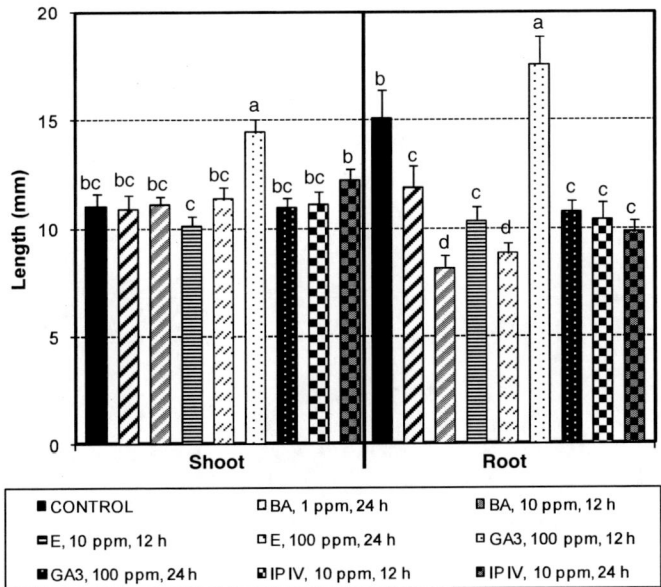

Figure 16. Root and shoot length on *Reseda complicata* after seed germination at 25°C in vermiculite. Vertical bars indicate ±1SE of the mean.

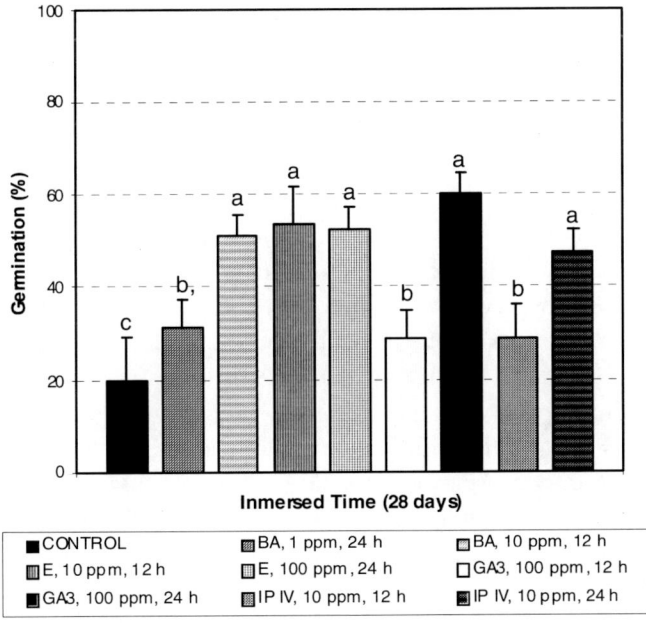

Figure 17. Effect of different treatments on the seed germination of *Reseda complicata* at 25°C in vermiculite.

Seeds of *G. versicolor* (Figure 12) and *R. complicata* (Figure 17) imbibited with E, BA and GA$_3$ and sown afterwards in vermiculite presented satisfactory results for seed germination in this study. In some cases, the effect of these PGRs on germination was reported to be very positive [91-92]. However, only GA$_{3;10-24}$ except for *H. spinosa* (Figure 18).

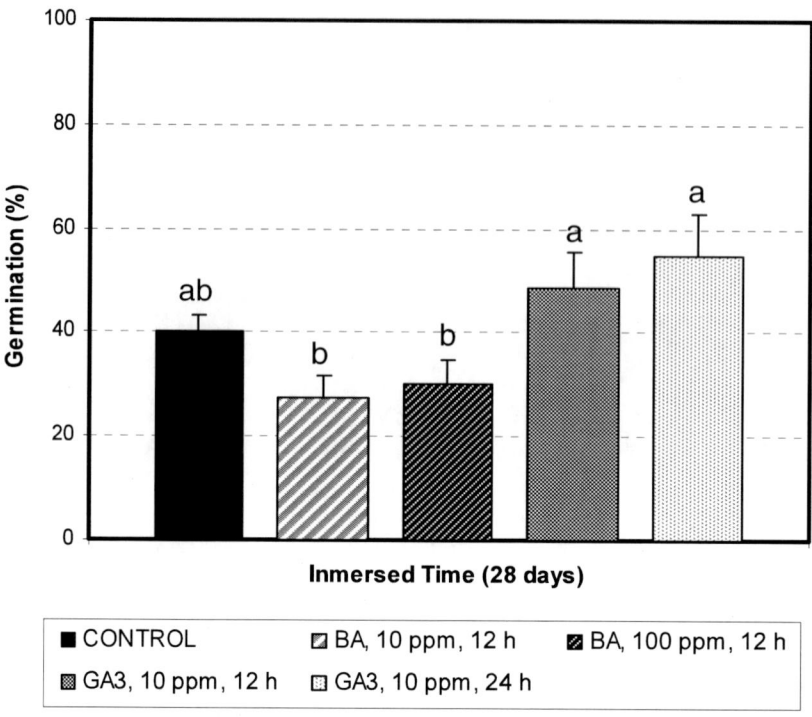

Figure 18. Effect of different treatments on the seed germination of *Hormathophylla spinosa* at 15°C in vermiculite.

Seedling growth after E application showed certain characteristic features in these species. It is well known that compounds that release ethylene can promote [93], inhibit [94], or have no effect on roots [95]. In the case of *G. versicolor* and *R. complicate*; higher concentrations of E promoted growth more than lower concentrations. Then, it might be stated that root elongation depended on the applied ethylene concentration [96].

Ethylene production in seeds begins immediately after the onset of imbibition, and it increases with time, but its development differs markedly between species. However, root protrusion from the seed coat is invariably associated with a peak ethylene release [97]. This implies a close relationship between the ability to produce ethylene and seed vigour in *G. versicolor* and *R. complicate*. This ability appears to be greater in the former than in the latter species. Also, CO_2 and ethylene may interact in the germination and growth processes [98]. This may occur by ethylene biosynthesis and axial root growth, especially in *R. complicata* seeds, furthermore breaking thermodormancy and subsequently raising the germination percentage [97]. This gaseous exchange, however, and its action on seeds, depends on sowing depth and thickness of the vermiculite cover, among other factors [98-99].

Also, the application of BA on germinated seedlings in vermiculite was positive for shoot but not root growth in both species. It is known that CKs can inhibit root growth [100]. Also, CKs can inhibit hypocotyl elongation and can act independently, and additively with light. This is in the sense that neither of the two factors can cause greater inhibition under conditions in which the other factor is saturated [101]. Su and Howell [101] also showed that CKs inhibit hypocotyl elongation in *Arabidopsis* due to the high levels of ethylene, while light acts as a mechanism that does not depend directly on ethylene [102]. This hypothesis states that the endogenous levels of ethylene present in *G. versicolor* and *R. complicata* seeds possibly increase in the presence of BA, confirming that ethylene may, at least in some cases, mediate the CKs action [100].

The response differences showed by the two species may be caused by alterations in one or more cellular processes (e.g., water uptake or metabolism) derived from tissues sensitivity to the exogenous application of BA. Cytokinins are rapidly metabolised by plants [103], but different studies have demonstrated the complexity of such metabolism [104]. This might partly account for the different effectiveness of BA in each of the two study species [105].

The nutrient solution added to the used substrate can influence the CK's synthesis [106]. Thus, Samuelson and Larsson [107] reported that high CK's levels in rye roots were owed to increased N concentrations in the medium. Also Wagner and Beck [108] noted that low CK's levels at low N concentrations, could also reduce the available quantities of CKs in leaves, buds, and roots [109]. On the other hand, a media with an excess of minerals or even with an optimal solution can diminish the CKs activity in roots, without provoking changes in other tissues [110].

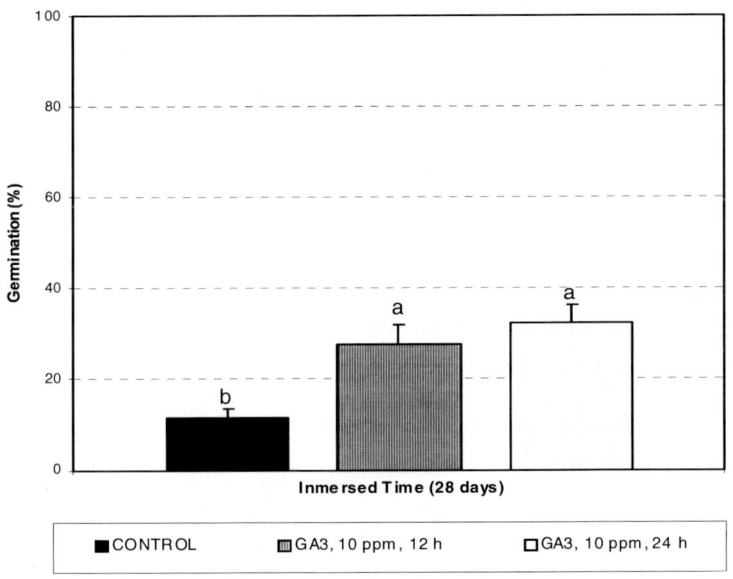

Figure 19. Effect of different treatments on the seed germination of *Hormathophylla spinosa* at 25°C in vermiculite.

Hormathophyla spinosa seeds treated with GA_3 at 15 and 25°C showed greater germination percentages than in control treatment (Figure 18 and 19). However, these results were lower than those obtained when the study was conducted in Petri dishes. Shoot and root lengths were also greater in the GA_3 treatments than in the controls in *H. Spinosa* (Figure 20 and 21). The only exception was in $GA_{3;10\text{-}24}$ treatment, where shoot and root lengths were greater in the control (Figure 21). Treatments with IP IV showed similar percentage germination than in the control at both study temperatures in *T. serpylloides* (Figure 22 and 23). Other studies have shown that low concentrations of this plant growth regulator help to restore the response to gravity of the roots [111].

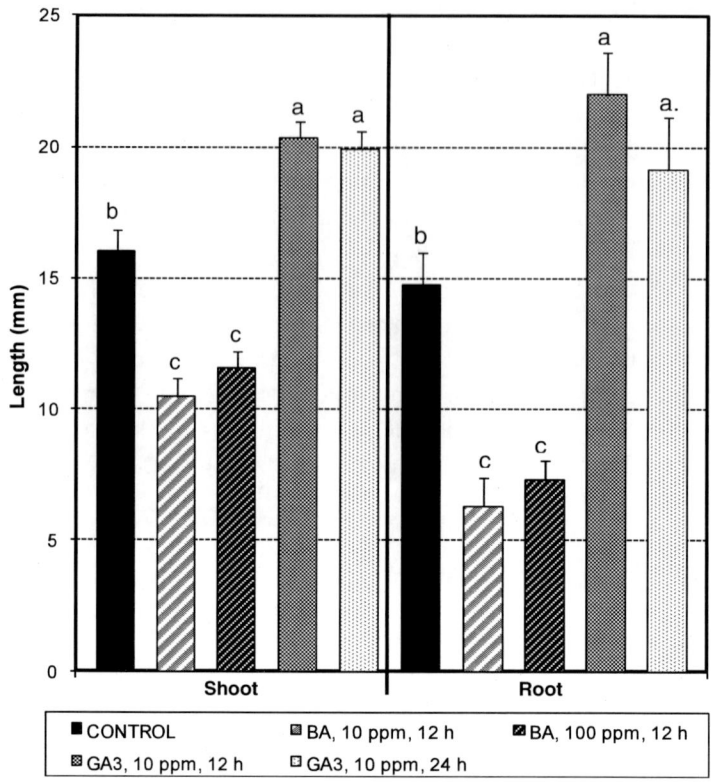

Figure 20. Root and shoot length in *Hormathophylla spinosa* sown in vermiculite, 15ºC.

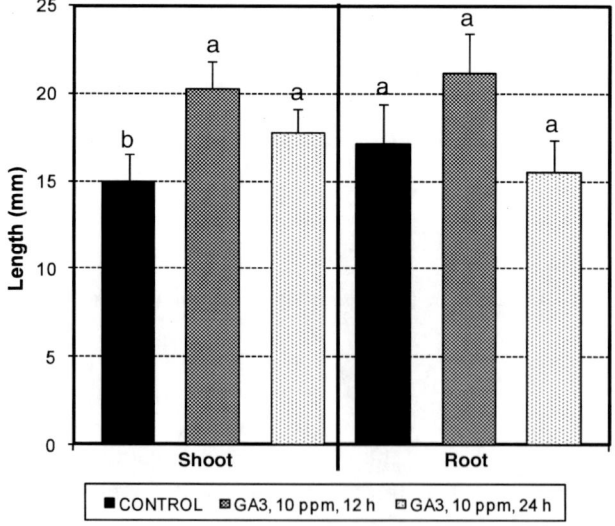

Figure 21. Root and shoot length in *Hormathophylla spinosa* sown in vermiculite, 25ºC.

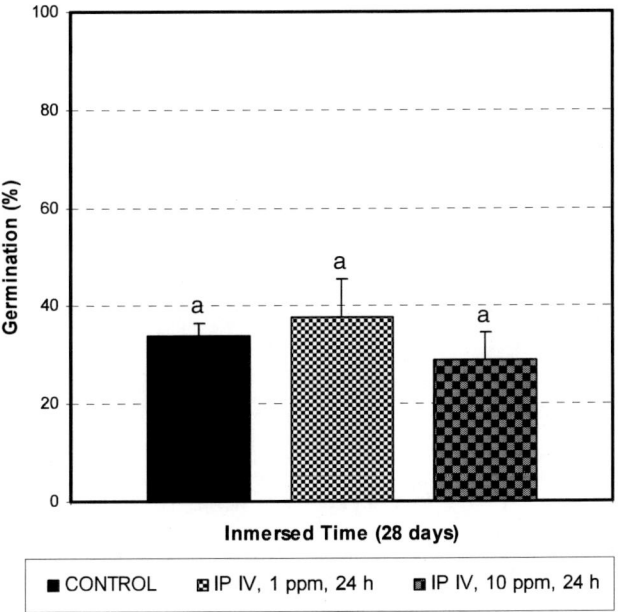

Figure 22. Effect of different treatments on the seed germination of *Thymus serpylloides* at 15°C in vermiculite.

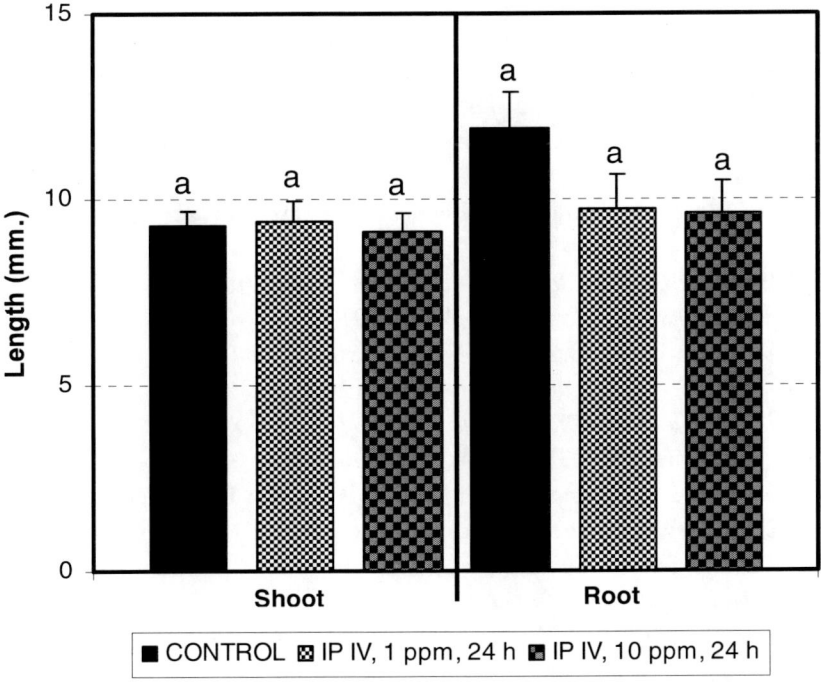

Figure 23. Root and shoot length in *Thymus serpylloides* sown in vermiculite, 25°C.

Conclusion

The study plant growth regulators improved eithr shoot or root growth (or both) in the study species. This suggests their suitability for plant cover restoration. The exogenous application of PGRs affected germination, rooting, and developmental processes in all of the study species most often in a positive manner. PGRs application combined with light affected the germination of all study species, but not always positively. Also, vermiculite seems to be a useful substrate in the laboratory for studying germination rate, rooting, and growth in the first stages of seedling development, as well as for testing the action of phytoregulators on these processes. This raises the expectaion that, when transferred to the field, these treatments might be an effective aid to help plant recruitment in recovery programmes.

References

[1]   A. Miyawaki, *Ecol. Res.* 19, 83 (2004).
[2]   K. M. Urbanska, N. R. Webb and P. J. Edwards. In: K. M. Urbanska,; N. R. Webb and P. J. Edwards (Eds.), *Restoration Ecology and Sustainable Development*, pp. 3-7. Cambridge Univ. Press. (1997).
[3]   C. G. Daily, *Science* 269, 350 (1995).
[4]   I. N. Vogiatzakis, A. M. Mannion and G. H. Griffiths, *Progr. Phys. Geog.* 30, 175 (2006).
[5]   L. Giménez-Benavides, A. Escudero and F. Pérez-García, *Ecol. Res.* 20, 433 (2005).
[6]   S. Rivas–Martínez. In: V. C. Duplessey, A. Poms and R. Fantecli (Eds.), *Climate And Global Change* (1990).
[7]   G. Blanca, *Flora amenazada y endémica de Sierra Nevada*. Junta de Andalucía. Ed. Univ. Granada. Granada. 410 p (2002).
[8]   CETURSA, *Datos de pistas y remontes de la estación de esquí de Sierra Nevada*, Available: http://www.sierranevadaski.es (2012).
[9]   T. M. Macyk. In: R. L. Barnhisel, R. G. Darmody and W. L. Lee Daniels, (Eds.), pp. 537-566. *Am. Soc. Agr. Publ.*, Madison, WI (2000).
[10] C. Körner. *Alpine plant life-Functional Plant Ecology of High Mountain Ecosystems.* Springer, Heidelberg (2003).
[11] J. C. Chambers, J. A. Macmahon and G. W. Wade. In: J. C. Chambers and G. W. Wade (Eds.), Evaluating reclamation success: the ecological considerations-Proc. Symp., pp. 59-72. USDA Forest Service, Northeastern Forest Experiment Station, General Technical Report NE-164 (1992).
[12] G. Grabherr, M. Gottfried, A. Gruber and H. Pauli. In: F.S. Chapin III and C. Körner (Eds.) *Arctic and alpine biodiversity: patterns, causes and ecosystem consequences*, pp. 167–181, Springer, Berlin (1995).
[13] J. C. Chambers, R.W. Brown and R.S. Jonhston. In: T. A. Colbert and R. L. Cuany, (Eds.), Proc. High Altitude Revegetaion Workshop, N° 6, 215. *Water Resour. Inst. Inf. Series* N° 53. Fort Collins. Colorado St. Univ. (1984).

[14] K. M. Urbanska. In: K. Urbanska, N. R. Webb and P. J. Edwards (Eds.), *Restoration ecology and sustainable development*, pp. 81–110. Cambridge: Cambridge University Press (1997).

[15] K.M. Urbanska, *Biodiv. Conserv.* 6, 1655 (1997).

[16] R. Delarze, *Bot. Helv.* 104, 3 (1994).

[17] R. Gros, L. J Monrozier, F. Bartoli, J. L. Chotte and P. Faivre, *Appl. Soil Ecol.* 27, 7 (2004).

[18] E. Ruth-Balaganskaya and K. Myllynen-Malinen. *Landscape Urban Plan.* 50, 259 (2000).

[19] M. Sánchez-Marañón, M. Soriano, G. Delgado and R. Delgado, *Soil Sci. Soc. Am. J.* 66, 948 (2002).

[20] S. Muller, T. Dutoit, D. Alard and F. Grevilliot, *Restor. Ecol.* 6, 94 (1998).

[21] J. P. Grime. *Plant strategies and vegetation processes.* John Wiley, Chichester (1979).

[22] W. D. Bowman, T. A. Theodose, J. C. Schardt and R. T. Conant, *Ecology* 74, 2085 (1993).

[23] F. F. Munshower. *Practical handbook of disturbed land revegetation.* CRC Press Inc., Boca Raton (1994).

[24] K. M. Urbanska, In: R. M. M. Crawford (Ed.), Disturbance and recovery in artic lands, pp. 481-501. Kluwer Acad., Dordrecht (1997).

[25] W. H. Drury and C. T. Nisbet, *J. Arnold Arbor.* 54, 331 (1973).

[26] A. D. Bradshaw, In: K. M. Urbanska, N. R. Webb and P. J. Edwards (Eds.), *Restoration Ecology and Sustainable Development*, pp. 8-13. Cambridge Univ. Press (1997).

[27] J. P. Kimmins. *Forest Ecology.* Prentice Hall, NJ (1997).

[28] I. Noble and R.O. Slatyer, *Vegetation* 43, 5 (1980).

[29] R. Vallejo, J. Cortina, A. Vilagrosa, J. P. Seva and J. A. Alloza, In: J. M. Rey Benayas, T. Espigares Pinilla and J. M. Nicolau Ibarra (Eds.), *Restauración de ecosistemas mediterráneos*, pp. 11-42. Univ. Alcalá. Madrid (2003).

[30] C. Mataix, In: C. López Jimeno (Ed.), Manual de estabilización y revegetación de taludes, pp. 595-633. Univ. Pol. Madrid. Madrid. (2002).

[31] J. Dorner. *An introduction to using native plants in restoration projects.* Plant Conservation Alliance Bureau of Land Management, US Department of Interior, U.S. Environmental Protection Agency (2002).

[32] M. Pezzi and M. E. Martín-Vivaldi, *Características del relieve residual sobre dolomías en la orla alpujárride de Sierra Nevada.* VI Coloquio de Geografía (1983).

[33] J. Galindo-Zaldívar, F. González-Lodeiro and A. Jabaloy, *Tectonophysics* 227, 105 (1993).

[34] J. A. Ortega-Olivencia, T. Devesa Rodríguez-Riaño, *Biología floral en Fabaceae.* RJB (CSIC), Madrid (1999).

[35] F. Ortega and Y. Jiménez, *El clima de Sierra Nevada.* Inéd. (1999).

[36] J. D. Bewley and M. Black, *Seeds: Physiology of development and germination.* 2nd ed. Plenum Press, New York (1994).

[37] J. D. Bewley, *Plant Cell*, 9, 1055 (1997).

[38] C. C. Baskin and J. M. Baskin. *Seeds: Ecology, Biogeography, and Evolution of Dormancy and Germination.* Acad. Press, San Diego, CA (1998).

[39] F. Besnier. *Semillas: Biología y Tecnología.* Ed. Mundiprensa. Madrid (1989).

[40] G. H. P. Arts and R. A. J. M Vand der Heijden, *Aqua. Bot.* 37, 139 (1990).

[41] G. Ne'eman, N. Henig-Server and A. Eshel, *Physiol. Plant.* 106, 47 (1999).

[42] ISTA (International Seed Testing Association), *International Rules for Seed Testing* (1985).

[43] C. Hackett, *J. Ecol.* 52, 159 (1964).

[44] G. D. Turner, R. R. Lau and D. R. Young, *Environ. Exp. Bot.* 30, 383 (1988).

[45] O. T. Okusanya, *Oikos* 30, 549 (1978).

[46] J. N. Maloof, J. O. Borevitz, D. Weigel and J. Chory, *Cell Develop. Biol.* 11, 523 (2000).

[47] H. A. Borthwick, S. B. Hendricks, M. W. Parker, E. H. Toole and V. K. Toole, *Proc. Natl. Acad. Sci.* 38, 662 (1952).

[48] R. E. Kendrick, *Sci. Prog.* 63, 347 (1976).

[49] C. C. Baskin and J. M. Baskin. *Am. J. Bot.* 75, 286 (1988).

[50] N. Colbach, B. Chauvel, C. Dürr and G. Richard, *Weed Res.* 42, 210 (2002).

[51] C. A. Thanos, K. Georghious and F. Skarou, *Ann. Bot.* 63, 121 (1989).

[52] A. J. Trewavas, *Physiol. Plant.* 55, 60 (1982).

[53]] M. C. Rodríguez-Gacio and A. J. Matilla, *Physiol. Plantarum* 112, 273 (2001).

[54] L. A. Voesenek, J. J. Benshop, J. Bou, M. C. Cox, H. W. Groeneveld, F. F. Millenaar, R. A. M. Vreeburg and A. J. M. Peeters, *Ann. Bot.* 91, 205 (2003).

[55] B. Białecka and J. Kępczyński, *Plant Growth Regul.* 51, 21 (2007).

[56] A. Ferrante, A. Mensuali-Sodi, G. Serra and F. Tognoni, *Plant Growth Regul.* 38, 119 (2002).

[57] J. Pospíšilová, M. Vágner, J. Malbeck, A. Trávníčková and P. Baťková, *Biol. Plantarum* 49, 533 (2005).

[58] C. A. Parks and T.H. Boyle, *HortScience* 37, 202 (2002).

[59] C. M. Karssen, S. Zagórski, J. Kepczyñski and S. P. C. Groot, *Ann. Bot.* 63, 71 (1989).

[60] F. B. Abeles, P. W. Morgan and M. E. Salveit Jr. *Ethylene in Plant Biology.* 2ª Ed. Acad. Press, New York (1992).

[61] H. Kawade, R. Imai, T. Sassa and Y. Kamiya, *J. Biol. Chem.* 272, 21706 (1997).

[62] A. J. Matilla, *Seed Sci. Res.* 10, 111 (2000).

[63] Z. M. Sawan, A. A. Mohamed, R. A. Sakr and A. M. Tarrad, *Environ. Exp. Bot.* 44, 59 (2000).

[64] F. Serrano-Bernardo, PhD Thesis, University of Granada, Spain (2005).

[65] S. Burés, *Sustratos,* Ediciones Agrotécnicas, Madrid (1999).

[66] AOSA, *Tetrazolium Testing Handbook* N° 29, Assoc. Offic. Seed Anal., Las Cruces, New Mexico (2000).

[67] J. L. Rosúa and J.C. Martín, In: J. Chacón, and J. L. Rosúa (Eds.), 1ª Conf. Int. Sierra Nevada: Conservación y Desarrollo Sostenible, Vol. V, pp. 77-87, Granada (1996).

[68] J. Rigaurd and A. Puppo, *J. Gen. Microbiol.* 88, 223 (1975).

[69] J. Lorite, F.B. Navarro and F. Valle, *Plant Biosystems* 141, 1 (2007).

[70] J.A. Hódar and R. Zamora, *Biodivers. Conserv.* 13, 493 (2004).

[71] T. F. Chanyenga, C. J. Geldenhuys and G. W. Sileshi, *S. Afr. J. Bot.* 81, 25 (2012).

[72] J. J. Casal, R. A. Sánchez and J. F. Botto, *J. Exp. Bot.* 49, 127 (1997).

[73] C. Vázquez-Yanes, M. Rojas-Arechiga, M. E. Sánchez-Coronado and A. Orozco-Segovia, *Tree Physiol.* 16, 871 (1996).

[74] J. J. Casal and R. A Sánchez, *Seed Sci. Res.* 8, 317 (1998).

[75] G. J. Blionis and D. Vokou, *Plant Ecol.* 178, 77 (2005).

[76] F. Ortega, In: J. Chacón, F. Ortega, A. Pulido and J. L. Rosúa, pp. 25-31, Univ. Granada (1989).

[77] T. Angosto and A. J. Matilla. *Seed Sci. Tech.* 22, 319 (1994).

[78] C. C. Baskin, J. M. Baskin, E. W. Chester and M. Smith, *Am. J. Bot.* 90, 620 (2003).

[79] J. L. García Martínez and J. Gil, *J. Plant Growth Regul.* 20, 354 (2002).

[80] H. W. M. Hillhorst and C. M. Karssen, *Plant Physiol.* 86, 591 (1988).

[81] C. Min-Seok, K. Soojung and P. Tae-Ho, *Afr. J. Agric. Res.* 6(32), 6720 (2011).

[82] T. H. Thomas, *Plant Growth Regul.* 11, 239 (1992).

[83] C. C. Chou, W. S. Chen, K. L. Huang, H. C. Yu and L. J. Liao, *Plant Physiol. Biochem.* 38, 309 (2000).

[84] T. H. Thomas, P. D. Hare and J. Van Staden, *Plant Growth Regul.* 23, 105 (1997).

[85] E. Barzilai and A. M. Mayer, *Aust. J. Biol. Sci.* 17, 797 (1964).

[86] J. Van Staden and P. F. Wareing, *Planta* 104, 126 (1972).

[87] K. J. Steadman, G. P. Bignell and A. J. Ellery, *Weed Res.* 43, 458 (2003).

[88] Q. Tian and J. W. Reed, *J. Plant. Growth. Regul.* 20, 274 (2001).

[89] P. Villar Salvador, In: J. M. Rey Benayas, T. Espigares Pinilla and J. M. Nicolau Ibarra (Eds.), *Restauración de ecosistemas mediterráneos*, pp. 65-86, Univ. Alcalá. Madrid (2003).

[90] F. Lloret, C. Casanovas and J. Peñuelas, *Funct. Ecol.* 13, 210 (1999).

[91] B. Kucera, M. A. Cohn and G. Leubner-Metzger, *Seed Sci Res.* 10, 281 (2005).

[92] M. Iqbal, M. Ashraf and A. Jamil, *Plant Growth Regul.* 50, 29 (2006).

[93] J. B. Zhen, Y. J. Lin and B. H. Pei, *Sci. Silv. Sinica.* 29, 19 (1993).

[94] M. Jusatis, *Sci. Hort.* 29, 77 (1986).

[95] K. W. Mudge and B. T. Swanson, *Plant Physiol.* 61, 271 (1978).

[96] R. Pan, J. Wang and X. Tian, *Plant Growth Regul.* 36, 135 (2002).

[97] F. Corbineau and D. Côme, In: J. Kigel and G. Galili (Eds.), *Seed Development and germination*, pp. 397-424, Marcel Dekker, Inc., New York (1995).

[98] H. S. Saini, P. K. Bassi, E. D. Consolación and M. S. Spencer, *Can. J. Bot.* 64, 2322 (1986).

[99] H. G. Edelman, G. Gudi and F. Kühnemann, *J. Exp. Bot.* 53, 1627 (2002).

[100] M. Brault and R. Maldiney, *Plant Physiol. Biochem.* 37, 403 (1999).

[101] W. Su and S. Howell, *Plant Physiol.* 108, 1423 (1995).

[102] M. J. Laskowski, E. Seradge, J. R. Shinkle and W. R. Briggs, *Plant Physiol.* 100, 95 (1992).

[103] M. Kamínek, R. Motyka and R. Vankova, *Physiol. Plant.* 101, 689 (1997).

[104] P. E. Jameson, In: D. W. S. Mok and M. C. Mok (Eds.), *Cytokinins. Chemistry, Activity and Function*, pp. 113-128. CRC Press Inc., Boca Raton (1994).

[105] J. Suttle, *Plant Growth Regul.* 35, 199 (2001).

[106] P. Battal and B. Tileklioglu, *Turk. J. Bot.* 25, 123 (2001).

[107] M. E. Samuelson and C. M. Larsson, *Plant Sci.* 93, 77 (1993).

[108] B. M. Wagner and E. Beck, *Planta,* 190, 511 (1993).

[109] A. Salama and P. F. Wareing, *J. Exp. Bot.* 30, 971 (1979).

[110] P. Gallego, J. Hernández-Nistal, L. Martín, G. Nicolás and, N. Villalobos, *Plant Sci.* 77, 207 (1991).

[111] T. Chunn, S. Taketa, S. Tsurumi and M. Ichii. *Plant Growth Regul.* 39, 161 (2003).

Tropical Forest Ecosystems

In: From Seed Germination to Young Plants
Editor: Carlos Alberto Busso

ISBN: 978-1-62618-653-8
© 2013 Nova Science Publishers, Inc.

Chapter 15

Early Growth Improvement on Endemic Tree Species by Soil Mycorrhizal Management in Madagascar

H. Ramanankierana[1], R. Baohanta[1], J. Thioulouse[2], Y. Prin[3],
H. Randriambanona[1], E. Baudoin[4], N. Rakotoarimanga[1],
A. Galiana[3], E. Rajaonarimamy[1], M. Lebrun[4]
and Robin Duponnois[4,5,]*

[1]Laboratoire de Microbiologie de l'Environnement. Centre National de Recherches sur l'Environnement, Antananarivo, Madagascar
[2]Université de Lyon, Lyon; Université Lyon 1; CNRS, UMR5558, Laboratoire de Biométrie et Biologie Evolutive, Villeurbanne, France
[3]CIRAD. Laboratoire des Symbioses Tropicales et Méditerranéennes (LSTM), UMR 113 CIRAD/INRA/IRD/SupAgro/UM2, Campus International de Baillarguet, TA A-82/J, Montpellier, France
[4]IRD. Laboratoire des Symbioses Tropicales et Méditerranéennes (LSTM), UMR 113 CIRAD/INRA/IRD/SupAgro/UM2, Campus International de Baillarguet, Montpellier, France
[5]Laboratoire Ecologie & Environnement (Unité associée au CNRST, URAC 32) Faculté des Sciences Semlalia, Université Cadi Ayyad, Marrakech, Maroc

Abstract

Mycorrhizal fungi are ubiquitous components of most ecosystems throughout the world and are considered key ecological factors in (1) governing the cycles of major plant nutrients and (2) sustaining the vegetation cover. Two major forms of mycorrhizas are usually recognized: the arbuscular mycorrhizas (AM) and the ectomycorrhizas (ECM). The lack of mycorrhizal fungi on root systems is a leading cause of poor plant

[*] E-mail address: robin.duponnois@ird.fr.

establishment and growth in a variety of forest landscapes. Numerous studies have shown that mycorrhizal fungi are able to improve the survival and early growth of various tree species in the field. Mycorrhizal association is estimated to occur in 95% of native undisturbed vegetation, whereas it occurs in less than 1% of vegetation from disturbed sites. Thereafter, mycorrhizal symbiosis has to be reestablished at these latter sites to benefit from the mycorrhizal effects on plant growth. This can be achieved by enhancing the mycorrhizal status of seedlings before they are transplanted to disturbed sites. It is necessary that nurseries produce tree seedlings associated with mycorrhizal fungi that are ecologically compatible with the tree species and the planting sites to make afforestation successful. According to these conditions that have to be taken into account, different methods of mycorrhizal inoculation have been identified to optimize fungal effects on plant growth. The main objective of this chapter was to describe some methods to obtain mycorrhizal seedlings at the nursery and to present some tree growth data resulting from the use of mycorrhization under such conditions in Madagascar.

Introduction

Land degradation is expanding around the world, and the (i) decline in soil fertility, and deterioration of soil physical and biological properties, and (ii) invasion by aggressive vegetation are serious concerns to forest regeneration. Particularly, tropical deforestation is of great concern worldwide for its impact on biological diversity and biochemical cycles, especially the global C cycle, which is known to affect climatic changes [1]. Trees play major ecological and functional roles within ecosystems. Also, they are regarded as a source of cash, savings and assets to the rural poor, and can help to meet the growing global demand for timber and other forest products [2]. However, forest cover continues to decrease over the world, and native species are particularly endangered especially on tropical ecosystems. Fostering of reforestation, formation of riparian woodlands and agroforestry programs have been undertaken to reverse this trend, especially in arid areas and deforested lands that have poor natural forest regeneration. Tree seedling mortality and development during the early growth stage are major factors influencing forest dynamics. Thus, research must aim at understanding how and why tree seedling either grow or die [3].

Among tropical forests, Madagascar's natural forest contains a diverse and highly endemic flora and fauna [4]. Composed by rainforest, and dry and spiny forests, Malagasy native forest cover was about 9.4 hectares in 2005, all of which was considered as highly exploited and endangered [5]. The recruitment processes are poorly known within these native forests, and they constitute a serious gap in our understanding of forest recovery processes and forest regeneration and conservation. Knowledge of seedling development and plant coexistence are not only important for our understanding of the forest recovery processes, but are also required for increasing the success and efficiency of restoration practices, and the performance of afforestation. In this way, it has been demonstrated that mycorrhizal fungi, an ubiquitous component of most ecosystems throughout the world, are an ecological key in improving seedling dynamics (development and mortality rate) by governing the cycle of major plant nutrients and by mitigating the attack of plant pathogens [6, 7, 8, 9].

Mycorrhizas constitute an important root symbiosis for approximately 92% of plant families and offer the potential to make a significant contribution to natural regeneration of vegetation communities [10]. Two major forms of mycorrhizas are usually recognized: the arbuscular mycorrhizas (AMs) and the ectomycorrhizas (ECMs). Arbuscular mycorrhizal symbiosis is the most widespread mycorrhizal association type and is fundamental in optimizing plant fitness and soil quality [11]. Particularly, the AM symbioses improve the resilience of natural plant communities against environmental stresses [12]. Some studies in Africa and in Central America have shown that most plant species found in rainforests are endomycorrhizal [13].

However, the impact of AM fungi on growth of individual plant species varies depending on the AM fungal taxa involved [7, 14]. The ectomycorrhizal symbiosis is phylogenetically restricted, and has evolved separately in several lineages of land plants [10]. ECMs are clearly younger than the ancient AMs, and occur in the forests of cool-temperate and boreal latitudes [11]. They also occur in an ecologically and economically important minority of tropical tree species belonging to the families and sub-families of Fagaceae, Caesalpinioideae, Betulaceae, Diptericarpaceae, Leptospermoideae in the Myrtaceae, Phyllanthaceae, Gnetaceae, Sapotaceae, Papilionoideae, Proteaceae, Asteropeiaceae, Sarcolaenaceae, Casuarianaceae and Acacieae [15, 16, 17, 18, 19, 20, 21].

For many decades, the importance of mycorrhizal fungi to terrestrial ecosystems has been recognized, and their potential use in forestry has been explored. In Madagascarian forest ecosystems, it has been illustrated that some endemic trees are associated with a high diversity of ectomycorrhizal fungi [22, 23]. Also, a large part of endemic forest trees have evolved with, at least, one type of mycorrhizal structure [24]. In this chapter, we address the implications of mycorrhiza on the early growth of some Madagascarian, endemic tree seedlings with emphasis on the importance of mycorrhizal fungi diversity and some pioneer plant species.

Mycorrhizal Status Description of Native Forest Tree Species in the Central and Eastern Parts of Madagascar

Although mycorrhizal structures are dominant within native tree and shrub species in the natural forests of Madagascar, little is known about the importance of these symbiotic structures on the regeneration strategies of forests or the ecological restoration of perturbed areas. The mycorrhizal status of dominant shrub and tree species within three Malagasy natural forest formations is indicated in Table 1. Surveyed sites were located along the eastern (Analalava and Ianjomara forest) and the central (Sclerophyllous forest of Arivonimamo) part of Madagascar. Analalava and Ianjomara forests are situated in well-preserved stands of coastal tropical rainforests. They are characterized by a high diversity of endemic trees. The sclerophyllous forest of Arivonimamo is mainly formed by a population of *Uapaca bojeri* with some shrub species of Sarcolaenaceae and Asteropeiaceae, two botanical families endemic to Madagascar.

Table 1. Mycorrhizal status of dominant shrub and tree species from Analalava, Ianjomara and Arivonimamo forests in the eastern and central parts of Madagascar

Genus/species[1]	Family	Sites[2]	Mycorrhizal status[3]
Amyrea sp. (?)	Euphorbiaceae	Ana.	AM
Anthostema madagascariense Baill. (E)	Euphorbiaceae	Ana.	AM
Breonia havilandiana Homolle (?)	Rubiaceae	Ana.	AM
Canarium madagascariense Engl. (E)	Burseraceae	Ana.	AM
Casearia nigrescens Tul. (E)	Salicaceae	Ana.	AM
Cynometra capuronii Du Puy et R. Rabev. (E)	Fabaceae	Ana.	AM
Clitoria lasciva Bojer ex Benth. (E)	Fabaceae	Ana.	AM
Colubrina sp. (?)	Rhamnaceae	Ana.	AM
Conchopetalum madagascariense Radlk. (E)	Sapindaceae	Ana.	AM
Croton lepidotus Aug. DC. (E)	Euphorbiaceae	Ana.	AM+
Cryptocarya acuminata Schinz (?)	Lauraceae	Ana.	AM
Dicoryphe sp. (?)	Hamamelidaceae	Ana.	AM
Dillenia triquetra (Rottb.) Gilg (?)	Dilleniaceae	Ana.	AM
Diospyros bernieri Hiern (?)	Ebenaceae	Ana.	AM+
Diospyros sp. (?)	Ebenaceae	Ana.	AM+
Dracaena reflexa Lam. (n)	Asparagaceae	Ana.	AM+
Dypsis sp. (?)	Arecaceae	Ana.	AM
Ellipanthus madagascariensis (G. Schellenb.) Capuron ex Keraudren (E)	Connaraceae	Ana.	AM
Erythroxylum sp. (?)	Erythroxylaceae	Ana.	AM+
Eugenia louvelii H. Perrier (?)	Myrtaceae	Ana.	AM+
Fernelia sp. (?)	Rubiaceae	Ana.	AM
Ficus cocculifolia Baker (n)	Moraceae	Ana.	AM+
Ficus lutea Vahl. (n)	Moraceae	Ana.	AM+
Gaertnera macrostipula Baker (?)	Rubiaceae	Ana.	AM
Harungana madagascariensis Lam. ex Poir. (?)	Hypericaceae	Ana.	AM+
Homalium involucratum (DC.) O. Hoffm. (E)	Salicaceae	Ana.	AM
Landolphia nitens Lassia (E)	Apocynaceae	Ana.	(AM)
Leptolaena multiflora Thouars (E)	Sarcolaenaceae	Ana.	AM&ECM
Macaranga cuspidata Boivin ex Baill (?)	Euphorbiaceae	Ana.	AM+
Macphersonia madagascariensis Blume (E)	Sapindaceae	Ana.	AM
Malleastrum minutifoliolatum J.-F. Leroy (E)	Meliaceae	Ana.	AM
Mascarenhasia arborescens A. DC. (n)	Apocynaceae	Ana.	NM
Memecylon xiphophyllum R. D. Stone (?)	Memecylaceae	Ana.	AM

Genus/species[1]	Family	Sites[2]	Mycorrhizal status[3]
Nesogordonia macrophylla Arènes (E)	Malvaceae	Ana.	AM
Paropsia madagascariensis (Mast.) H. Perrier (E)	Passifloraceae	Ana.	AM
Psiadia sp. (?)	Asteraceae	Ana.	AM+
Psidium cattleianum Sabine (n)	Myrtaceae	Ana.	(AM)
Psorospermum lanceolatum (Choisy) Hochr. (E)	Hypericaceae	Ana.	AM
Ravenala madagascariensis Sonn. (E)	Strelitzaceae	Ana.	(AM)
Ravenea julietiae Beentje (E)	Arecaceae	Ana.	AM
Rhodocolea racemosa (Lam.) H. Perrier (E)	Bignoniaceae	Ana.	AM
Rhopalocarpus thouarsianus Baill. (E)	Sphaerosepalaceae	Ana.	AM
Saldinia proboscidea Hochr. (E)	Rubiaceae	Ana.	AM
Suregada boiviniana Baill. (?)	Euphorbiaceae	Ana.	AM
Symphonia tanalensis Jum. & H. Perrier (E)	Clusiaceae	Ana.	AM
Syzygium emirnense (Baker) Labat & G. E. Schatz (?)	Myrtaceae	Ana.	AM
Tabernaemontana coffeoides Bojer ex A. DC. (n)	Apocynaceae	Ana.	AM
Tambourissa purpurea (Tul.) A. DC. (E)	Monimiaceae	Ana.	AM
Tina fulvinervis Radlk. (E)	Sapindaceae	Ana.	AM
Uapaca louvelii Denis (E)	Euphorbiaceae	Ana.	AM&ECM
Vepris sp. (?)	Rutaceae	Ana.	AM
Zanthoxylum tsihanimposa H. Perrier (E)	Rutaceae	Ana.	AM
Aphloia theiformis (Vahl) Benn. (n)	Aphloiaceae	Ian.	AM+
Aristida similis Steud. (?)	Poaceae	Ian.	AM+
Burasaia madagascariensis DC. (E)	Menispermaceae	Ian.	AM
Cinnamumum camphoratum Blume (n)	Lauraceae	Ian.	AM+
Cinnamomum zeylanicum Blume (n)	Lauraceae	Ian.	AM+
Clidemia hirta (L.) D. Don (n)	Melastomataceae	Ian.	(AM)
Colubrina decipiens (Baill.) Capuron (n)	Rhamnaceae	Ian.	AM
Commelina sp. (?)	Commelicaceae	Ian.	AM+
Dactyloctenium sp (?)	Poaceae	Ian.	AM
Dalbergia madagascariensis Vatke (E)	Fabaceae	Ian.	AM+
Dombeya dolichophylla Arènes (?)	Malvaceae	Ian.	AM
Dracaena reflexa Lam.(n)	Asparagaceae	Ian.	(AM)
Dichapetalum leucosia (Spreng.) Engl. (E)	Dichapetalaceae	Ian.	AM
Dypsis sp. (?)	Arecaceae	Ian.	(AM)
Dypsis nodifera Mart. (E)	Arecaceae	Ian.	(AM)
Agelaea pentagyna (Lam.) Baill. (?)	Connaracea	Ian.	AM
Gaertnera macrostipula Baker (?)	Rubiaceae	Ian.	AM

Table 1. (Continued)

Genus/species[1]	Family	Sites[2]	Mycorrhizal status[3]
Gaertnera obovata Baker (?)	Rubiaceae	Ian.	AM
Grevillea banksii R. Br. (n)	Proteaceae	Ian.	(AM)
Harungana madagascariensis Lam. Ex Poir.	Hypericaceae	Ian.	AM+
Hugonia sp (?)	Linaceae	Ian.	AM
Landolphia myrtifolia (Poir.) Markgr. (E)	Apocynaceae	Ian.	AM
Landolphia sp. (?)	Apocynaceae	Ian.	NM
Landolphia gummifera (Poir.) K. Schum. (E)	Apocynaceae	Ian.	(AM)
Macaranga cuspidata Boivin ex Baill (?)	Euphorbiaceae	Ian.	AM+
Macarisia lanceolata Baill. (?)	Rhizophoraceae	Ian.	AM
Machaerina flexuosa (Boeckeler) J. Kern (?)	Cyperaceae	Ian.	AM
Macphersonia madagascariensis Blume (E)	Sapindaceae	Ian.	AM
Merremia tridentata (L.) Hallier f. (n)	Convolvulaceae	Ian.	AM
Noronhia emarginata (Lam.) Thouars (E)	Oleaceae	Ian.	AM+
Osmunda regalis L. (?)	Osmondaceae.	Ian.	AM
Ouratea sp. (?)	Ochnaceae	Ian.	AM
Panicum luridum Hack. (?)	Poaceae	Ian.	AM
Phyllanthus amarus Schumach. & Thonn. (n)	Phyllanthaceae	Ian.	(AM)
Poupartia chapelieri (Guillaumin) H. Perrier (E)	Anacardiaceae	Ian.	AM+
Psidium cattleianum Sabine (n)	Myrtaceae	Ian.	(AM)
Psorospermum fanerana Baker (E)	Clusiaceae	Ian.	AM
Ravenala madagascariensis Sonn. (E)	Strelitzaceae	Ian.	(AM)
Rubus sp. (?)	Rosaceae	Ian.	AM+
Sauvagesia erecta L. (n)	Ochnaceae	Ian.	AM
Scolopia maoulidae S. Hul, Labat & O. Pascal (?)	Salicaceae	Ian.	AM
Streblus dimepate (Bureau) C.C. Berg (?)	Moraceae	Ian.	AM
Symphonia fasciculata (Noronha ex Thouars) Vesque (E)	Clusiaceae	Ian.	AM
Tacca leontopetaloides (L.) Kuntze (?)	Discoreaceae	Ian.	AM
Trema orientalis (L.) Blume (?)	Cannabaceae	Ian.	AM+
Tristemma virusanum Juss. (n)	Melastomataceae	Ian.	AM
Trophis montana (Leandri) C.C. Berg (?)	Moraceae	Ian.	AM
Uapaca ferruginea Baill. (E)	Euphorbiaceae	Ian.	AM &ECM
Urena lobata L. (n)	Malvaceae	Ian.	AM
Voacanga thouarsii Roem. & Schult. (n)	Apocynaceae	Ian.	NM
Uapaca bojeri L. (E)	Euphorbiaceae	Ariv	AM&ECM
Leptolaena bojeriana (E)	Sarcolaenaceae	Ariv	AM&ECM

Genus/species[1]	Family	Sites[2]	Mycorrhizal status[3]
Trema sp (n)	Ulmaceae	Ariv	AM
Aphloia theaeformis (Vahl.) Benn. (n)	Flacourtiaceae	Ariv	AM+
Rhus taratana (Baker.) H. Perrier (n)	Anacardiaceae	Ariv	AM+
Helychrysum rusillonii Hochr. (?)	Asteraceae	Ariv	AM+
Psiadia altissima (D.C.) Drake. (?)	Asteraceae	Ariv	AM+
Rubus apetalus Poir. (n)	Rosaceae	Ariv	AM

[1]Plant species: following the genus, species, and authority names, available data on endemicity are indicated: (**E**): endemic, (**n**): nonendemic, (**?**): not fully established. www.mobot.org/phillipson/catalogue/catalogue.htm

[2]Collection sites: **Ana** : Analalava, **Ian.** Ianjomara, **Ariv**: Arivonimamo

[3]Mycorrhizal status: **AM**: arbuscular mycorrhiza, (**AM**): lightly infected, **AM+**: heavily infected, **AM&ECM**: co-existence of arbuscular mycorrhizas and ectomycorrhizas, **NM**: nonmycorrhizal

Mycorrhizal results were obtained by examining 30 randomly chosen root fragments of 10 mm length each using a light microscope, for each plant species. Roots were considered AM when intracellular arbuscules and/or hyphal coils and/or vesicles were observed. The degree of AM infection was assessed according to four classes:

(i) nonmycorrhizal (termed "NM"), when no fragments presented any trace of AM infection;

(ii) lightly infected [termed "(AM)"], when only one to three fragments presented AM intracellular structures;

(iii) AM infected (termed "AM"), when four to 29 fragments presented AM structures, and

(iv) heavily infected (termed "AM+"), when all 30 fragments presented abundant AM structures.

Only four tree species (*Mascarenhasia arborescens* and *Tabernaemontana coffeoïdes* in Analalava forest, *Landolphia sp* and *Voacanga thouarsii* in Ianjomara forest) were identified as nonmycorrhizal. These four nonmycorrhizal tree species belong to the botanical family of Apocynaceae. Among the 111 study plant species, 12 were lightly infected; 62 species presented typical, well-developed AM infections; 27 species were heavily infected, and 6 species were found with both AM and ECM. In the family of the Sarcolaenaceae, all the examined species had both ECM and AM. These results illustrated the massive occurrence of mycorrhizal structures within the Malagasy flora, particularly within the endemic flora. All 42 endemic species presented mycorrhizal structures. Moreover, results of table 1 showed that more than 95% of the examined species in the three different forest ecosystems were associated with mycorrhizal fungi.

Importance of Mycorrhizal Symbionts on Seedling Development under Controlled Conditions

In addition to the high diversity observed within the flora of Madagascar, Malagasy natural forests are well known by their high rate of endemicity [25]. However these native tree species which have economical and ecological value were rarely used by the national program of reforestation. This is because of the little knowledge on the conditions of early development of their seedlings. The success of an outplanted nursery –grown tree seedlings depends on their ability to rapidly access nutrients and water held within the soil matrix [26]. In nature, this process is enhanced by the formation of symbiotic mycorrhizal associations. However, on many disturbed sites (e.g., mine spoils or abandoned agricultural lands), suitable mycorrhizal fungi are lacking, and this might limit seedling establishment and growth [27]. In this part of the chapter, we describe research activities relative to the effect of soil symbiotic microorganisms, especially of mycorrhizal fungi on seedling development of Malagasy native tree species. These activities affected particularly some forest tree species for which socio-economical and/or ecological values have already been illustrated.

Effects of Arbuscular Mycorrhizal Native Strains on Seedling Development of *Adansonia za* (Jum & H. Perrier) H. Perrier

Among eight species of *Adansonia* (baobab) all over the world, six species (*A. grandidieri, A. madagascariensis, A. perrieri, Rubrostipa, A. suarezensis, A. za*) are endemic to Madagascar. Another species (*A. digitata*) develops in the western, central and Eastern part of Africa, and the last species (*A. gobossa*) is endemic to North-western Australia. Depending on the species, baobabs develop in a wide range of ecosystems, including arid zones and savannahs, as well as dry and wet forests.

Adansonia za constitutes a well known Baobab in the western part of Madagascar because of its different use in everyday life of Malagasy people in this region of the island. However, ecosystems of *A. za* have been highly disturbed by deforestation. Large parts of these ecosystems have been transformed to agriculture lands, especially to rice lands, which really threatens the population of that tree species. Moreover, seedlings of *A. za* have been rarely observed within these ecosystems, where baobab's populations are particularly constituted by adult trees. This species of baobab belongs to the *Longitubae* section which makes their seeds with water-impermeable coats [28]. Thus, severe treatments are needed to remove the physical dormancy to allow seed germination.

Controlled mycorhization of *A. za* was undertaken by Razafimiaramanana in 2010 by using *Glomus intraradices* as a reference mycorrhizal strain, and three native strains of arbuscular mycorrhizas (*Glomus* sp., *Scutellospora* sp. and *Entrophospora* sp.) [29]. They were isolated from a baobab ecosystem of Kirindy forest in the western part of Madagascar. After 6 months of culturing under greenhouse conditions, the native strain *Glomus* sp. stimulated the development of *A. za* seedlings more than the other strains did (Table 2). Compared to the control, shoot growth of plants inoculated with *Glomus* sp., *Glomus intraradices, Scutellospora* sp. or *Entrophospora* sp. was stimulated 4.6, 3.7, 1.9 or 2.4 times, respectively. Shoot and root dry weights of all inoculated plants were significantly higher than values in the control treatment. These results showed a high degree of mycorrhizal dependency of *A. za* seedlings, and particularly the importance of native strains on the development of mycorrhiza on this plant. Thus, the establishment program of *A. za* seedlings in these original areas and/or in others degraded soils requires a preliminary management of soil mycorrhizal communities. Under natural conditions, the germination of baobab seeds constitutes a limiting factor to plant regeneration [30]. In this case, the development of regeneration or multiplication technologies is an important option to increase seedling performance of Baobab, and to preserve this genetic resource of great economic and medicinal value.

Table 2. Shoot and root growth, mycorrhizal dependency and mycorrhizal development and of *A. za* seedlings after 6 months inoculation with *G. intraradices* or native mycorrhizal strains in pot cultures

	Treatments				
	C*	GI	GL	SC	EN
Shoot biomass (g dry weight plant^{-1})	0.19a**	0.71c	0.88d	0.37b	0.46b
Root biomass (g dry weight plant^{-1})	0.32a	0.95c	1.11d	0.59b	0.65b
Mycorrhizal dependency (%)	0a	72.6d	77.7d	48.4b	58.2c
AM colonization (%)	0a	66.4d	73.19e	37.59b	50.2c

*C: Control; GI: Glomus intraradices; GL: Glomus sp.; SC: Scutellospora sp.; EN: Entrophospora sp.
** Data in the same row followed by the same letter are not significantly different (p>0.05) after one-way analysis of variance.

Effects of Arbuscular Mycorrhiza Native Strains on Seedling Development of *Dalbergia trichocarpa* Baker

Malagasy species of *Dalbergia* are characterized by an undeniable wood quality. As a result, they have a great socio-economical, environmental or commercial value all over the world. Among the 125 described species of *Dalbergia*, 42 out of the 48 found in Madagascar are endemic [31]. A large part of these endemic tree species is scarce due to its overexploitation in many natural forest regions of Madagascar. As an example, 52,000 tones of wood from 100,000 individual trees of rosewood (*Dalbergia* spp.) and ebony trees were logged in north-east of Madagascar [32]. During the last decade, illegal logging and export of rosewood was undertaken even within protected areas [33, 34]. In this situation, efforts should focus in forest preservation, and if possible, in increasing the population of these valuable forest tree species.

The potentiality of the plant-soil-microorganism association was explored to optimize both growth and regeneration of the endemic species of *Dalbergia* [35]. These studies illustrated that all 8 studied species formed symbiosis structures with nitrogen-fixing bacteria. Since then, little information was available related to the importance of soil microorganisms on the growth stimulation of *Dalbergia* seedlings. Recently, the presence of arbuscular and vesicular mycorrhizal structures was reported on the root systems of the two endemic species of *Dalbergia* (*Dalbergia maritima* R. Vig) [24]; *Dalbergia trichocarpa* Baker [36]. Then, the first study was conducted exploring the importance of both arbuscular and vesicular mycorrhizas and nitrogen-fixing bacteria on the growth of *D. trichocarpa* [37]. Three strains of arbuscular and vesicular mycorrhizas were used including two native strains (*Glomus* sp1-ME and *Glomus* sp2-ME; isolated from undisturbed stand of *D. trichocarpa*) and one exotic strain of *Glomus* [(*Glomus intraradices*; provided by the Laboratoire Commun de Microbiologie (IRD/UCAD/ISRA) Dakar-Senegal]. A strain of nitrogen-fixing bacteria [from the strain collection of the Laboratory of Environmental Microbiology (CNRE), Antananarivo-Madagascar] was co-inoculated with a single or a multiple strain of arbuscular and vesicular mycorrhizas. This strain of nitrogen-fixing bacteria was isolated from the root system of *D. trichocarpa* collected in an undisturbed stand of this tree.

The results of these experiments illustrated the great importance of native mycorrhiza strains on the development of *D. trichocarpa* seedlings (Table 3). Compared to the control, the total root and shoot growth of seedlings were stimulated 3.5 or 5.8 times, respectively, after inoculation with the nitrogen-fixing bacteria STM 609 and *Glomus sp1*-ME or *Glomus sp2*-ME. At the same time, total growth of roots and shoots was 2.9 times on plants inoculated by the exotic strain *Glomus intraradices*. Shoot and root dry weights of all inoculated plants were significantly higher than valves in the control treatment (with single or multiple strains of arbuscular and vesicular mycorrhizas). Shoot and root development of seedlings was stimulated more in the multiple strain of arbuscular and vesicular mycorrhiza than in control or a single strain of arbuscular and vesicular mycorrhiza treatments. For these treatments, the importance of native strains on the stimulation of seedling development was illustrated. Indeed, the high levels of shoot and root developments were observed on plants inoculated by the two native strains of arbuscular and vesicular mycorrhizas with or without the exotic strain of this group of mycorrhizas (Table 3). Similar results were observed on each plant for the mycorrhizal and nodule developments in the root system. The highest levels of mycorrhizal colonization, mycorrhizal dependency and nodule number were registered on plants inoculated by the multiple strains of arbuscular and vesicular mycorrhizas and the nitrogen-fixing bacteria strain.

Effects of Ectomycorrhizal Symbionts Diversity on Seedling Development of *Intsia bijuga* (Colebr.) O. Kuntze

Intsia bijuga is found in its native range of Madagascar, the Seychelles, Indonesia, Malaysia, Thailand, Philippines, Papua New Guinea and Australia. This is in addition to its primary distribution in the western Pacific and Indo-Malaysian regions, from New Guinea and Palau in the west to Fiji, Tonga and Samoa in the Southeast, and to the Mariana Caroline and Marshall Islands in the north and northeast in the Pacific. A spreading tree of up to 40 m tall, *I. bijuga* is undoubtedly one of the most highly valuable trees in these regions, both in terms of its traditional cultural and commercial timber values. In Madagascar, *I. bijuga* occurs frequently in the eastern coastal rainforest, in primary or old secondary forests, and in open forests from 0 to 800 m.a.s.l. Trees of *I. bijuga* are in very high demand and permanently decreasing in abundance because of their overexploitation for house posts, canoe making and due to its indiscriminate modern commercial logging.

Table 3. Shoot and root growth, mycorrhizal and nodule development and mycorrhizal dependency of *D. trichocarpa* seedlings inoculated with the nitrogen-fixing bacteria STM 609 and a single or a multiple strain of arbuscular and vesicular mycorrhizas in sterilized soil after 4 months culturing

Treatments	Number of nodule plant^{-1} (STM 609)	Mycorrhizal colonization (%)	Mycorrhizal dependency (%)	Shoot biomass (g dry weight plant^{-1})	Root biomass (g dry weight plant^{-1})
T*	0 a**	0 a	0 a	0.127 a	0.070 a
A	61 c	47.59 c	29.2 d	0.419 c	0.286 c
B	43 b	34 b	23.2 b	0.359 b	0.229 b
C	85 e	68.19 e	59.5 f	0.722 e	0.422 e
A+B	46 b	36.79 b	26.1 c	0.388 b	0.234 b
A+C	91 f	73.40 f	74.6 g	0.873 f	0.486 f
B+C	66 d	54.20 d	39.6 e	0.523 d	0.393 d
A+B+C	112 g	88.80 g	84 h	0.967 g	0.685 g

*T: Control; A: Glomus sp1-ME; B: Glomus intraradices; C: Glomus sp1-ME

**Data in the same column followed by the same letter are not significantly different (p>0.05) following one-way analysis of variance.

Belonging to the family of the Fabaceae, subfamily Caesalpinoideae, *I. bijuga* is not a nodulated tree, and it has been found forming exclusively ectomycorrhizas [24, 38]. There is no evidence to date that this tree species associates with vesicular-arbuscular mycorrhiza fungi [11]. *Intsia bijuga* associates with a few groups of ectomycorrhiza fungi [39] despite the exceptional diversity of the ectomycorrhizas fungi associated with native or endemic trees of Madagascar [22, 40]. In natural stands of the Seychelles, only Tedersoo et al (2007) identified 15 species of ectomycorrhiza fungi associated with *I. bijuga* by using DNA sequencing of mycorrhizal root tips [38].

In Madagascar, mycorrhizal inoculation of *I. bijuga* seedlings was initiated by Rakotoarimanga in 2010 by using single or multiple strains of ectomycorrhizal fungi [36]. Four strains of ectomycorrhizas fungi were used in their studies. Two strains of *Scleroderma* (SC02-ME and SC03-ME) were isolated from two fruiting bodies that were collected under (1) *Uapaca bojeri* within the sclerophyllous forest of the Madagascarian highland, and (2) an *Intsia bijuga* stand in the eastern littoral forest of Madagascar, respectively. One strain of *Pisolithus* (*Pisolithus* sp. Pis02-ME) was isolated from a sporophore collected under *Pinus* and *Eucalyptus* plantations in the central highland of Madagascar. The last isolated strain was a species of *Boletus* (*Boletus sp* BO01-ME), obtained from a sporophore collected under *I. bijuga* in the eastern rainforest of Madagascar. After 4 months of culturing in pots, the effects of each inoculation treatment on seedling growth and mycorrhizal development were as shown in Table 4. Compared to the control, a significant development of shoot seedling biomass was found on all treatments with *Pisolithus* sp. Pis02-ME on single and multiple treatments. However, no significant root development was found between the control and all treatments. For the mycorrhizal dependency and ectomycorrhizal colonization, each type of inoculation (single or multiple) had variable effects depending on the strain used. Generally, high levels of ectomycorrhizal colonization were observed on treatments with Pis02-ME, except on single inoculation with SC03-ME. For this last treatment, no effect of high levels of ectomycorrhizal colonization was recorded on seedling growth (shoot and root biomass). These results illustrated that ectomycorrhizal symbionts associated to exotic trees were able to stimulate the development of *I. bijuga* seedlings.

Effects of Dual Mycorrhization (Endo and Ectomycorrhization) on Seedling Development of *Uapaca bojeri* L. (Euphorbiaceae)

Some plant species such as *Uapaca bojeri* [22] may contain the two forms of mycorrhizal symbiosis (endomycorrhizae and ectomycorrhizae), in their root system. The importance of each association depends on the developmental stage of the plant [41]. In general, endomycorrhizal (AM) fungi colonize seedlings initially, and then are replaced by ectomycorrhizas through a process of competition after a few months [42].

A native tree species, *Uapaca bojeri*, of the sclerophyllous forest in Madagascar, is highly dependent on both types of mycorrhiza (Table 5).

Table 4. Shoot and root growth, mycorrhizal dependency and mycorrhizal development of *I. bijuga* seedlings after 4 months of culturing and inoculation by a single or multiple ectomycorrhizal strains in pot culture

Treatments	C*	SC02	SC03	Pis02	BO01	SC02+SC03	SC02+SC03+Pis02	SC02+SC03+Pis02+BO01
Shoot biomass (g dry weight plant^{-1})	2.85a	3.12a	3.02a	4.39b	3.01a	3.07a	4.50b	4.13b
Root biomass (g dry weight plant^{-1})	0.82a	0.91a	0.77a	1.18b	0.83a	0.82a	0.94a	0.85a
Mycorrhizal dependency (%)	-	8.02a	1.48a	33.2b	0.91a	5.67a	31.67b	4.21a
Ectomycorrhizal colonization (%)	0.00a	67e	20.6cd	24.6d	7.50b	9.20b	19.09c	18c

*C: Control; SC02: *Scleroderma sp* SC 02-ME ; SC03: *Scleroderma sp* SC 01-ME ; Pis02: *Pisolithus sp* Pis 02-ME ; BO01: *Boletus sp* BO01-ME

** Data in the same row followed by the same letter are not significantly different after a one-way analysis of variance (p>0,05).

A high occupancy of AM fungi appeared first on young seedlings (3-month-old roots) followed by ECM colonization (Figure 1) [22]. Chen et al. (2000) [43] described, after studying *Eucalyptus urophylla* growth that these fungi interact mainly during the first four months of plant growth. AM species colonized first and had little effect on ECM colonization. The succession of these two types of mycorrhizas did not compromise plant development. This was because the greatest growth response was seen on plants colonized by both types of mycorrhiza [41, 44, 45].

Table 5. Shoot growth, mycorrhizal development, and relative mycorrhizal dependency of *U. bojeri* seedlings 5 months after *G. intraradices* and/or *Scleroderma sp* SC 02-ME inoculation insterilized soil

Treatments	Shoot biomass (mg plant⁻¹)	Ectomycorrhizal colonization (%)	Arbuscular colonization (%)	RMD* (%)
Control	91.1 a	0a	0a	-
Scleroderma sp SC 02-ME	181.2 b	8.7b	0a	47.6a
G. intraradices	160.1b	0a	77.5b	42.7a
Scleroderma sp SC 02-ME + *G. intraradices*	360.3c	11.5b	82.5b	70.7b

*RMD: Relative mycorrhizal dependency
**Data in the same column followed by the same letter are not significantly different (p<0,05) after a one-way analysis of variance

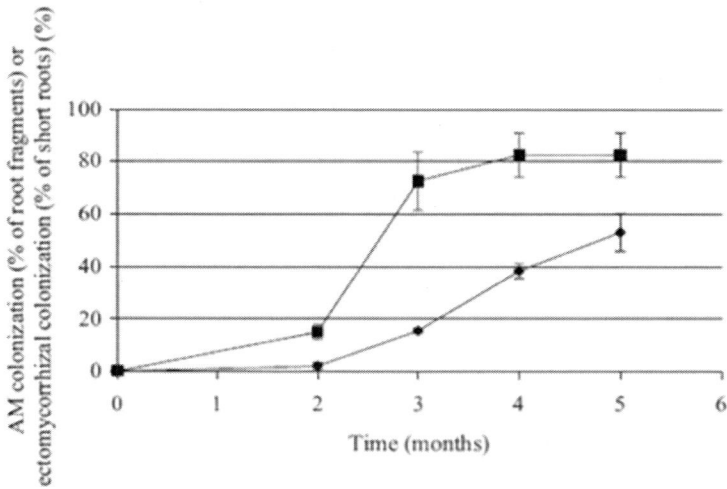

Figure 1. Sequence of mycorrhizal colonization on *U. bojeri* seedlings (■: AM colonization; ♦: total ectomycorrhizal colonization).

Indeed, positive effects of the dual inoculation were shown for seedling growth and root mycorrhizal colonization of *Uapaca bojeri* (Table 5) in comparisson to the non-inoculated control treatment under greenhouse conditions. This co-occurrence of AM with ECM in the same root system might determine the success of plant species to colonize a wide range of habitats and allow plant establishment (i.e. forest restoration) on degraded areas [46].

Nurse Plant Phenomenon and its Importance on Late Successional Plant Regeneration and on Forest Restoration

Following perturbation, it is well known that some plant species (e.g., pioneer or perennial plants) can associate with beneficial soil microorganisms which could have positive effects on late successional plant species [47, 48, 49].

Within two disturbed forest ecosystems of *Uapaca bojeri* (an endemic tree species with high socio-economical value), located at Arivonimamo (Region of Itasy) and Ambatofinandrahana (*Region of Amoron'I Mani*a) in the Central part of Madagascar, another kind of facilitation through shared mycorrhizal fungi was observed. It was first found that the degraded areas, previously occupied by *Uapaca bojeri*, were colonized by shrub species which in most cases were associated with mycorrhizal fungi (Table 6). Some of these shrub species were associated with endo- and ectomycorrhizal fungi like *U. bojeri* as it was described by Baohanta (2011) [50]. This characteristic might help to explain their ability to establish on poor soils. This is because of the improvement of water and mineral acquisition and plant protection throughout the mycorrhizal symbiosis [11, 51, 52].

Table 6. Mycorrhizal status of pioneer shrub species within the degraded area of two study sites

Plant species [1]	Family	Mycorrhizal status [2]
Leptolaena pauciflora Baker. (E)	Sarcolaenaceae	ECM & MVA
Leptolaena bojeriana (Baill.) Cavaco. (E)	Sarcolaenaceae	ECM & MVA
Sarcolaena oblongifolia Cavaco. (E)	Sarcolaenaceae	ECM
Trema sp. (n)	Ulmaceae	MVA
Vaccinium emirnense Hook. (n)	Ericaceae	Endo
Aphloia theaeformis (Vahl.) Benn. (n)	Flacourtiaceae	MVA
Rhus taratana (Baker.) H. Perrier (n)	Anacardiaceae	MVA
Helychrysum rusillonii Hochr. (?)	Asteraceae	MVA
Psiadia altissima (D.C.) Drake. (?)	Asteraceae	MVA
Rubus apetalus Poir. (n)	Rosaceae	MVA
Erica sp. (n)	Ericaceae	Endo

[1]Plant species: following the genus, species and authority names; available data on endemicity are indicated: (**E**): endemic at the genus level; (**n**): nonendemic; (**?**): not fully established (http://www.mobot.org/phillipson/ catalogue/catalogue.htm).

[2]Mycorrhizal status: AM arbuscular mycorrhiza; ECM, ectomycorrhiza; AM&ECM, co-existence of arbuscular mycorrhiza and ectomycorrhiza; Endo, endomycorrhizal.

Pioneer species, which often reflect the stage of degradation of forest soils, are among the most studied "nurse plants". Many studies have been conducted to determine their impact on soil biological and chemical functioning, and on plant succession [49, 53, 54]. In arid ecosystems, seedling establishment and survival have been greater underneath the canopies of shrubs than in the open interspaces [55]. The ability of such species to persist or to re-establish on disturbed sites might also allow the survival of mycorrhizal fungi propagules in the soil, even though woody mycorrhizal host plants are absent. In turn, the presence of established mycorrhizal fungi in the soils may facilitate the establishment or the re-establishment of mycorrhizal tree seedlings following disturbance [56, 57, 58, 59], potentially contributing to plant succession. As a result, nurse plants might be able to i) resist various environmental stresses, ii) create microclimates or "fertile microclimates" that could facilitate the establishment of other species, iii) be less competitive compared to the target species [60].

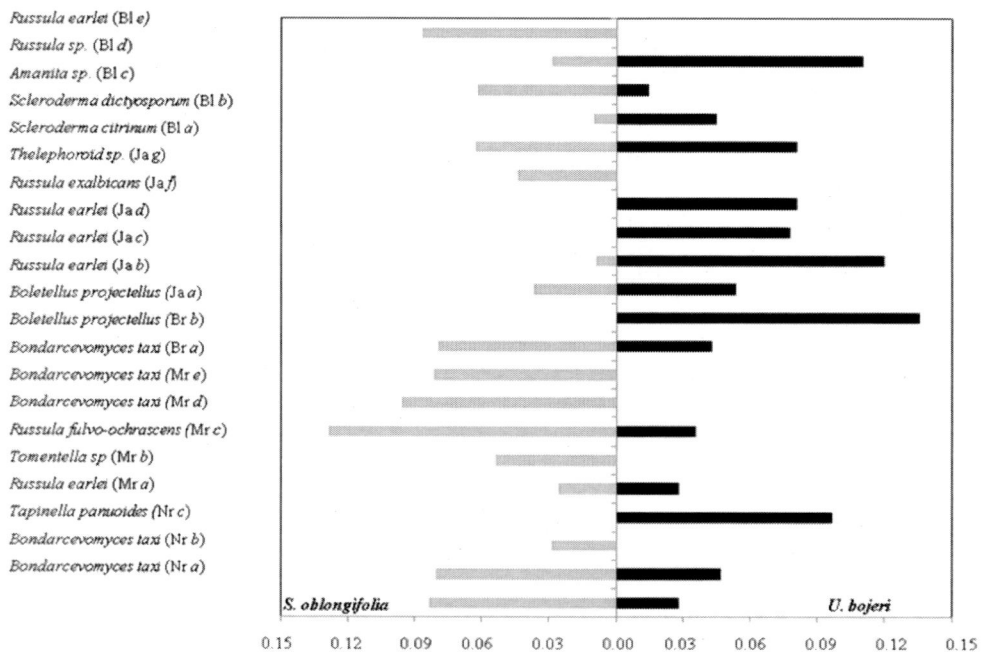

Figure 2. Relative frequency of identified RFLP types based on ITS region sequences on roots of *Uapaca bojeri* and *Sarcolaena oblongifolia*. Relative frequency was calculated as the number of occurrences of each RFLP type divided by the total number of occurrences of all RFLP types.

Shared mycorrhizal symbionts between two plant species within the same environment and belonging to the same genus, family or different families is an important positive interaction [61, 62, 63]. This kind of association was observed between the two shrubs species, *Leptolaena bojeriana* or *Sarcolaena oblongifolia*, and *Uapaca bojeri* (the native tree species) within the two study sites. Indeed, some ectomycorrhizal species were associated with both shrub species and with *Uapaca bojeri*. This was after the comparison of RFLP-type of ectomycorrhizas collected from harvested *Sarcolaena oblongifolia* roots with those associated with *Uapaca bojeri* by using restriction fragment length polymorphism (RFLP) (Figure 2).

Table 7. Effect of *L. bojeriana* / *U. bojeri* succession and dual-cultivation of *L. bojeriana* / *U. bojeri* seedlings on growth and ectomycorrhizal colonization of *U. bojeri*

Treatments	*U. bojeri* Shoot biomass (Dry weight in g)	*U. bojeri* ECM Colonization [4] (%)
Bulk soil		
Control [1]	0.1205 [a] ±0.03[5]	29.33 [a] ±9.61
L. bojeriana [2]	0.2769 [b] ±0.02	30.33 [a] ±4.16
L. bojeriana WA [3]	0.3085 [b] ±0.05	65.33 [b] ±2.52
Pinus patula soil		
Control	0.0855 [a] ±0.02	16.33 [a] ±4.16
L. bojeriana	0.2327 [b] ±0.02	65.33 [b] ±5.69
L. bojeriana WA	0.3331 [b] ±0.11	79.33 [c] ±7.02
Eucalyptus sp. soil		
Treatments	*U. bojeri* Shoot biomass (Dry weight in g)	*U. bojeri* ECM Colonization [4] (%)
Control	0.0832 [a] ±0.02	36.00[a] ±3.61
L. bojeriana	0.2331 [b] ±0.07	42.00[a] ±10.39
L. bojeriana WA	0.2501 [b] ±0.07	90.33[b] ±5.51

Data in the same column within each soil type followed by the same letter are not significantly different (p>0.05) according to the Newman-Keuls test
[1]*U. bojeri* without pre- or dual cultivation with *L. bojeriana*.
[2]*U. bojeri* with *L. bojeriana* seedlings without the aerial parts
[3]*U. bojeri* after dual-cultivation with *L. bojeriana* seedlings with aerial parts
[4]Root Ectomycorrhizal colonization (%)
[5] Standard error of the mean.

In a glasshouse study, *Uapaca bojeri* seedlings were grown near established *Leptolaena bojeriana* seedlings (dual cultivation) on soils collected either under exotic species (disturbed soil) or distant from any ectomycorrhizal host (bare soil). Results showed that the presence of the pioneer shrub species enhanced seedling development and root mycorrhizal colonization of the native species *Uapaca bojeri* in all soil samples, in comparison to the control without the shrubs species (Table 7). Increased mycorrhizal colonization of *Uapaca bojeri* seedlings near *Leptolaena bojeriana*, and the consequent increase in seedling nutrient uptake and growth potential, are the possible implications of inter-specific sharing of mycorrhizal fungi [64]. Indeed, sharing of mycorrhizal fungi may allow *U. bojeri* and *L. bojeriana* to form links into a common mycelial network, without initial constrains to establish mycorrhizal colonization [65]. This would also give seedlings a more rapid access to a potentially extensive, established mycelial network [58, 63]. It is also possible that nutrients may be transferred among plants *via* mycorrhizal linkages, fostering seedling development [11].

Facilitation Phenomenon for Native Tree Species Establishment: Are Exotic Plant Species Involved?

Most of the forest plantations in the world are carried out with exotic species [66] because of the lack of ecology and sylviculture knowledge of the native species. During 2000 to 2005, plantations in the world showed an expansion of approximately 2.8 million hectares per annum [67] due to the increasing demand for paper pulp, timber and fuelwood [68, 69, 70]. Other objectives of these plantations were to reduce of the pressures on the natural forest ecosystems, and the need for sequestering carbon to meet obligations under the Kyoto Protocol. However, the invasion of exotic plant species constitutes a threat for conservation and restoration of the natural ecosystems [71, 72]. The beginning of the 90s was marked by a new trend, which regarded the forest plantations as a catalyst for regeneration of the native species [73, 74, 75].

In Madagascar, classified among the first ten countries of hot spot of biodiversity with a rate of very high endemisms [25], little importance was granted to exotic plant species. It was considered that these species had a strong capacity of adaptation to hard ecological conditions in comparison to the insular, fragile Malagasy flora [76, 77]. For a few years, this threat became a reality. Binggeli (2003) reported a list of 38 invading exotic species (*Opuntia* spp., *Psidium cattleianum*, *Grevillea banksii*) endangering the Malagasy flora [78]. During a few decades, Madagascar did not have a clear plantation policy [79], despite the advantages and roles of the plantations on the environment (e.g., protection against erosion, production of firewood and paper pulp) [80]. As a result, plantations account for only 2% of the forest cover in the island [81], and the majority of plantation forests are planted with *Pinus* and *Eucalyptus*.

Light in the understory is an important factor to forest regeneration [82]. Tree planting may facilitate the process of forest succession by providing a nurse effect to colonizing native species. Facilitation, the positive effect of plants on the establishment or growth of others, has long been recognized as an important driving force for secondary succession [83]. It was defined by van Andel (2006) as an interaction between individuals of different species, where one of species changes the environment in such a way that is beneficial to the other [84].

In the southern center of Madagascar (commune of Androy, located 400 km south of Antananarivo; 21°22'S, 47°18'E; 1100–1200 m.a.s.l.) the pine plantations (*Pinus patula*) are located near the forest corridor which connects the national parks of Ranomafana and Andringitra. In this context of vicinity of the natural forest and plantations, native species were regenerated in the plantations which underwent various disturbances (wood extraction, cyclones, fires, culture).

We assessed the diversity of naturally regenerated native species (trees, shrubs, herbs and lianas) in the disturbed, exotic tree plantations (*Pinus patula*). Transects were used with this purpose (40 transects, 205 plots of 10m x 10m). The following hypothesis was formulated: gaps in the plantation facilitate native species regeneration. Use of correspondence analysis (CA) allowed identification of three vegetation groups, which corresponded to various stages of succession (Figure 3): (i) herbaceous vegetation, (ii) mixed herbaceous-woody vegetation and (iii) woody vegetation (forest regrowth). Understory species richness (S), Shannon diversity index (H'), and woody density (D) were studied within 10 plots randomly selected per vegetation group.

One hundred and twenty five (125) species divided into 46 families were inventoried, including 34 endemic species. The most common plant families found under the plantation were Asteraceae (19 species), Poaceae (14 species) and Rubiaceae (14 species). By growth form, there were 58 tree (46%), 55 herb (44%) and 12 liana (10%) species. Mean values for stem density (D) and basal area were 6843 individuals per hectare and 6.29 m^2 ha^{-1}, respectively, in the woody vegetation (Table 8).

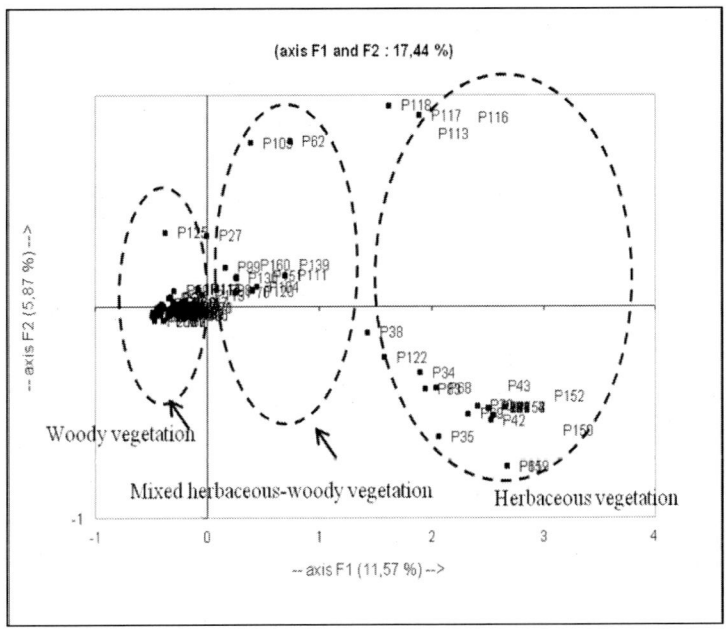

Figure 3. Correspondence analysis for all plots (based on presence/absence of species in the pine plantation, 125 species/205 plots).

Table 8. Mean values for vegetation group identified by COA of the pine plantation (standard errors in parenthesis), n: number of plots [87]

Floristic parameters	Vegetation group		
	Herbaceous vegetation n = 10	Mixed herbaceous-woody vegetation n = 10	Woody vegetation n = 10
Species richness S	5.6 (1.89) a*	29 (6.4) b	32.6 (9) b
Shannon H'	1.46 (0.23) a	3.31 (0.78) b	4.01 (0.67) b
Stem density D (No ha^{-1})	-	-	6843 (2276)
Basal area G (m^2 ha^{-1})	-	-	6.29 (4.62)

*Different letters within the same row indicate significant differences (p<0.05) following the Tukey HSD test).

Our results further provide information on the dynamic nature of vegetation. The first stages of succession are characterized by herbaceous vegetation which is replaced by mixed formations and finally by woody formations (forest regrowth). Floristic richness (S) and Shannon Wiener index (H') increased during succession: values were lower in the herbaceous vegetation (S<10 and 1.24<H'<1.46) than in the forest regrowth (27<S<33 and 3.95<H'<4.01).

Although monocultures are deemed to be "biological deserts" by some researchers [85], our results suggest that exotic plantations help to restore native species by stabilizing soil and creating favorable site conditions for plant recolonization. In our study site, the composition of the soil seed bank, and the availability of recent seed sources (forest corridor) in the vicinity of the plantation are important.

Pinus species are dependent on symbiosis to develop optimally under natural conditions [86]. Particularly, *Pinus patula* has the ability to symbiotically fix nitrogen with the help of certain species of actinomycetes. This relationship causes an increase in soil nitrogen content with time, and facilitates the regeneration of Malagasy native species. Likewise, exotic tree plantations potentially may greatly improve physical and biological site conditions catalyzing subsequent succession processes towards a natural forest [73].

Understanding the process of understory succession might contribute to conserve native biodiversity in Madagascar. The reproduction of the natural regeneration observed in the pine plantation can be used as a model to restore the degraded ecosystems of the region.

Conclusion

Results presented in this chapter show that mycorrhizal symbioses have a real potential to improve the performance of seedlings, especially of endemic trees, and could be used in afforestation programs or in ecological restoration processes of degraded areas in many forest ecosystems in Madagascar. Soil mycorrhizal communities can be managed by (i) using isolated strains in the framework of controlled mycorrhizal inoculation, or (ii) exploring the capacity of pioneer shrub species to stimulate the potentiality of residual mycorrhizal propagules that might facilitate the establishment of others tree seedlings. This second technology would allow to design multispecific reforestations or a two-phase reforestation strategy, mimicking the natural succession process, as soon as most shrub species are able to facilitate the early growth and survival of young forest tree seedlings. However, the development of these technologies is suggested from studies conducted under nursery and /or greenhouse conditions. Further experiments have to be carried out to test the positive effect of each technology both on a longer period of plantation and in an ecological restoration process under field conditions.

Management of native mycorrhizal strains proved to be more interesting than using introduced fungal strains to improve growth of Malagasy endemic tree seedlings. Moreover, use of sun-tolerant shrubs (which can have a positive effect on soil mycorrhizal communities) can be of great importance to the plantation program of endemic trees or to the forest ecosystem regeneration.

References

[1]   B. Bolin and R. B. Cook (Eds), *The major biochemical cycles and their interactions.* Scope 21, Wiley, Chichester (1983).

[2]   W. Bentley and M. Gowen (Eds), *Forest resources and wood-based biomass energy as rural development assets.* Winrock International and Oxford and IBH Publishing, New Delhi (1994).

[3]   A. Sims, A. Kiviste, M. Hordo, D. Laarman and K. V. Gadow, *Ann. Bot. Fen.* 46(4), 336–352 (2009).

[4]   S. M. Goodman and J. P. Benstead, *The natural history of Madagascar.* University of Chicago Press, Chicago (2003).

[5]   MEFT, USAID and CI. *Evolution de la couverture de forêts naturelles à Madagascar, 1990 – 2000 – 2005* (2009).

[6]   E. Seccon, S. Sanchez and J. Campo, *Plant Ecol.* 170, 277–285 (2004).

[7]   M. G. A. Van der Heijden, J. N. Klinomoros, M. Yrsic, P. Moutoglis, R. Streitwolf-Engel, T. Boller, A. Wiemken and I. R. Sanders, *Nature* 396, 69–72 (1998).

[8]   N. Requena, E. Perez-Solis, C. Azcon-Aguilar, P. Jeffries and J. M. Barea, *Appl. Environ. Microb.* 67, 495–498 (2001).

[9]   R. P. Schreiner, K. L. Mihara, K. L. McDaniel and G. J. Bethlenfalvay, *Plant Soil* 188, 199–209 (2003).

[10]   B. Wang and Y. L. Qiu, *Myvorrhiza* 16, 299–363 (2006).

[11]   S. E. Smith and D. J. Read, *Mycorrhizal symbiosis.* 3rd edition. Academic Press. Ltd. Cambridge (2008).

[12]   J. M. Barea, J. Palenzuela, P. Cornejo, I. Sanchez-Castro, C. Navarro-Fernandez, A. Lopéz-Garcia, B. Estrada, R. Azcon, N. Ferrol and C. Azcon-Aguilar, *J. Arid Environ.* 75, 1292–1301 (2011).

[13]   M. Härkönen, T. Saarimäki, L. Mwasumbi and T. Niemela, *Aquilo, Ser. Bot.* 31, 99-105 (1993).

[14]   T. R. Scheublin, R. S. P. Van Logtestijn and M. G. A. van der Heijden, *J. Ecol.* 95, 631–638. Available: 111/j.1365-2745.2007.01244.x (2007).

[15]   R. Watling and S. S. Lee, *J. Trop. For. Sci.* 7, 657–669 (1995).

[16]   M. Béreau, M. Gazel and J. Garbaye, *Can. J. Bot.* 75: 711–716 (1997).

[17]   B. Moyersoen, P. Becker and I. J. Alexander *New Phytol.* 150: 591–599 (2001).

[18]   T. W. Henkel, J. Terborght and R. Vilgalys, *Mycol. Res.* 106, 515–531 (2002).

[19]   Haug, M. Weiss, J. Homeier, F. Oberwinkler and I. Kottke, *New Phytol.* 165, 923–936 (2004).

[20]   Natarajan, G. Senthilarasu, V. Kumaresan and T. Rivière, *Curr. Sci.* 88, 1893–1895 (2005).

[21]   G. Peay, P. G. Kennedy, S. J. Davies, S. Tan and T. D. Bruns, *New Phytol.* 185, 529–542 (2010).

[22]   H. Ramanankierana, M. Ducousso, N. Rakotoarimanga, Y. Prin, J. Thioulouse, E. Randrianjohany, L. Ramaroson, M. Kisa, A. Galiana and R. Duponnois, *Mycorrhiza* 17, 195–208 (2007).

[23] Ducousso, G. Bena, C. Bourgeois, B. Buyck, M. Eyssartier, M. Vincelette, R. Rabevohitra, L. Randrihasipara, B. Dreyfus and Y. Prin, *Mol. Ecol.* 13, 231–236 (2004).

[24] Ducousso, H. Ramanankierana, R. Duponnois, R. Rabevohitra, L. Randrihasipara, M. Vincelette, B. Dreyfus and Y. Prin, *New Phytol.* 178, 233–238 (2008).

[25] R. A. Mittermeier, O. Langrand, P. P. Lowry II, G. Schatz, J. Gerlach, S. Goodman, M. Steininger, F. Hawkins, N. Raminosoa, O. Ramilijaona, L. Andriamaro, H. Randrianasolo, H. Rabarison and Z. L. Rakotobe, in: R. A. Mittermeier, P. R. Gil, M. Hoffman, J. Pilgrim, T. Brooks, C. G. Mittermeier, J. Lamoureux and G. A. B. Da Fonsesca (Eds), *CEMEX SA. De CV,* Mexico City, pp. 138-144 (2004).

[26] Dunabeitia, N. Rodriguez, I. Salcedo and E. Sarrionandia. *For. Ecol. Manage.* 195, 129–139 (2004).

[27] U. Ortega, M. Dunabeitia, S. Menendez, C. Gonzalez-Muria and J. Majada, *Tree Physiol.* 24, 64–73(2004).

[28] Danthu, D. Ravelomanana, J. Razanameharizaka and M. Grouzis, *Seed Sci. Res.* 161, 83–88 (2005).

[29] H. Razafimiaramanana, *Gestion durable des baobabs de Madagascar: Importance des symbioses mycorhiziennes.* Mémoire de DEA de Biochimie. Faculté des Sciences Université d'Antananarivo, Madagascar, 58 p. (2010).

[30] S. Sugandha, R. Shashi and K. Shagufta, *Nanobiotech. Univ.* 1(2), 107–112 (2010).

[31] J. Ballet, P. Lopez and N. Rahaga, in: N. Andriananirina, N. Rabevohitra, J. Ballet and F. Rasolofo (Eds.), L'Harmattan, Paris, pp. 119-135 (2010).

[32] H. Randriamalala and Z. Liu, *Madag. Cons.Develop.* 5(1), 11-22 (2010).

[33] D. Schuurman and P. P. Lowry II, *Madag. Cons.Develop.* 4(2), 98-102 (2009).

[34] M. A. Barrett, J. L. Brown, M. K. Morikawa, J. N. Labat and A. D. Yoder, *Science* 328, 1109-1110 (2010).

[35] A. Munive, *Diversité génétique de souches de Bradyrhizobium d'arbres de forêts tropicales humides en Guyane, en Guinée et à Madagascar.* Thèse de Doctorat. Univ. Claude Bernard-Lyon I. 148 p. (2002).

[36] N. C. Rakotoarimanga, *Structure et fonctionnement de la microflore rhizosphérique associée à la symbiose mycorhizienne en faveur de Intsia bijuga, plante autochtone malgache de la forêt de Tampolo* – Thèse de Doctorat. Univ. d'Antananarivo. 202 p. (2010).

[37] H. Rajaonarimamy, *Influence de la diversité mycorhizienne sur la symbiose Dalbergia trichocarpa – rhizobia et sur la structure de la microflore tellurique.* Mémoire de DEA, Univ. d'Antananarivo. 60 p. (2010).

[38] L. Tedersoo, T.Suvi, K. Beaver and U. Kôljag, *New Phytol.* 175, 321–333 (2007).

[39] J. Nugroho, I. Mansur, A. Purwito and S. Suhendang, *J. Biosci.* 17(2), 68–72 (2010).

[40] B. Buyck, A. Verbeken and U. Eberhardt, *Mycol. Res.* 111, 787–798 (2007).

[41] N. S. Aggangan, H. K. Moon and S. H. Han, *New For.* 39(2), 215–230 (2010).

[42] V. L. Dos Santos, R. M. Muchovej, A. C. Borges, J. C. L. Neves and M. C. M. Kasuya, *Braz. J. Microb.* 32, 81-86 (2001).

[43] Y. L. Chen, M. Q. Gong, F. Wang, Y. Chen, M. Zhang, B. Dell and N. Malajczuk, in: M.Q. Gong, D. Xu, C. Zhong, Y.L. Chen, B. Dell and Brundrett MC. (Eds.), *China For. Publ. House*, Beijing, pp. 21-28 (2000).

[44] R. Duponnois, S. Diédhiou, J. L. Chotte and M. O. Sy, *Can. J. Microbiol.* 49(4), 281–287 (2003).

[45] M. C. Pagano and M. R. Scotti, *Mycoscience* 49, 379–384 (2008).

[46] D. J. Read, *Experientia* 67, 367-391 (1991).

[47] L. Gómez-Aparicio, J. M. Gómez, R. Zamora and J. L. Boettinger, *J. Veg. Sci.* 16(2), 191-198 (2005).

[48] L. Gómez-Aparicio, F. Valladares, R. Zamora and J. L. Quero, *Ecography* 28, 757-768 (2005).

[49] K. M. Holl, *J. Ecol.* 90, 179–187 (2002).

[50] R. H. Baohanta, *Facilitating the regeneration of Uapaca bojeri by managing mycorrhizal community associated with pioneer shrub species within the degraded area of the sclerophyllous forest of Arivonimamo.* PhD Thesis. Life Sciences (Biochemistry). Dept. Fundam. Appl. Biochem. Fac. Sci., Univ. Antananarivo, Madagascar, 190 p. (2011).

[51] S. E. Smith and D. J. Read, *Mycorrhizal Symbiosis*, 2nd edition. Academic Press, Ltd. Cambridge (1997).

[52] M. Cardoso and T. W. Kuyper, *Ecosys. Environ.* 116, 72–84 (2006).

[53] Choler, R. Michalet and R. M. Callaway, *Ecology* 82, 3295-3308 (2001).

[54] L. A. Cavieres, M. T. K. Arroyo, A. Peñaloza, M. A. Molina-Montenegro and C. Torres, *J. Veg. Sci.* 13, 547–554 (2002).

[55] I. Yeaton and K. J. Esler, *Vegetatio* 88, 103-113 (1990).

[56] B. R. Kropp and J. M. Trappe, *Mycologia* 74, 479–488 (1982).

[57] Molina and J. M. Trappe, *New Phytol.* 90, 495–509 (1982).

[58] E. I. Newman, *Adv. Ecol. Res.* 18, 243–270 (1988).

[59] D. A. Perry, M. P. Amaranthus, J. G. Borchers, S. L. Borchers and R. E. Brainerd *Bioscience* 39, 230–237 (1989).

[60] M. Callaway, *Bot. Rev.* 61, 306–349 (1995).

[61] F. Richard, S. Millot, M. Gardes and M. A. Selosse, *New Phytol.* 166, 1011–1023 (2005).

[62] Nara, *New Phytol.* 169, 169–178 (2006).

[63] M. Moora and M. Zobel, in: F. I. Pugnaire (Ed.), CRC Press, Boca Raton, FL, pp. 79-98 (2010).

[64] M. G. A. Van der Heijden and T. R. Horton, *J. Ecol.* 97, 1139–1150 (2009).

[65] R. Horton and M. G. A. Van der Heijden, in: M. A. Leck, V. T. Parker and R. L. Simpson (Eds.), Cambridge University Press, Cambridge, pp. 189–214 (2008).

[66] A.E. Lugo, *For. Ecol. Manage.* 99, 9-19 (1997).

[67] ONF, *Forêt et Carbone.* Dossier N° 5, 1-4. Available: onf.fr/rp/files/Foretcarbone.pdf (2006).

[68] R. Sedjo, *New For.* 17, 339–259 (1999).

[69] L. Bowyer, *Wood Fibre Sci.* 33(3), 318–333 (2001).

[70] E. W. S. Lee, B. C. H. Hau and R. T. Corlett, *For. Ecol. Manage.* 212, 358-366 (2005).

[71] P. M. Vitousek, C. M. D'Antonio, L. L. Loope and R. Westbrooks, *Am. Scient.* 84, 468-478 (1996).

[72] D. Simberloff, in: P. H. Raven (Ed.), National Academy Press, Washington, DC. pp 325-334 (2000).

[73] J. A. Parrotta, *Agric. Ecosyst. Environ.* 41, 115-143 (1992).

[74] A.E. Lugo, in: M. K. Wali (Eds.), SPB Academic Publishing, The Hague, (1992).

[75] J. A. Parrotta, *J. Veg. Sci.* 6, 627-636 (1995).

[76] H. Perrier de la Bathie, *Rev. Bot. App. Agric. Trop.* 11(Bull. 121), *Etudes et dossiers*, pp.719-729 (1931).

[77] P. P. Lowry II, G. E. Schatz and P. B. Phillipson, in: S. M. Goodman and B. D. Patterson (Eds.), Smithsonian Inst. Press, Washington D.C., pp. 93-123 (1997).

[78] P. Binggeli, in: S. M. Goodman and J. P. Benstead (Eds.), The University of Chicago Press, Chicago pp. 257-268 (2003).

[79] C. A. Kull, J. Tassin and H. Rangan, *Mount. Res. Develop.* 27(3), 224–231 (2007).

[80] S. Carrière and H. A. Randriambanona, *Bois For. Trop.* 292(2), 5-21 (2007).

[81] M. Dufils, in: S. M. Goodman and J. P. Benstead (Ed.), The University of Chicago Press, Chicago and London, pp. 88-96 (2003).

[82] C. J. Geldenhuys, *For. Ecol. Manage.* 99, 101-115 (1997).

[83] **J. H. Connell and R. O. Slatyer,** *Am. Nat.* 111*, 1119–1144* (1977).

[84] J. Van Andel, in: J. van Andel and J. Aronson (Eds.), Blackwell Publishing, Oxford, pp. 58-69 (2006).

[85] S. G. Newmaster, F. W. Bell, C. R. Roosenboom, H. A. Cole and W. D. Towill, *Can. J. For. Res.* 36, 1218-1235 (2006).

[86] R. Molina, H. Massicote and J. M. Trappe, in: A. M. F. Routledge (Ed.), Chapman and Hall Inc., New York, pp. 357-423 (1992).

[87] H. A. Randriambanona, *Successions écologiques dans les plantations de Pinus, d'Acacia et dans les forets naturelles de la région Nord-Ouest du corridor de Fianarantsoa (Madagascar).* Thèse de Doctorat. Univ. Antananarivo, Madagascar. Ed. Universitaires Européennes. 124 p. (2011).

In: From Seed Germination to Young Plants   ISBN: 978-1-62618-653-8
Editor: Carlos Alberto Busso   © 2013 Nova Science Publishers, Inc.

Chapter 16

Tropical Seedling-Insect Interactions: The Importance of Herbivory from the Individual to the Community Level in Natural Forests and Restoration

Tara Joy Massad[*]

Program on the Global Environment
University of Chicago, Chicago, IL, US

Abstract

The study of plant-insect interactions constitutes a large and growing branch of ecology that includes the two most diverse groups of higher organisms. This field of research addresses issues ranging from species-specific evolutionary arms races to the immense diversity of tropical plant communities to the effects of insect herbivores on reforestation. The time in plant ontogeny when these relationships are most influential is likely the vulnerable seedling stage because stored resources are scarce, effectively increasing negative effects of herbivory on plant growth and survival. This chapter examines the effects of seedling herbivory in the tropics, demonstrating the importance of herbivores to individual plant species as well as plant community composition; it closes by discussing the application of this knowledge to tropical forest restoration.

Many, but not all, studies of seedling herbivory conclude that herbivore damage limits plant performance. In addition, abundant examples of plant adaptations to tolerate or avoid herbivory exist. For example, even young plants have strong chemical antiherbivore defenses, supporting the hypothesis that herbivory is costly to seedlings. The potential for herbivores to have positive effects on forest diversity is described in the Janzen-Connell hypothesis. According to this model, herbivores lead to enhanced diversity by negatively affecting seedling survival, particularly when conspecific seedlings are growing at high densities (negative density dependence) or near their parent plant (negative distance dependence). Few studies have explicitly tested the assumption that density dependent effects are mediated by herbivores, and even fewer have

---

[*] E-mail address: tmassad@uchicago.edu; tmassad77@gmail.com.

demonstrated resultant increases in diversity. However, new empirical support is emerging for the importance of herbivores in negative density and negative distance dependence.

Herbivores also play a significant role in both actively managed and natural forest regeneration. For example, the growth of seedlings planted in reforestation is negatively affected by herbivory damage, and herbivores impact seedling survival and diversity in forests recovering from disturbances such as fire. Restoration projects could therefore benefit from explicitly considering ecological measures to manipulate herbivory.

Introduction

Insect herbivory is recognized has having the potential to exert significant ecological [1, 2] and evolutionary [3, 4, 5, 6] effects on plants, eliciting both positive [7, 8, 9, 10,11] and negative [12, 13, 14] performance responses. In addition to the influence herbivores have at the level of the individual or species, one of the most important hypotheses addressing plant community diversity ascribes herbivory a causal role in increasing species evenness. This idea, known as the Janzen-Connell hypothesis, proposes herbivores have the potential not only to limit plant performance but to do so to such an extent that survival is reduced and community diversity is ultimately increased [15, 16, 17]. Janzen's contribution to this hypothesis was based on his observations in tropical forests, and since that time, it has become generally recognized that herbivore pressure and plant defensive responses are particularly strong in the tropics [13, 18], in spite of the fact that the vast majority of plant-herbivore studies come from temperate ecosystems. The point in ontogeny when herbivores are most likely to influence plants is the vulnerable seedling stage. This could be because of a lack of stored reserves to replace lost tissues [14, 19], less leaf area to compensate for lost tissue, or weaker antiherbivore defenses, making seedlings more appealing to herbivores [20, 21]. Because herbivore pressure and plant defenses are both strong in the tropics and because hypothesized effects of herbivory on diversity were conceived of in the tropics, this chapter will focus on tropical herbivore-seedling interactions, examining the effects of herbivory from the level of the individual through the community.

Herbivory and the Individual—Ecological and Evolutionary Responses of Seedlings to Herbivore Damage

Among the first tropical studies to demonstrate herbivory can, in fact, have negative effects on plant growth and reproduction, particularly when plants are small, is work with *Piper arieianum* C. DC. (Piperaceae). Results showed artificial herbivory (within the range of natural herbivory levels) can limit reproduction even over more than one fruiting season [22]. This important finding, along with landmark studies of herbivory and plant defenses [23, 24, 25], stimulated the study of insect herbivore-plant interactions in tropical forests.

Cotyledon stage plants can be especially vulnerable to herbivory. A study from French Guiana explored the relationships between seed mass, cotyledon type, and tolerance to artificial herbivory, and the authors found the growth form of the cotyledon was more important than seed mass in predicting a seedling's reaction to herbivory. Responses to stem cutting were measured eight months after damage, and seedlings with hypogeal cotyledons better tolerated damage than seedlings with epigeal cotyledons because their meristems were not always lost in the cutting treatment. Nonetheless, the survival of seedlings with both types of cotyledons was often negatively affected by the removal of 50% of their cotyledons. This result was light dependent such that all seedlings were affected when less than 1% full sunlight was available, but only seedlings with cryptocotylar, hypogeal cotyledons were negatively affected in more abundant light. In contrast, the removal of 50% of the first true leaves was unimportant for survival across all groups [26]. These findings suggest a loss of reserves can be very damaging to seedling survival. Further strengthening this conclusion is work from seven species on Barro Colorado Island, Panamá (BCI) which examined the effects of a complete loss of foliar and photosynthetic cotyledon tissue two weeks after the expansion of the first photosynthetically active leaf or cotyledon. Seedlings characterized as being more shade tolerant and having more total non-structural stem and root carbohydrates (TNC) were better able to tolerate photosynthetic tissue loss than less shade tolerant seedlings with higher early relative growth rates (RGR). Those seedlings with more TNC had higher survival one year after damage, and there was an overall positive correlation between the TNC pool and the pace of leaf regrowth post-herbivory [27]. These inherently slower growing, shade tolerant species with high TNC pools are also those predicted by the growth rate / resource availability hypothesis to invest more resources in chemical antiherbivore defenses [28].

Data do show storage and carbon-based defenses can be positively related. Light induced increases in concentrations of TNC, condensed tannins, and phenolic glycosides in *Populus tremuloides* Michx. (Salicaceae [29]). Pools of TNC and condensed tannins were also positively related in both the shade tolerant *Quercus mongolica* var. *grosseserrata* (Fagaceae) and the shade intolerant *Castanea crenata* (Fagaceae), but *Q. mongolica* allocated more carbon to defense as opposed to storage than *C. crenata* did. So, allocation priorities can also differ between species and functional groups [30]. In the case of *Q. mongolica* and *C. crenata*, allocation differences were manifested in the slower growing species investing relatively more carbon in defending its tissues rather than in carbohydrates that may be used for regrowth [30]. Interesting future work could therefore investigate how natural herbivory levels vary with TNC concentrations and examine the relationships between TNC, defenses, and the outcome of herbivory in tropical seedlings. Such data would clarify whether allocation priorities are optimized among species with both storage and defense capabilities as well as expand knowledge of plant functional traits by providing more information regarding growth rates, susceptibility to herbivory, and investments in defense or tolerance mechanisms.

The full spectrum of fast growth/low investment in reserves or defense and slow growth/high investment in reserves or defense also encompasses the fast growing species with very limited reserves [26] and defenses [23]. These species generally follow an 'escape' rather than tolerance or defense strategy, growing more quickly to shorten the period of heightened vulnerability to herbivores [31]. Other work with herbivory on seedlings along this gradient also demonstrates that reserves are very important for surviving early damage. Dalling et al. (1997) [32] and Dalling and Harms (1999) [33] studied several large-seeded species on BCI and found seedlings of *Prioria copaifera* Griseb. (Fabaceae) and *Gustavia superba* (H.B.K.) Berg (Lecythidaceae) developed effectively even after the loss of over half their cotyledon tissue. In addition, seedlings with intact or damaged cotyledons were able to resprout multiple times following shoot removal. Several other large-seeded species with hypogeal cotyledons were likewise able to resprout after multiple bouts of simulated herbivory [34]. Similar results were obtained for the Australian rainforest species, *Idiospermum australiense* (Diels) S. T. Blake (Calycanthaceae [35]. Green and Juniper (2004) [36] extended this work with 15 Australian rainforest species with hypogeal cotyledons and found positive correlations between seed and cotyledon mass and the number and mass of resprouts developed after the excision of stems that had a single set of leaves. These findings support the hypothesis that large cotyledons evolved as reserves for times of negative carbon balance. The positive mediation of herbivore damage by cotyledonary reserves declines once roots, which can serve a similar carbohydrate provisioning function, are established [37]. Data from two dipterocarp species, however, suggest not all large-seed species mobilize root reserves for resprouting [38].

Emphasizing the necessity of evolved mechanisms to tolerate or minimize herbivory are other studies documenting the prevalence and negative effects of early herbivory. For example, work from BCI shows insect herbivory is quite common early in ontogeny; 30.6% of seedlings studied experienced foliar damage during their first year of growth, and 22-100% of these seedlings subsequently died, depending on the species in question. In contrast to previous studies, however, no distinct patterns correlating seed or cotyledon characteristics with mortality due to leaf herbivory emerged. Both heavy and light seeded species succumbed to or resisted mortality following leaf damage, and species with both reserve and foliar cotyledons had 100% mortality following herbivory, although only one species with foliar cotyledons was included in the study. This work was conducted after the development of true leaves, however, so the stage at which any cotyledonary benefits exist may have passed [39]. A similar study from BCI directly compared the effects of cotyledon or leaf removal on three species with cotyledons of differing functions: photosynthetic, semi-photosynthetic and reserve provisioning, or only reserve provisioning. Cotyledon removal reduced survival by about 50% across species, indicating that all cotyledon types are important while the first leaves are developing. True leaf removal had a negative effect on the survival of species with storage cotyledons, while survival of the species with only photosynthetic cotyledons was basically equal with or without leaf removal because the leaf area removed was small in comparison to the photosynthetic area of the cotyledons [40]. Data on propagule herbivory from three species of mangroves in Panamá also illustrate early herbivory is particularly damaging; mortality was much higher and growth much reduced with predispersal herbivory [41]. A study of *Dipteryx panamensis* (Fabaceae) in Costa Rica showed seedlings with even just one percent of their leaf area consumed by herbivores during their first month of growth were not able to survive through the year [42]. Further supporting the heightened vulnerability of very young seedlings are data from simulated herbivory on *Metrodorea pubescens* (Rutaceae). In this species, only the smallest size class of seedlings tested (5-50 cm) had their relative growth rate reduced by simulated herbivory, and only individuals < 20 cm tall had their survival affected by herbivory [43]. Together, these data indicate seedlings are very vulnerable to tissue losses during their establishment, and the relative impact of the type of tissue loss changes as provisioning roles shift from cotyledons to true leaves. Herbivory may therefore be an important selective pressure on the evolution of both cotyledon and foliar characteristics. As the importance of cotyledons as an energy source (via reserve or photosynthate provisioning) is transferred to true leaves, it may be that investments in antiherbivore defenses also change. Increased research into defense properties of cotyledons and true leaves should therefore be undertaken to help complete our understanding of the relationship between herbivory and plant characteristics throughout ontogeny [21].

The evolution of antiherbivore defenses at this very early stage is likely important despite tradeoffs predicted between growth and defense [20]. A few cases of strong investments in defenses have already been found early in ontogeny; a study of cyanogenic glycosides in the Australian rainforest species, *Ryparosa kurrangii* B.L.Webber (Achariaceae), indirectly demonstrated the importance of herbivore pressure by showing high concentrations of cyanogenic glycosides were common in seedlings, particularly in cotyledon stage plants. These concentrations of cyanogenic glycosides required a high investment of nitrogen, and it can therefore be hypothesized that protection from herbivores is of great importance to these seedlings, outweighing potential benefits of investing nitrogen directly into photosynthesis or growth. Cyanogenic glycoside concentrations decreased throughout ontogeny, further indicating herbivores most negatively affect seedlings and cotyledon stage plants. However, this study also revealed increases in leaf mass per unit area with ontogeny, and because herbivory was not quantified, it cannot be conclusively determined whether a shift in defense investment (from chemical to physical barriers) or simply a shift in defense strategy occurs with ontogeny [44]. Quantification of phenolics in seedlings and saplings of *Inga vera* Willd. (Fabaceae) also showed seedlings invest more in hydrolyzable tannins than older plants [45].

Defenses are also important before germination; the potential for large seeds to contain high quantities of secondary defenses was among other hypothesized benefits of large seed size in tropical moist forest species [46]. Early work by Janzen [47] demonstrated that larger Fabaceae seeds were not attached by Bruchidae (pea weevils) while small seeded species almost invariably were, and he concluded that deterrent concentrations of secondary metabolites such as alkaloids and saponins present in larger seeds served as important defenses against bruchid predation. Later work showed defenses are actually very common in tropical legume seeds [48, 49, 50], and a study of seed defenses in tropical pioneer species demonstrated that small seeded pioneers can also be well defended. The fortification of these seeds against fungal pathogens and arthropods is related to their ability to persist in the soil seed bank for extended periods [51].

The important possibility that defense compounds are transferred from the seed to the seedling has not received adequate study, although it was argued that defense compounds are likely passed from seeds to germinating plantlets [52]. This hypothesis is based on data showing anthocyanins and flavonoids were higher in the first emerging leaves than in the mature leaves of the marama bean, *Tylosema esculentum* (Burch.) Schreiber (Fabaceae [52]). In addition, the alkaloid profile of the pioneer species, *Bocconia frutescens* L. (Papaveraceae), showed both qualitatively and quantitatively that more alkaloids were present in seeds than leaves [51]; the possibility then exists that some of these alkaloids may be translocated and diluted in the growing leaves. Work in annual plants also suggests transgenerational defense induction is possible [53, 54, 55], although such maternal effects may be less likely in long-lived species. Suggestive evidence for the ability of *Mimosa bimucronata* (DC.) Kuntze (Fabaceae) seeds to induce defenses in response to bruchid predation exists, although it is correlative in nature, as phenolic levels were only compared between naturally attacked and intact seeds [56]. Further investigations of maternal effects on early plant defense should be emphasized, and resulting data would help clarify how young plants may simultaneously grow and defend themselves.

Negative effects of herbivory on seedlings continue past the cotyledon stage, but interestingly, growth and survival are not always affected in the same way. Results from forests in Malaysia [57], Panamá [58], and two Aracaceae species in the Amazon [59] showed seedling survival, but not growth, was negatively affected by even low rates of herbivory. Causes of these differential effects are not always clear, and even within the same species herbivory was found to decrease survival in one experiment but only decrease shoot biomass in another [60].

Distinctions should be made between studies of natural versus simulated herbivory. A study by Jackson and Bach simulated 50% herbivory, which was much higher than the roughly five percent herbivory measured naturally on seedlings of the same seedlings [60]. Blundell and Peart also studied effects of artificial herbivory in Borneo, and seedling survival was most reduced with 90% leaf area removal [61]. Nonetheless, ample evidence exists for negative effects of natural levels of herbivory on seedling survival and growth [57, 58, 62, 63]. A study including eight co-occurring species of Dipterocarpaceae in Malaysia showed natural herbivory in excess of 50% leaf area removed doubled mortality relative to seedlings with less herbivory [64]. Relevant to effects of climate change was the finding that low levels of herbivory ($< 10\%$) combined with drought tripled mortality rates of seedlings in Borneo [64]. An earlier study at the same site on a subset of the same species produced contrary results when water availability was experimentally reduced. In this case, effects of 50% leaf removal, apical meristem removal, and fine root damage were compared (all damage was artificial), and survival was only reduced by root damage in the low water treatment. Relative monthly height increment was limited by all the damage treatments but only in plots with natural water availability [65]. This speaks to the question of whether herbivory is most damaging when growth is vigorous or stressed, and it is likely that multiple factors affect the overall outcome.

Effects of herbivory should be evaluated in conjunction with as many environmental parameters as possible, as light [61, 66, 67] and water availability [64, 65, 68] can mediate the effect size of changes in growth after herbivore damage. A Namibian study crossing the effects of nutrient and water availability with herbivory in a nursery setting found that seedling growth was inhibited (number of branches produced) by herbivory when water was abundant. At low water levels, herbivory did not affect branch production. Interestingly, this study, which used a species from a pan-tropical genus, *Terminalia* (Combretaceae), as its focal organism showed 15.5 month old seedlings were capable of simultaneously investing in growth and tannin production, particularly under low resource conditions [68].

In terms of light, the survival of *Shorea quadrinervis* Sloot (Dipterocarpaceae) individuals less than one centimeter in diameter was much more inhibited by high, simulated herbivory in the understory than in nearby gap habitats, although high herbivory reduced the growth in height of seedlings in gaps more than seedlings in the understory [61]. This study also demonstrated the variable nature of responses to herbivory of differing degrees; low levels of herbivory or meristem damage on gap individuals induced a positive growth response, while when herbivory was greater than 10% leaf area removed, overcompensation was not possible [61]. Similarly, five month old *Swietenia macrophyla* King (Meliaceae) seedlings in gaps were also much more resistant to negative effects of herbivory. Half of the individuals planted in the understory died when simulated herbivory was greater than 50% whereas only five percent of individuals planted in gaps died with the same amount of damage. Despite the strong reductions in survival of understory seedlings, growth was again more limited by herbivore damage in gaps. In the understory, growth was maintained at a constant, low level across degrees of leaf damage while growth in gaps was much higher but significantly reduced with over 50% leaf area removal [66]. In another forest and using natural rather than artificial herbivory, *S. macrophyla* and *Cedrella odorata* L. (Meliaceae) mortality also increased when damage exceeded 50% leaf area removal, and this result was also stronger for seedlings planted in unthinned (lower light) as compared to thinned canopy (higher light) sites. Interestingly, the effect of herbivory changed through time such that compensatory growth was observed for some seedlings after one year, although it was unclear why some individuals were able to respond positively to herbivory while others were not [69]. Overall, when resources are already limited, herbivory seems to induce strong reductions in survival, while under ample resource conditions, growth is affected before mortality, and plants seem to have more flexibility in responses depending on the degree of damage. The variable nature of the relationship between herbivory and growth requires long-term study to fully understand the lasting effects of herbivory, and this may be of particular importance in applied work.

Positive mediation of herbivory at high light is also species dependent. A factorial study comparing effects of light and herbivory levels found seedlings of *Brosimum alicastrum* Sw. (Moraceae) could compensate more for herbivory at higher light levels in terms of leaf area production. In the same study, seedlings of the vine, *Vitus tiliifolia* Humb. and Bonpl. ex Schult. (Vitaceae), increased their net assimilation rate more in high light after herbivory, but their growth was not affected [67]. In addition, herbivore exposure (seedlings were enclosed in mesh cages or planted in the open) and gap size interacted to reduce the relative growth rate of two of three pioneer species. Specifically, a small gap size and exposure to herbivores limited the growth of the most light-demanding species, *Trema micrantha* (L.) Blume (Ulmaceae), and although the growth of of the least light demanding species tested, *Miconia argentea* DC. (Melastomataceae), was limited by herbivory, effects were independent of gap size [70].

These results demonstrate the interesting point of how environmental variation interacts with herbivory to influence seedling growth and survival, which could ultimately be manifested in changes in diversity. High light conditions seem to often favor herbivores [70, 71, 72, 73, 74, 75] in spite of the often higher levels of carbon-based defenses present in plants grown at high light [72]. For example, natural herbivory from specialist caterpillars (*Steniscadia poliophaea*, Noctuidae) on *S. macrophyla* was shown to be much higher in gaps, which may be due to increased plant vigor or simply an environmental preference of the herbivores for high light environments [66]. In a study of the recovery of forest diversity after repeated fires in the southern Amazon, herbivory was also higher in burned forest, where the canopy was more open, than in undisturbed forest [14].

Other experimental factors that are important to consider when assessing the effects of herbivory are the method of measuring herbivory and the length of the study. For example, leaf abscission can follow herbivore damage [64, 76], and it is recognized this must be carefully dealt with when measuring herbivory [66]. Studies must also be of appropriate monitoring frequency and duration to detect effects of herbivory. Denslow found that recently germinated Bombacaceae seedlings were able to recover from foliar herbivory within their first six weeks of growth; she attributed this to their ability to rely on their cotyledonary reserves [77]. After two years, she found only three surviving seedlings from the original cohort of 93 seedlings, but she could not correlate this with early herbivory, probably because such a long gap existed between monitoring periods.

The characteristics of specific herbivores are also highly relevant to the outcome of their feeding. The most interesting aspect of the above mentioned study of Bombacaceae herbivory was the observation that foliar damage did not show negative distance dependence while apical meristem homopteran damage did [77]. Herbivore identity is particularly important to consider in applied work, especially with invasive plants where herbivores can provide effective biocontrol. The shrub, *Mimosa pigra* L. (Fabaceae), is one of Australia's most problematic invasives, and a study with a geometrid caterpillar from its native range, *Macaria pallidata,* demonstrated that growth of approximately two month old seedlings was reduced one week after caterpillar damage [78]. These results are encouraging for biocontrol, but such studies should, again, span longer time periods, particularly when the goal is to determine relevant applied effects of herbivory. Another study with a highly invasive species in Australia, *Acacia nilotica* (L.) Willd. ex Delile (Fabaceae), showed seedlings responded negatively to artificial defoliation and stem damage, especially to repeated bouts of damage, suggesting multivoltine species may be successful in its control [79]. The growth and survival of *Melaleuca quinquenervia* (Cav.) Blake (Myrtaceae), a tree native to Australia but invasive in the United States, can be limited by a host specific psyllid, although a host specific leaf feeding curculionid did not reduce growth. It is interesting to note that these negative effects became apparent at lower insect densities in the field than in the lab [80].

Not all studies of herbivory result in negative effects on plant performance, however. Herbivory had minimal effects on the growth of the liana, *Connarus turczaninowii* Triana (Connaraceae), on BCI, reducing growth in only one of three years documented and having no effect on the survival of seedlings and saplings [81]. The growth and survival of seedlings of *I. vera* were likewise unaffected by herbivory [45]. A second study using the same three pioneer species as mentioned above [70] found relationships between natural herbivory and growth or survival were not significant [74]. In addition, the aforementioned study in which growth and survival of young *M. pubescens* individuals were negatively related to herbivory also showed larger saplings (100-150 cm) actually increased their relative growth rates in response to high amounts of leaf damage [43]. Saplings (0.5-1.5 m tall) of *Casearia nitida* (L.) Jacq. (Salicaceae) also displayed compensatory growth in response high levels of simulated herbivory, a response they were capable of due to mobilization of their stored carbohydrates [19]. Another study that failed to detect negative effects of herbivory on growth focused on the relationship between *Manilkara bidentata* (A. DC.) A. Chev. (Sapotaceae) and its specialist leaf mining herbivores. Although herbivory was on average 24.5% in plots of low conspecific densities and up to 38.% in plots of high densities, no significant effect of herbivory on growth was detected [75]. This work examined saplings between 20 and 250 cm tall, however, and in conjunction with results from the study of *M. pubescens* [43] and *C. nitida* [19], it is reasonable to suspect that herbivory is not as damaging to larger plants.

To summarize the effects of herbivory on juvenile woody plants, a meta-analysis was performed on all the studies of herbivory at the cotyledon, seedling, or sapling stage indexed in the Web of Science between 2000 and 2010. Means, sample sizes, and standard error/deviation were collected from up to three experiments per paper and used to calculate the standardized effect size (Hedges' d, d_{si}) of herbivory on growth, photosynthesis, or reproduction. When both growth and reproduction were measured, the response variable was chosen at random. Within the growth response, biomass was preferentially selected when available; if it was not measured, a response was selected randomly. Effect sizes were examined to determine the overall importance of herbivory on plant performance, and they were also compared between temperate and tropical studies using the between class variable statistic, Q_B [82]. All statistics were performed with SAS 9.1.3 (SAS Institute, Inc.).

A previous meta-analysis showed herbivory has stronger negative effects in the tropics than in the temperate zone [13]. In spite of this, the new dataset used in this meta-analysis which compared the effects of herbivory on only seedlings and saplings of woody plants showed the effect size of herbivory was equivalent in the two areas (temperate zone d_{si} = 0.84, 95% CI = 0.31, n = 169; tropical zone d_{si} = 0.85, 95% CI = 0.54, n = 56). In both latitudinal zones, herbivory significantly inhibited plant performance, demonstrating an overall negative effect of early herbivory.

Scaling up – How Differential Herbivory Can Shape a Community

Although herbivory has overall negative effects on plant performance, not all plants are equally susceptible to herbivore damage. As described above, both plant tolerance and defense can moderate the outcome or occurrence of herbivory. Slow growing species invest more heavily in phenolics and leaf toughness, and their mature leaves suffer less herbivory than those of fast growing pioneer species [24, 83, 84]. Fine et al examined these patterns in terms of their influence on species distributions within the landscape, and range restrictions were clearly related to differences in vulnerability to herbivore-induced mortality [84]. This work showed white sand specialist species grow slowly but are well defended whereas clay specialists grow faster but are less well defended [84]. When these two groups were reciprocally transplanted across soil types in a factorial design manipulating herbivore access, it was shown clay specialists could not compensate for herbivory on nutrient poor white sands and suffered much higher mortality when exposed to herbivores than their sister species that evolved in white sand habitat [85]. This work strongly suggests herbivorous insects play an important role in determining the composition of plant communities across habitats.

Within a habitat, herbivores have long been hypothesized to affect species evenness and diversity [15, 16]. These ideas, formally known as the Janzen-Connell hypothesis, were thoroughly reviewed by Carson et al. [17], so only a few more recent examples will be discussed here. Carson et al. [17] concluded evidence was overall supportive of the negative density and negative distance predictions of the hypothesis—that herbivory and pathogen infection would be greatest when conspecific seedlings were found at high densities or at a short distance from their parent plant. This concentrated damage can result in higher seedling mortality where conspecifics are densely aggregated, effectively opening space for the recruitment of other species and leading to a positive effect on species evenness. However, in spite of the hypothesis' long history, it is only recently being fully explored—by documenting that survival-limiting herbivory opens space on the forest floor in dense patches of conspecific seedlings or in patches of seedlings germinating near to their parent tree and measuring resultant increases in diversity and evenness in natural communities.

Two important recent contributions using data from the 50 ha forest dynamics plot on BCI showed density dependent effects exist in the seedling bank; conspecific neighbors were negatively related to seedling survival locally (in 1 m^2 plots), although not all species were affected in the same way—29 out 59 species examined were differentially affected by con- as opposed to heterospecific neighbors [86]. When the scale was increased to encompass the entire 50 ha plot, the negative relationship held between seedling survival and conspecific densities with two important caveats—density was relevant when measured as basal area but not the number of stems, and species had to be separated into functional groups related to their shade tolerance for the pattern to appear. The survival of light-demanding seedlings was most negatively related to adult basal area; shade tolerant species' survival was also negatively related to basal area but to a lesser degree [86]. This could be related to stronger chemical defenses in shade tolerant species, providing greater resistance to herbivory. Basal area is likely strongly correlated with leaf area, so plant apparency to herbivores may be more related to basal area than stem density *per se*. The second study demonstrated that rare species are actually more negatively affected by conspecific neighbors than common species [87], which speaks to an empirical difficulty in testing the Janzen-Connell hypothesis noted by Carson et al. [17]: the distribution of rare species may be more likely to follow predictions of the Janzen-Connell hypothesis, but these species are also more difficult to study by virtue of their very rarity. In sum, these new studies make a strong contribution to understanding patterns of diversity and seedling survival, but they do not provide the mechanistic data needed to connect herbivory to negative density dependence. Such data were gathered in an Amazonian forest, however. Alvarez-Loayza and Terborgh carefully monitored the causes of seedling mortality in highly dense 'seedling carpets' of simultaneously germinating conspecifics. [63]. Their forensic-like approach to examining agents of mortality allowed to conclude that seedlings in these dense patches were overwhelming dying due to specialist arthropod or pathogen damage whereas seedlings of the same species surrounded by a variety of heterospecific neighbors were surviving at much higher levels.

Although it was reported seedlings of rare species were more likely to demonstrate Janzen-Connell effects [87], most work has found Janzen-Connell effects in common species. In the Peruvian Amazon, there were highly pronounced negative distance and negative density dependent patterns of recruitment as common species transitioned from seedlings to saplings [91]. Work from the Ecuadorian Amazon also showed a community compensatory trend for common species of the Myristicaceae [92], meaning seedling mortality was greater for common as opposed to rare species [93]. 'Seedling' was defined differently in each study, however, suggesting, as mentioned above, that the life-stages at which patterns are sought may produce very different results. Seedlings in the work on rare and common species on BCI [87] were at least 20 cm tall whereas those in the study from Ecuador [92] were as small as 1 cm in height, a likely more vulnerable stage. Negative density dependence was also important for changes in diversity between seeds and seedlings of commonly occurring species [94]. Studying only older seedlings may therefore omit critical data from the puzzle. However, saplings that showed negative distance and density dependence in Perú [95] were at least 1 m tall, so more factors are involved than plant age. Another important consideration in evaluations of the Janzen-Connell hypothesis is the study scale. For example, 60 x 60 m plots revealed a significant correlation between conspecific density and specialist herbivory in Puerto Rican forests, whereas no such relationships were detected in smaller 20 x 20 m plots [75]. Work in gap and understory environments also demonstrated that negative-density effects operate on a larger scale in gaps versus forest understory [66]. On the other hand, negative density dependence has also been recorded to diminish in strength with scale, and herbivore satiation has been mentioned as potential explanation for this pattern [86]. This idea is based on a study of mammalian seed predation [88], but because of the r-selected nature of insect populations, it may be less likely that insect numbers and feeding would plateau before their slower growing food supply. Examples of herbivore satiation involving insects are most often focused on seed predators [89, 90], and it makes sense that insects which rely on a food source with a short window of temporal availability, such as seeds, may not be capable of increasing their populations in time to fully exploit the resource.

Overall, strong evidence exists for negative density dependence and the role of herbivorous insects in structuring this pattern. Support for specialized herbivores operating in a negative distance dependent fashion by increasing mortality on very young seedlings near conspecific adults has also recently been found [62] (although new studies indicate pathogens and soil-borne organisms also contribute to negative density [96, 97] and distance [95] dependence). The final step needed to support the Janzen-Connell hypothesis is to show that the end result of this herbivory and mortality is increased local diversity. A study in the seasonally dry forests of the southern Amazon excluded crawling insects from naturally recruiting communities of seedlings and found the effects of herbivory on recruitment diversity differed between intact and recently burned forest [14]. The presence of herbivores was related to increases in diversity in burned, but not unburned, forest when nitrogen was added to plots; herbivory was higher on seedlings in the burned forest, and nitrogen favored seedling growth (Massad, *unpublished data*). This combination of properties—high herbivore pressure and strong growth—enabled herbivores to have a positive effect on recruitment diversity. Data from the following year at the same site show contrasting patterns, indicating that herbivory increases diversity in unburned but not burned forest. These differences are likely due to variation in the availability of propagules to recruit into the forest and replace individuals lost to herbivores (Massad, *unpublished data*).

Resistance to negative effects of herbivory in the form of tolerance or antiherbivore defenses may contribute to unusually high dominance of individual species in some tropical forests and patterns contrary to Janzen-Connell effects [98]. For example, *Eperua grandiflora* (Aublet) Bentham (Fabaceae) was very tolerant of herbivory and had high recruitment rates under the seed shadow of its parent tree [99]. However, this lack of distance dependence was observed between seven and 19 months post-germination, while survival before seven months was negatively related to distance from the parent tree [99]. This re-emphasizes important ontogenetic differences in effects of herbivory. In addition, when there is high intraspecific variability in chemical defenses within a single species, such as is found in members of the genus *Hymenaea* (Fabaceae [25]) or in *Nectandra ambigens* (Blake) C.K. Allen (Lauraceae [100], or even among species of *Bursera*, the dominant genus in Mexican dry forests [101], taxa may show clumped distributions, resulting in patches with high local dominance rather than the evenness associated with Janzen-Connell effects [15].

In sum, evidence is accumulating in support of the contribution of negative density and distance dependence to tropical forest diversity, although more studies of naturally recruiting seedling communities are needed to demonstrate the entire chain of events from herbivore related mortality at high conspecific densities or near parent plants to resulting increases in diversity and evenness.

Applications to Restoration

The response of seedlings to herbivory, while not uniform, is generally negative, making an evaluation of plant-herbivore interactions highly relevant to reforestation work. Experimental factors that may influence herbivory, such as planting diversity and seedling defenses, have only recently been incorporated into the design of tropical forest restoration studies [102, 103]. A small number of reforestation studies examining timber production also tested the potential for herbivory to be limited in polycultures as opposed to monocultures [104] or by combinations of polycultures and insecticides [105]. Results varied by species, but herbivory was often reduced in polycultures [102, 103, 104], and the chemical control of herbivores was correlated with enhanced growth [105]. Several earlier studies, while not directly designed to measure the effects of herbivory in reforestation, also noted that herbivores limited seedling survival or growth. Leaf-cutter ants (*Atta* spp), for example, are especially damaging in neotropical reforestation [106, 107] and areas undergoing natural regeneration [108, 109]. Among the best known 'pest' species affecting commercial tropical reforestation is the shootborer, *Hypsipyla grandella*, which attacks members of the Meliaceae, causing pronounced damage particularly in monoculture plantings [110, 111, 112, 113]. Monocultures often have the effect of facilitating herbivory by increasing plant apparency to insects. This was outlined in the resource concentration hypothesis developed in agricultural systems [114], but it is equally applicable to reforestation. The idea that herbivores will cause greater damage when their preferred host plants are highly concentrated is also the basis of the Janzen-Connell model, demonstrating the similarity of cause and effect in plant-herbivore interactions in natural and applied settings.

Species and planting designs most often used in reforestation may therefore actually favor herbivores and inadvertently limit seedling growth. According to the growth rate hypothesis [28], the fast growing species often planted in reforestation may be weak in secondary defenses and consequently more attractive to herbivores. In addition, monocultures are still a common reforestation method even though they simplify insect host-searching and can increase herbivory. Consideration of these ecological factors in future restoration work may improve reforestation naturally by decreasing herbivore damage and promoting seedling growth. Polycultures alone may not be enough to reduce herbivory, but careful attention to the defense characteristics of the species comprising a polyculture can help limit herbivory and enhance growth [103], and species specific neighborhood effects are recognized as being important across ecosystems [115].

Attention to herbivory is even required while seedlings are still in the nursery. *Parkia pendula* (Willd.) Benth. ex Walp. (Fabaceae) is planted in Amazonian rehabilitation sites, but it is commonly attacked by a gall-forming cecidomyiid. Gall damage in a nursery reached 35% of all leaflets on attacked individuals, and aboveground biomass of these plants was significantly reduced [116].

These results further highlight the importance of considering susceptibility to herbivory when selecting species for restoration as planting particularly vulnerable species may be an unproductive use of the limited resources available for restoration.

Conclusion

The field of plant-insect interactions has grown dramatically since the publication of *Insects on Plants: Community Patterns and Mechanisms* in 1984, the seminal work formalizing the topic [117], and thousands of interesting studies have explored relationships between plants and their herbivores.

Work on the effects of herbivory on tropical seedlings illustrates the intricacies of these interactions, and many questions remain to be explored. Herbivory seems to have its strongest negative effects on very young seedlings, so more work should focus on this, albeit difficult to document, life-stage.

Environmental factors have important effects on plant resource budgets and alterations in their investments in storage, defense, and growth. Further studies of interactions between abiotic variables and herbivory would therefore be instructive, particularly to understand how these relationships will be altered with the multiple manifestations of global change. Scaling up to connect herbivory to plant diversity is a provocative avenue for research, and work is now needed to directly connect herbivory to increased evenness in natural communities. Finally, improvements in tropical forest restoration methods can be made by incorporating knowledge of plant-herbivore dynamics, and more applied work should integrate consideration of these biotic interactions.

While many factors make it difficult for researchers to dedicate themselves to long-term studies, it is clear that the most relevant and revealing results in the future will come from work that addresses the impact of herbivory through time—illustrating the ultimate effects of herbivory on plant performance, forest diversity, restoration success, and responses to global change.

References

[1] W. F. Morris, R. A. Hufbauer, A. A. Agrawal, J. D. Bever, V. A. Borowicz, G. S. Gilbert, J. L. Maron, C. E. Mitchell, I. M. Parker, A. G. Power, M. E. Torchin and D. P. Vazquez, *Ecology* 88(4), 1021–1029 (2007).

[2] A. M. O. Oduor, J. M. Gomez and S. Y. Strauss, *Biol. Invas.* 12(2), 407–419 (2010).

[3] P. R. Ehrlich and P. H. Raven, *Evolution* 18(4), 586–608 (1964).

[4] J. B. Harborne, *Biochemical Aspects of Plant and Animal Coevolution*. Academic Press, New York (1978).

[5] D. J. Futuyma and A. A. Agrawal, *Proc. Nat. Acad. Sci.U.S.A.* 106(43), 18054–18061 (2009).

[6] E. Garrido, G. Andraca-Gomez and J. Fornoni, *New Phytol.* 193(2), 445–453 (2012).

[7] S. J.McNaughton, *Amer. Nat.* 113(5), 691–703 (1979).

[8] K. N. Paige and T. G. Whitham, *Amer. Nat.* 129(3), 407-416 (1987).

[9] J. Maschinski and T. G. Whitham, *Am. Nat.* 134(1), 1-19 (1989).

[10] J. Escarre, J. Lepart and J. J. Sentuc, *Oecologia* 105(4), 501–508 (1996).

[11] K. Poveda, M. I. Gomez Jimenez and A. Kessler, *Ecol. Applic.* 20(7), 1787–1793 (2010).

[12] J. Bergelson, T. Juenger and M. J. Crawley, *Amer. Nat.* 148(4), 744–755 (1996).

[13] L. A. Dyer and P. D. Coley, in T. Tscharntke and B. A. Hawkins (Eds.), Cambridge University Press, Cambridge, pp. 67-88 (2001).

[14] T. J. Massad, J. K. Balch, E. A. Davidson, P. M. Brando, C. Lahís Mews, P. Porto, R. Mota Qintino, S. A. Vieira, B. H. Marimon Junior, S. E. Trumbore, *Oecologia* DOI 10.1007/s00442-012-2482-x (2012).

[15] D. H. Janzen, *Amer. Nat.* 104(940), 501–528 (1970).

[16] J. H. Connell, in P. J. den Boer and G. Gradwell (Eds.), Wageningen, Pudoc, pp. 298-312 (1971).

[17] W. P. Carson, J. T. Anderson, E. G. Leigh and S. A. Schnitzer, in W. P. Carson and S. A. Schnitzer (Eds.), Blackwell Publishing, Malden, pp. 210-241 (2008).

[18] P. D. Coley and J. A. Barone, *Ann. Rev. Ecol. Syst.* 27, 305–335 (1996).

[19] K. Boege and R. J. Marquis, *Trends Ecol. Evol.* 20(10), 526–526 (2005).

[20] D. A. Herms and W. J. Mattson, *Quart. Rev. Biol.* 67(3), 283–335 (1992).

[21] K. E. Barton and J. Koricheva, *Amer. Nat.* 175(4), 481–493 (2010).

[22] R. J. Marquis, *Science* 226(4674), 537–539 (1984).

[23] P. D. Coley, *Ecol. Monogr.* 53(2), 209–233 (1983).

[24] P. D. Coley, *Ecology* 64(3), 426–433 (1983).

[25] J. H. Langenheim and W. H. Stubblebine, *Biochem. Syst. Ecol.* 11(2), 97–106 (1983).

[26] C. Baraloto and P. M. Forget, *Am. J. Bot.* 94(6), 901–911 (2007).

[27] J. A. Myers and K. Kitajima, *J. Ecol.* 95(2), 383–395 (2007).

[28] P. D. Coley, J. P. Bryant and F. S. Chapin, *Science* 230(4728), 895–899 (1985).

[29] J. D. C. Hemming and R. L. Lindroth, *J. Chem. Ecol.* 25(7), 1687–1714 (1999).

[30] A. Imaji and K. Seiwa, *Oecologia* 162(2), 273–281 (2010).

[31] T. A. Kursar and P. D. Coley, *Biochem. Syst. Ecol.* 31(8), 929–949 (2003).

[32] J. W. Dalling, K. E. Harms and R. Aizprua, *J. Trop. Ecol.* 13, 481–490 (1997).

[33] J. W. Dalling and K. E. Harms, *Oikos* 85(2), 257–264 (1999).

[34] K. E. Harms and J. W. Dalling, *J. Trop. Ecol.* 13, 617–621 (1997).

[35] W. Edwards and P. Gadek, *J. Trop. Ecol.* 18, 943–948 (2002).

[36] P. T. Green and P. A. Juniper, *J. Ecol.* 92(3), 397–408 (2004).

[37] I. M. Barberis and J. W. Dalling, *J. Trop. Ecol.* 24, 607–617 (2008).

[38] T. Ichie, I. Ninomiya and K. Ogino, *J. Trop. Ecol.* 17, 371–378 (2001).

[39] S. Alvarez-Clare and K. Kitajima, *Biotropica* 41(1), 47–56 (2009).

[40] K. Kitajima, *Biotropica* 35(3), 429–434 (2003).

[41] W. P. Sousa, P. G. Kennedy and B. J. Mitchell, *Oecologia* 135(4), 564–575 (2003).

[42] D. A. Clark and D. B. Clark, *Amer. Nat.* 124(6), 769–788 (1985).

[43] M. T. Nascimento and J. D. Hay, *J. Trop. Ecol.* 10, 611–620 (1994).

[44] B. L. Webber and I. E. Woodrow, *J. Ecol.* 97(4), 761–771 (2009).

[45] R. W. Myster, *For. Ecol. Manage.* 169(3), 231–242 (2002).

[46] S. A. Foster, *Bot. Rev.* 52(3), 260–299 (1986).

[47] D. H. Janzen, *Evolution* 23(1), 1–27 (1969).

[48] S. S. Rehr, D. H. Janzen and P. P. Feeny, *Science* 181(4094), 81–82 (1973).

[49] D. H. Janzen, C. A. Ryan, I. E. Liener and G. Pearce, *J. Chem. Ecol.* 12(6), 1469–1480 (1986).

[50] K. P. Modi, N. M. Patel and R. K. Goyal, *Chem. Pharmac. Bull.* 56(3), 357–359 (2008).

[51] J. W. Veldman, K. G. Murray, A. L. Hull and M. Garcia-C, *Biotropica* 39(1), 87–93 (2007).

[52] P. A. Ndakidemi and F. D. Dakora, *Func. Plant Biol.* 30(7), 729–745 (2003).

[53] J. Lammerink, D. B. MacGibbon and A. R. Wallace, *New Zeal. J. Agric. Res.* 27(1), 89–92 (1984).

[54] A. A. Agrawal, C. Laforsch and R. Tollrian, *Nature* 401(6748), 60–63 (1999).

[55] A. A. Agrawal, *Ecology* 83(12), 3408–3415 (2002).

[56] D. Kestring, L. C. C. R. Menezes, C. A. Tomaz, G. P. P. Lima and M. N. Rossi, *J. Plant Biol.* 52(6), 569–576 (2009).

[57] M. P. Eichhorn, R. Nilus, S. G. Compton, S. E. Hartley and D. F. Burslem, *Ecology* 91(4), 1092–1101 (2010).

[58] T. Brenes-Arguedas, P. D. Coley and T. A. Kursar, *Ecology* 90(7), 1751–1761 (2009).

[59] M. A. W. Pacheco, *J. Ecol.* 89(3), 358–366 (2001).

[60] R. V. Jackson and C. E. Bach, *Austr. J. Ecol.* 24(3), 278–286 (1999).

[61] A. G. Blundell and D. R. Peart, *J. Ecol.* 89(4), 608–615 (2001).

[62] J. M. Norghauer, J. Grogan, J. R. Malcolm and J. M. Felfili, *Oecologia* 162(2), 405–412 (2010).

[63] P. Alvarez-Loayza and J. Terborgh, *J. Ecol.* 99(4), 1045-1054 (2011).

[64] D. P. Bebber, N. D. Brown and M. R. Speight, *J. Trop. Ecol.* 20, 11–19 (2004).

[65] D. Bebber, N. Brown and M. Speight, *J. Trop. Ecol.* 18, 795–804 (2002).

[66] J. M. Norghauer, J. R. Malcolm and B. L. Zimmerman, *J. Ecol.* 96(1), 103–113 (2008).

[67] H. S. Ballina-Gómez, S. Iriarte-Vivar, R. Orellana and L. S. Santiago, *J. Trop. Ecol.* 26(2), 163–171 (2010).

[68] M. L. J. Katjiua and D. Ward, *J. Chem. Ecol.* 32(7), 1431–1443 (2006).

[69] K. Gerhardt, *Trees-Struct. Funct.* 13(2), 88–95 (1998).

[70] T. R. H. Pearson, D. F. R. P. Burslem, R. E. Goeriz and J. W. Dalling, *J. Ecol.* 91(5), 785–796 (2003).

[71] S. Harrison, *Oecologia* 72(1), 65–68 (1987).

[72] C. M. Nichols-Orians, *Oecologia* 86(4), 552–560 (1991).

[73] B. E. Howlett and D. W. Davidson, *J. Trop. Ecol.* 17, 285–302 (2001).

[74] T. R. H. Pearson, D. F. Burslem, R. E. Goeriz and J. W. Dalling, *Oecologia* 137(3), 456–465 (2003b).

[75] P. Angulo-Sandoval and T. M. Aide, *J. Trop. Ecol.* 16, 447–464 (2000).

[76] A. G. Blundell and D. R. Peart, *Am. J. Bot.* 87(11), 1693–1698 (2000).

[77] J. S. Denslow, *Biotropica* 12(3), 220–222 (1980).

[78] L. A. Wirf, *Biol. Control* 37(3), 346–353 (2006).

[79] K. Dhileepan, C. J. Lockett, M. Robinson and K. Pukallus, *Ann. Appl. Biol.* 154(1), 97–105 (2009).

[80] S. J. Franks, A. M. Kral and P. D. Pratt, *Environ. Ent.* 35(2), 366–372 (2006).

[81] T. M. Aide and J. K. Zimmerman, *Ecology* 71(4), 1412–1421 (1990).

[82] J. Gurevitch and L. V. Hedges, in S. M. Scheiner and J. Gurevitch (Eds.), Oxford University Press, New York, pp. 347-369 (2001).

[83] P. D. Coley, *Rev. Biol. Trop.* 35, 151–164 (1987).

[84] P. V. A. Fine, Z. J. Miller, I. Mesones, S. Irazuzta, H. M. Appel, M. H. Stevens, I. Sääksjärvi, J. C. Schultz and P. D. Coley, *Ecology* 87(7), S150–S162 (2006).

[85] P. V. A. Fine, I. Mesones and P. D. Coley, *Science* 305(5684), 663–665 (2004).

[86] L. S. Comita and **S**. P. Hubbell, *Ecology* 90, 328–334 (2009).

[87] L. S. Comita, H. C. Muller-Landau, S. Aguilar and S. P. Hubbell, *Science* 329(5989), 330–332 (2010).

[88] E. W. Schupp, *Amer. Nat.* 140(3), 526–530 (1992).

[89] M. J. Crawley and C. R. Long, *J. Ecol.* 83(4), 683–696 (1995).

[90] S. Raghu, C. Wiltshire and K Dhileepan, *Austral Ecol.* 30(3), 310–318 (2005).

[91] V. Swamy, J. Terborgh, K. G. Dexter, B. D. Best, P. Alvarez and F. Cornejo, *Ecol. Let.* 14(2), 195–201 (2011).

[92] S. A. Queenborough, D. F. R. P. Burslem, N. C. Garwood and R. Valencia, *Ecology* 88(9), 2248–2258 (2007).

[93] J. H. Connell, J. G. Tracey and L. J. Webb, *Ecol. Monogr.* 54(2), 141–164 (1984).

[94] K. E. Harms, S. J. Wright, O. Calderón, A. Hernández and E. A. Herre, *Nature* 404(6777), 493–495 (2000).

[95] V. Swamy and J. Terborgh, *J. Ecol.* 98(5), 1096-1107 (2010).

[96] R. Bagchi, T. Swinfield, R. E. Gallery, O. T. Lewis, S. Gripenberg, L. Narayan and R. P. Freckleton, *Ecol. Let.* 13(10), 1262–1269 (2010).

[97] S. A. Mangan, S. A. Schnitzer, E. A. Herre, K. M. L. Mack, M. C. Valencia, E. I. Sanchez and J. D. Bever, *Nature* 466(7307), 752–755 (2010).

[98] J. H. Connell and M. D. Lowman, *Amer. Nat.* 134(1), 88–119 (1989).

[99] P. M. Forget, *Biotropica* 24(2), 146–156 (1992).

[100] M. E. Sanchez-Hidalgo, M. Martinez-Ramos and F. J. Espinosa-Garcia, *Funct. Ecol.* 13(5), 725–732 (1999).

[101] J. X. Becerra, in *Proc. Nat. Acad. Sci. U.S.A.* 104(18), 7483–7488 (2007).

[102] T. J. Massad, J. Q. Chambers, S. G. Rolim, R. M. Jesus and L. A. Dyer, *Restor. Ecol.* 19(SI), 257-267 (2010).

[103] T. J. Massad, *Appl. Veg. Sci.* 5(2), 125-139 (2012).

[104] F. Montagnini, E. Gonzalez, C. Porras and R. Rheingans, *Commonw. Forest Rev.* 74, 306–314 (1995).

[105] M. Plath, K. Mody, C. Potvin and S. Dorn. *Forest Ecol. Manage.* 261(3), 741–750 (2011).

[106] F. Tilki and R. F. Fisher, *Forest Ecol. Manage.* 108(3), 175–192 (1998).

[107] A. Moulaert, J. P. Mueller, M. Villarreal, R. Piedra and L. Villalobos, *Agroforestry Syst.* 54(1), 31–40 (2002).

[108] D. C. Nepstad, C. Uh., C. A. Pereira and J. M. Cardoso da Silva, *Oikos* 76(1), 25–39 (1996).

[109] H. L. Vasconcelos and J. M. Cherrett, *J. Trop. Ecol.* 13, 357–370 (1997).

[110] A. S. Guimarães Neto, J. M. Felfili, G. Fernandes da Silva, L. Mazzei, C. W. Fagg and P. E. Nogueira, *Revista Árvore* 28(6), 777-784 (2004).

[111] D. Piotto, E. Víquez, F. Montagnini and M. Kanninen, *Forest Ecol. Manage.* 190(2-3), 359-372 (2004).

[112] D. R. Perez-Salicrup and R. Esquivel, *Forest Ecol. Manage.* 255(2), 324–327 (2008).

[113] F. D. Menalled, M. J. Kelty and J. J. Ewel, *Forest Ecol. Manage.* 104(1-3), 249–263 (1998).

[114] R. B. Root, *Ecol. Monogr.* 43(1), 95–120 (1973).

[115] P. Barbosa, J. Hines, I. Kaplan, H. Martinson, A. Szczepaniec and Z. Szendrei, *Ann. Rev. Ecol. Evol. Syst.* 40, 1–20 (2009).

[116] G. W. Fernandes, J. C. Santos, F. M. C. Castro and A. Castilho, *Rev. Bras. Entom.* 51(4), 471–475 (2007).

[117] D. R.Strong, J. H. Lawton and T. R. E. Southwood, *Insects on Plants: Community Patterns and Mechanisms*. Cambridge, Massachusetts (1984).

Cool-Temperate, Deciduous, Broad-Leaved Forest Ecosystems

In: From Seed Germination to Young Plants
Editor: Carlos Alberto Busso

ISBN: 978-1-62618-653-8
© 2013 Nova Science Publishers, Inc.

Chapter 17

Environmental Conditions after Seed Dispersal of Spring Ephemeral Species at the Floor of Deciduous, Broad-Leaved Forests

Kojiro Suzuki[*]

Department of Landscape Architecture, Faculty of Regional Environmental Science,
Tokyo University of Agriculture, Tokyo, Japan

Abstract

The vegetation in Japan is classified into four zones from north to south; evergreen coniferous forests at the subarctic zone; deciduous, broad-leaved forests at the cool-temperate zone; evergreen, broad-leaved forests at the warm-temperate zone; and subtropical rain forests at the subtropical zone. Spring ephemerals (e.g., *Erythronium japonicum*, *Anemone debilis*, *Trillium apetalon*, *Epimedium grandiflorum* var. *thunbergianum*, *Disporum sessile* and *Hepatica nobilis* var. *japonica*) are distributed at the floor of the deciduous, broad-leaved forests, or substitution forests in evergreen, broad-leaved forests, kept by thinning. The seeds of *Trillium apetalon,* a representative spring ephemeral, are dispersed in the litter layer with the myrmecochory after seeds are produced in early summer. They can germinate only under the litter in winter, and thereafter the first leaf (plumule) emerges above the leaf litter in spring. The relationship between the seed germination habit of *Trillium apetalon* (Trilliaceae) and the litter layer is discussed under the particular temperature, light, and water conditions at the floor of a deciduous, broad-leaved forests.

---

[*] E-mail address: kojiros@nodai.ac.jp.

Introduction

Satoyama is the village forest where a semi-natural area is nearby a populated area at a deciduous broad-leaved forest in Japan (Figures1 and 2). Trees in the overstory of Satoyama include *Carpinus japonica, C. laxiflora, C. tschonoskii, Quercus crispula, Q. acutissima, Q. variabilis, Q. dentata, Q. serrata, Castanea crenata, Zelkova serrata, Magnolia obovata, M. praecocissima, Prunus jamasakura, Rhus succedanea, Acer amoenum* var. *matsumurae, A. mono* var. *marmoratum* f. *dissectum* and *Styrax japonica. Pourthiaea villosa, Callicarpa mollis, Viburnum erosum* var. *punctatum* and *Lonicera gracilipes* var. *glabra* are shrub species in the understory. Wood of these trees is useful for architecture or furniture.

Satoyama village forests were kept almost intact by human life or activity until half a century ago. However, people used the forest trees with several purposes. They include (i) Periodic forest thinning; cut woods were used as fuel and the production of shiitake mushroom (*Lentinus edodes*); (ii) Fallen leaves were used for making compost (leaf mold) to produce vegetation in the field; (iii) Some trees were used for making furniture (*Abies firma, Torreya nucifera, Acer mono* var. *marmoratum* f. *dissectum* and *Paulownia tomentosa*, etc.), glazing (*Rhus verniciflua*), materials for candles (*Rhus succedanea*) and for wrapping food (*Magnolia obovata* and *Idesia polycarpa*). These tree species are specially protected and kept, and planted in the coppice forest; (iv) Plants at the forest floor were often used as edible plants, medicinal herbs or ornamental plants (i.e. *Erythronium japonicum, Lilium auratumn, Polygonatum odoratum* var. *pluriflorum* and *Allium grayi*) and (v) Sine a lot of insects live in the coppice forest, children used to catch and keep *Oryzias latipes, Misgurnus anguillicaudatus*, a stag beetle (Lucanidae) and a Japanese rhinoceros beetle (*Allomyrina dichotoma*) for pleasure until several years ago. However, the Satoyama area has decreased recently. This was because the land was used with other purposes: housing and building of offices, buildings, factories, hospitals or shopping malls. This was the result of adaptation to the modern life style, which mainly uses oil energy.

Figure 1. Landscape of Satoyama, a village forest of Yokohama city, Kanagawa Prefecture, central Japan.

Figure 2. Vegetation of Satoyama, *Quercus crispula*, *Q. acutissima*, *Q. variabilis*, *Q. dentata* and *Q. serrata*, in Fukushima Prefecture, Northern Japan.

Spring Ephemerals

Many plant species can be found in the forest floor under the deciduous, broad-leaved trees (i.e., *Ophiopogon japonicus*, *Liriope platyphylla*, *Lilium auratum*, *Polygonatum odoratum* var. *pluriflorum*, *Disporum sessile*, *D. smilacinum* and *Heterotropa nipponica*). Among them, there are many spring ephemeral species (i.e., *E. japonicum*, *T. apetalon*, *T. tschonoskii* –LILIACEAE-, *Cephalanthera falcata*, *C. longibracteata*, *C. erecta* – ORCHIDACEAE-, *Coptis japonica*, *Hepatica nobilis* var. *japonica*, *Anemone flaccida*, *A. raddeana*, *A. pseudo-altaica* and *Adonis amurensis*; Figure 3).

Life History of Spring Ephemerals

Spring ephemerals are mainly distributed at the forest floor of deciduous, broad-leaved trees in cool-warm temperate zones, or on substitution forests in evergreen, broad-leaved forests, kept by thinning.

Adult plants which undergo dormancy as underground bulbs or roots during winter grow leaves with scape and produce flowers, before deciduous, broad-leaved trees break dormancy and open the leaves in early spring. Each plant finishes blooming until deciduous, broad-leaved trees open the leaves in spring. Seeds are dispersed by early summer right after the fruits are produced, and shoots (leaves and scape) are dispersed highly at that time.

The dormant stage apparently occurs in summer, lasting for at least five months, from May to September. The sprouting of new roots in occurs underground in October. New aerial shoots are formed within the bulb. They remain underground during winter, until spring comes [1].

Erythronium japonicum   *Erythronium japonicum*

Trillium apetalon   *Trillium tschonoskii*

Hepatica nobilis var. *japonica*   *Hepatica nobilis* var. *japonica*

Anemone pseudo-altaica   *Anemone pseudo-altaica*   *Cephalanthera falcata*

Figure 3. Some flowers of spring ephemerals at the forest floor.

Dispersal and Seed Germination Strategy of *Trillium apetalon* (Trillaceae)

Trillium apetalon, a representative spring ephemeral, has unique dispersal and seed germination strategies in its life history (Figure 4).

a) The shoot (scape and leaf) appears from underground and blooms, just after shoot appearance in early spring, when snow melts. At this time, there is no canopy of deciduous, broad-leaved trees. Shoots of *T. apetalon* disappear from the ground when the deciduous, broad-leaved trees develop a canopy (Figure 4).

b) Seeds of *T. apetalon* are dispersed by myrmecochory in late spring or early summer (around June). *Trillium kamtschaticum* and *T. tschonoskii* have *Aphaenogaster smythiesi japonica* and *Mymica ruginodis* [2] as main agents of seed dispersal. *Erythronium japonicum* is also a representative spring ephemeral. Ants (*Polyrhachis lamellidens* and *Aphaenogaster smythiesi japonica*) are strongly attracted to seeds with elaiosomes, and carry them into their nests; however, all such seeds are thereafter carried out from their nests and abandoned in the area, around the nests [1]. Thus, seeds are moved to the soil surface, under the leaf litter layer.

c) Dispersed seeds remain under the leaf litter layer until autumn. When over the leaf litter layer the temperature is beyond 25°C that one under the leaf litter layer is around 20°C during summer.

d) When leaf falling occurs from the deciduous, broad-leaved trees in autumn, from late September to October, dispersed seeds germinate. Radicles protrude through the seed coats under the leaf litter layer. However, some seeds do not germinate in autumn; they do so in the next spring or late summer.

e) Seeds placed under the leaf litter layer with snow in winter (from November to March) produce a radicle, and thereafter a plumule (first leaf) until early spring. At this time, temperatures reach subzero values above the snow cover or leaf litter layer. However, temperatures around the dispersed seeds under the leaf litter layer with snow cover keep from 0 to 5°C.

f) When leaves grow from adult plants and stems from underground, and bloom in early spring, the plumule (first leaf) of *T. apetalon* seedlings appear after breaking through the fallen leaves (Figure 5). However, it may represent the cotyledon of *E. japonicum*, which has broken the leaf litter layer. Incidentally, the leaf apex of adult plants in *E. japonicum* is cuspidate enough to break the fallen leaves (Figure 6).

Temperature, Light and Water Effects on the Germination of *T. apetalon* (Trilliaceae), a Representative Spring Ephemeral

Seeds of spring ephemerals need special conditions during their germination (defined as the appearance of the radicle after breaking through the seed coat) and seedling stage (appearance of plumules, first leaf, from under the leaf litter layer to above it) periods.

(1) & (6) Early Spring: late March to early April

Adult plants bloomed flower before the deciduous tree develope the leaves.
Seedling: the first leaf is prolonged breaking trough leaves.

(1)   (6)

 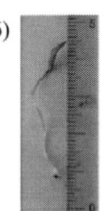

The seeds are dispersed with Myrmecochory just after fruit is fructified.

(2) Early Summer: early June

(3) Summer: June to September

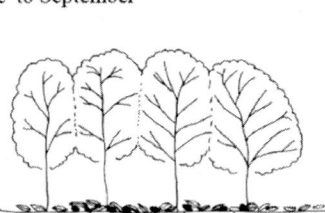

Above litter : 25-30℃
Under litter : around 20℃

The seeds are put under leaf litter layer.
The temperature of above litter show from 25 to 30℃, under litter is around 20℃.

(4) Autumn: Late September to October

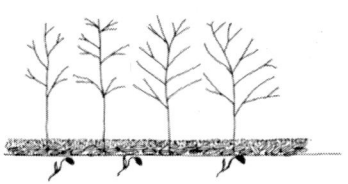

The seed germinate under leaf litter layer, when the leaves fall off from deciduous - leaves trees.

(5) Winter: November to February

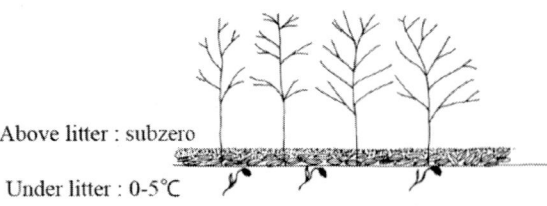

Above litter : subzero

Under litter : 0-5℃

The first leaf is appeared during winter, existing cotyledon. The temperature of above litter show subzero, but under litter is from zero to 5℃.

Figure 4. Seed germination strategy of *Trillium apetalon*, a representative species of spring ephemerals.

Temperature Conditions for Seed Germination

The *T. apetalon* seeds germinated well under 20°C, after exposure to 20°C for about two months, 10°C for three months, and 5°C during three months (Figure 7). However, germination was low at 15°C, after exposing seeds at constant temperatures of 5°C for two months, 10°C for three months, and 5°C for three months (Figure 7). Constant temperatures of 20°C have no effect on the germination of *T. apetalon* seeds collected after seed dispersal (Figure 7).

Figure 5. Seedlings of *Erythronium japonicum* breaking through the fallen litter in early spring.

Light Condition for Seed Germination

Fallen leaves are cut off from the sunlight [3]. *Trillium apetalon* seeds are dispersed to under the leaf litter layer by myrmecochory. It means they need to germinate under dark conditions. In fact, *T. apetalon* seeds can germinate under both light and dark conditions. Thus, seeds of *T. apetalon* can germinate even though they are under the leaf litter layer.

Water Condition for Seed Germination

The leaf litter layer, where seeds are dispersed to the soil surface water contents. Snow cover on the leaf litter layer also maintains the high water contents. However, *T. apetalon* seeds may lose seed germinability by drying. Thus, seeds of *T. apetalon* dispersed under the leaf litter layer by myrmecochory are able to keep germinability by the covering leaf litter layer, or sometimes because of snow cover.

Figure 6. The leaf shape of adult plants, and adult plants of *Erythronium japonicum* breaking through the fallen litter in early spring.

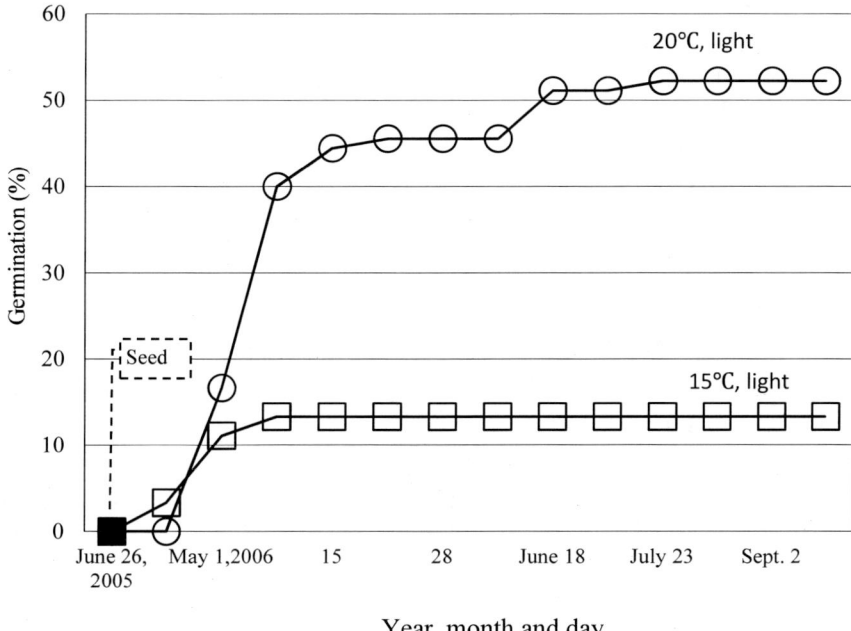

Figure 7. Seed germination of *Trillium apetalon* under different temperature conditions. Seeds were collected on June 26, 2005 in Fukushima prefecture, northern Japan. Seed germination tests were done with elaiosomes in an artificially controlled environment. ○: The seed lots were exposed to 20°C under light conditions after storage to 20°C for two months, 10°C for three months, and 5°C for three months under wet conditions. :The seed lots were exposed to 15°C under light conditions after storage to 5°C for two months, 10°C for three months, and 5°C for three months under wet conditions. No germination occurred when temperature was maintained constant at 20°C (unpublished).

Conclusion

There have been many arguments on the significance of myrmecochory for seed dispersal and establishment [4]. The maximum and mean distance of dispersal reported for *T. tschonoskii* seeds have been 2.7 m and 0.64 ± 0.41 m (mean ± 1SD), respectively [5]. *Aphaenogaster smythiesi japonica* and *M. ruginodis* were the main agents for the dispersal of *Trillium kamtschaticum* and *T. tschonoskii* seeds; the dispersal distance was either 3-30 m (maximum) or 0-60 m (average). Only about 15% of seeds were removed by ants and the other 85% were left near the mother plants, though all of them lost elaiosomes. Ingestion of elaiosomes by ground beetles may cause a previously reported clumping of seedlings near adult plants [2]. The ant *Myrmica kotokui* frequently carries *E. japonicum* seeds. However, the frequency of seed removal has been reported to be low, and most seeds are dispersed as far as 1 m or less. The spatial distribution of *E. japonicum* individuals was alomost random, and most seedlings were established 5-20 cm away from the fertile plants. However, even this small scale of seed dispersal contributes to avoid crowding of seedlings [6]. Ants (*P. lamellidens* and *A. smythiesi japonica*) are strongly attracted to seeds with elaiosomes and carry them into their nests; however, all such seeds might soon be abandoned in the area surrounding the nests [1]. All seeds that fall from fruits to above the leaf litter layer go down to such litter layer by wind or finger touch directly.

Myrmecochory, thereafter appears essential for the germination of *Trillium*. It allows seeds of the spring ephemerals to be deposited below a thick, mesic leaf-litter layer on the forest floor after seed dispersal. There is little effect of putting away the second generation (seedling) from the mother plant for myrmecochory. Ohkawara et al. [6] argued that this response may facilitate the germination and establishment of *E. japonicum* seedlings. Myrmecochory has the role of carryinging seeds under the leaf litter layer.

Leaf litter has various roles [7]. Litter helps to prevent (i) the freezing of soils, or (ii) make the freezing thinner [8]. Litter also protects the fern apices from freezing damage during winter [9, 10]. The air temperature has been higher than that of the soil surface under the litter [11]. Other studies found that both burning and hand litter removal increased the soil temperatures in a glass land by 5°C during the entire growing season [12].

The importance of the leaf litter layer on seed germination habit of *T. apetalon* has been reported in other studies [13]. Leaf litter is one of the important elements of the forest. After dispersal, it allows the basic germination conditions of *T. apetalon* seeds. In short, (i) leaf-litter cover on the forest floor maintains a moderate temperature of approximately 20°C, even during the summer in central Honshu; (ii) seed germination rates decrease under moist and high temperatures of 25–30°C, almost the same temperatures than above the leaf litter; (iii) *T. apetalon* seeds are able to germinate under dark conditions: the thick leaf-litter cover block sunlight; and (iv) seeds are covered with a thick leaf-litter layer that maintains a remarkably high water content and germinability, thus reducing germinability losses by drying.

Vegetation has to be increased in urban or suburban areas by planting forest plant species which provide a forest floor, where spring ephemerals may grow. This is because the green space in the suburb has been decreased by human activities. The plant technique of using leaf litter will help to keep the current conditions in the deciduous, broad-leaved forest floor.

Acknowledgments

I thank Professor Hiroto Iwanaga, of Tokyo University of Agriculture, for correcting the English in this paper.

References

[1]   S. Kawano, *Plant Sp.Biol.* 20, 67–74 (2005).
[2]   M. Ohara and S. Higashi, *J. Ecol.* 75, 1091-1098 (1987).
[3]   K. Suzuki, *J. Landsc. Architect.* Extra Issue 10, 1–127 (2006).
[4]   C. C. Baskin and M. J. Baskin, *Seeds*. Academic Express, NewYork. 666pp (2001).
[5]   S. Higashi, S. Tsuyuzaki, M. Ohara and F. Ito, *Oikos* 54, 389-394 (1989).
[6]   H. Ohkawara, S. Higashi and M. Ohara, *Oecologia* 106, 500-506 (1996).
[7]   J. M. Facelli and S. T. A. Pickett, Bot. Rev. 57, 1-32 (1991).
[8]   A. L. McKinney, *J. Range Manage.* 40, 119-121 (1929).
[9]   A. S. Watt, *New Phytol.* 55, 369-388 (1956).

[10] A. S. Watt, *New Phytol.* 69, 431-449 (1970).

[11] J. E. Weaver and N. W. Rowland, *Bot. Gaz.* 114, 1-19 (1952).

[12] L. C. Hulbert, *Ecology* 50, 874-877 (1969).

[13] K. Suzuki and S. Kawano, *Plant Sp. Biol.* 25, 231-239 (2010).

Mine Tailing Ecosystems

In: From Seed Germination to Young Plants
Editor: Carlos Alberto Busso

ISBN: 978-1-62618-653-8
© 2013 Nova Science Publishers, Inc.

Chapter 18

Mycorrhizal Fungi from Mine Sites and Rehabilitation of Marginal and Mine Sites in Surigao, Philippines

Nelly S. Aggangan[1,*], *Ma. Ruth R. Edradan*[2], *Grace B. Alvarado*[2], *PrexyPearl C. Macana*[2], *Lafayette Kirsi S. Noel*[2] *and Debra Ruth R. Edradan*[2]

[1]National Institute of Molecular Biology and Biotechnology (BIOTECH), University of the Philippines Los Baños, College, Laguna, Philippines
[2]Caraga Regional Science High School, Surigao Sur, Philippines

Abstract

Most mining companies are unsuccessful in the rehabilitation of their mine tailing areas due to the unfavorable environmental conditions particularly the elevated concentrations of metals. Copper and nickel are the top metals being mined in the Philippines. This study was conducted to assess the mycorrhizal diversity in two mine sites in Surigao del Norte, Philippines, to compare the plant growth promoting effect of locally developed mycorrhizal inoculants with the mycorrhizal fungi native in the mine sites, and to select the best mycorrhizal fungi-reforestation tree species combination for the rehabilitation of marginal and Cu and Ni mine sites in Surigao.

Seedlings of mangium (*Acacia mangium*), falcata (*Paraserianthes falcataria*) and mangkono (*Xanthostemon verdugonianus*) were inoculated with a mix spores of arbuscular mycorrhizal (AM) fungi native in the mine sites or commercially available mycorrhizal inoculants with brand name Mykovam and Mycogroe and grown in marginal or mine waste soils. Mykovam is a soil-based biofertilizer consisting of AM fungi: *Gigaspora margarita, Glomus macrocarpum, G. etunicatum, G. fasciculatum* and other unidentified *Glomus* species. Mycogroe, on the other hand, consists of spores of ectomycorrhizal (ECM) fungi *Scleroderma* and *Pisolithus*. Height and diameter were measured periodically and biomass production, mycorrhizal root colonization and heavy

---

* corresponding author.

metal accumulation were assessed at harvest. The native ECM fungi observed in plants growing in the mine sites were *Scleroderma* and *Thelepora* while AM fungi were species under the genera: *Glomus*, *Gigaspora*, *Acaulospora* and *Entrophospora*.

After three months in the nursery, mycorrhiza infected roots were higher in marginal than in mine soil. Mycorrhizal plants grew better in marginal than in mine soil. Mycorrhiza inoculated plants outperformed the uninoculated counterpart. The native mycorrhizal fungi promoted the best growth of mangium and mangkono. On the other hand, inoculation did not promote growth of falcata seedlings. Mycorrhizal plants contained higher P than the uninoculated counterpart. The highest Cu concentration was observed in mangium, inoculated with the native mycorrhizal fungi. In Cu and Ni mine tailing sites, mangium inoculated with the native promoted higher seedling survival and improved plant growth comparable with the commercial mycorrhizal inoculants. In conclusion, mangium inoculated with the native mycorrhizal fungi can be planted to rehabilitate marginal and mine sites in Surigao and perhaps in other mine tailing sites in the Philippines with high Cu and Ni deposits.

Keywords: Arbuscular mycorrhizal fungi, ectomycorrhizal fungi, heavy metals, copper

Introduction

Most mining companies are unsuccessful in the rehabilitation of their mine tailing areas due to the unfavorable environmental conditions particularly the elevated concentrations of metals. Copper (Cu) and nickel (Ni) are the top metals being mined in the Philippines. The toxic effects of heavy metals and the infertile state of the soil do not favor plant growth and survival. Thus, mine out soils are normally devoid of plants. The use of plants is considered as the most viable, safe and cheap method of rehabilitating mine tailing dump sites where toxic metals are absorbed by the plants and thus, prevented from leaching to the underground water.

Mycorrhizal fungi can improve growth and survival in infertile soil and can increase tolerance of plants to environmental stresses such as heavy metals. It is therefore envisioned that the use of mycorrhizal fungi and plants native or indigenous in the mine tailing sites can be tapped for a successful rehabilitation. Plants can absorb more metals in the presence of mycorrhizal fungi in their roots and keep the metals as long as the plant lives (Aggangan 1996, Aggangan et al. 2007).

Mycorrhizal inoculants with brand name Mykovam and Mycogroe have been developed at the National Institute of Molecular Biology and Biotechnology (BIOTECH), University of the Philippines Los Baños, and are now commercially available in the country. Mykovam consists of arbuscular mycorrhizal (AM) fungi which are effective in promoting growth, survival and yield of a variety of agricultural crops, forestry crops, fruit crops, ornamentals and forage crops. Mycogroe contains ectomycorrhizal (ECM) fungi which are specific to pines, eucalypts, dipterocarps and acacias though eucalypts and acacias are also associated with AM fungi (Brundrett et al. 1996).

The beneficial effects of mycorrhizal fungi in increasing plant tolerance and the capability of extracting heavy metals have been shown in many reports. For example, Reyes et al. (2006) inoculated narra seeds with Mykovam or MineVAM (mixture of native AM fungi in mine sites in Luzon island) during seed sowing in garden soil, and three weeks after, the treated seedlings were transplanted in non-sterilized garden or mine waste soils of Antamok, Benguet (Luzon island) and Toledo, Cebu (Visayas island), Philippines. Reyes et al. (2006) reported that total biomass of mycorrhizal plants was significantly reduced in mine waste soils. By contrast, percent mycorrhizal infection and nutrient uptake were significantly increased through inoculation with MineVAM. Phosphorus nutrition of narra was greatly improved by mycorrhizal inoculation thus, in spite of the high concentration of Cu and Pb in the plant tissues, toxicity symptoms were not observed. Most of the Cu and Pb were concentrated in the roots of mycorrhizal plants. The effect of native MineVAM inoculation was significantly greater than MYKOVAM inoculation in promoting growth and heavy metal uptake of narra grown in mine waste soil. Similarly, Landim (2003) reported greater growth increases when native mixed inoculum from plantation or forest plots were used than with pure inoculum of *Glomus clarum*. She suggested that native inoculum can be successfully used in restoration of deforested regions in Northeast Brazil.

To date, there are no studies conducted on the utilization of mycorrhizal fungi for the reforestation of Cu rich sites specifically in Manila Mining Corporation (MMC), Placer, Surigao del Norte in the island of Mindanao, Philippines. It was envisioned that the use of indigenous mycorrhizal fungi and plants surviving in such sites will contribute to the success of rehabilitation or reforestation of denuded mining areas and mine tailing dumpsites.

This study was conducted to assess the mycorrhizal diversity in two mine sites in Surigao del Norte, Philippines, to compare the plant growth promoting effects of locally developed and commercially available mycorrhizal inoculants (Mykovam and Mycogroe) with the mycorrhizal fungi native in mine sites on growth and survival of mangium (*Acacia mangium*), falcata (*Paraserianthes falcataria*), and mangkono (*Xanthostemon verdugonianus*) in marginal or mine waste soils of MMC, Placer, Surigao del Norte, Philippines, and to select the best mycorrhizal inoculant-reforestation tree species combination for the rehabilitation of marginal and Cu and Ni mine sites in Surigao.

Methodology

Experimental Design

The experimental design used was a three factor (tree species, soil and inoculation treatments) in a Randomized Complete Block Design (RCBD). The experiment was conducted at the Department of Environment and Natural Resources Provincial Environment and Natural Resources Office-Community Environment and Natural Resources Office (DENR-PENRO-CENRO) in Surigao City.

Reconnaissance Survey at the Mine Tailing

Field survey was conducted to assess the mycorrhizal diversity and plants (native and reforestation species) thriving in the marginal and mine waste sites of the Manila Mining Corporation (MMC), Placer, Surigao del Norte (Mindanao island), Philippines. Soil information, site details, morphological details and identity of plants and mycorrhizal species were recorded.

Roots and rhizosphere soil (0-20cm depth) of five dominant plant species and fruit body of ECM fungi were collected. In the laboratory, root samples were cleared and stained with trypan blue (Philips and Hayman, 1970) where mycorrhizal colonization was counted. Likewise, mycorrhizal spore count was obtained from ten grams dry rhizosphere soil samples.

The soils for pot experiments were collected at the top 20 cm depth in a marginal (covered with talahib grass, *Saccharum spontaneum*) and mine waste soil area (adjacent to each other) in Suyoc dumpsite of MMC, Placer, Surigao del Norte. The soils were air-dried, pulverized and passed through a 5 mm screen. Two hundred fifty grams of dry marginal or mine waste soils were dispensed into plastic pots with dimensions of 6" (top diameter) x 4" (bottom diameter) and 4" height and with a hole at the bottom for drainage.

The physical and chemical analyses were done at the Bureau of Soil based at UP Los Baños and sample solutions were read at the Bureau of Soil and Water Management, Diliman, Quezon City using an Atomic Absorption Spectrophotometer. Analytical techniques were based from PCARRD (1991). Soil samples collected from Suyoc dumpsite (mine waste soil) has the following characteristics: pH 5.0, 0.03% nitrogen, 5.18 ppm phosphorus, 0.22 exchangeable K [cmol](+)/kg soil, 1.7% OM, 18.24 [cmol] Ni/kg soil, 106 ppm Cu, and silty clay soil texture.

Biological Materials

Mangium *(Acacia mangium,* Fabaceae*)* falcata *(Paraserianthes falcataria,* Leguminosae), mangkono *(Xanthostemon verdugonianus,* Myrtaceae*)* were used as test plants. These exotic (mangium and falcata, found dominant in MMC) and endemic (mangkono) forest species are being used for reforestation in abandoned mines and mine tailing sites in Mindanao.

The soil inoculants used in the experiments were collected in the rhizosphere of mangium growing in mine tailings in MMC, Placer, Surigao del Norte. The selection was based on the spore count and diversity of mycorrhizal fungi it contained. The roots of mangium were found to be associated with both ecto and endomycorrhizal (or AM) fungi. It consisted of *Glomus, Gigaspora, Acaulospora* and *Entrophospora.* Sporocarps of ECM fungi *Scleroderma* and *Thelepora* were observed and the roots were prevalently colonized with *Scleroderma* (white fungus). The effect of the native inoculants (from mangium) was compared with the commercial mycorrhizal inoculants (Mykovam and Mycogroe) available in the Philippines produced at the National Institute of Molecular Biology and Biotechnology (BIOTECH), University of the Philippines Los Baños, College, Laguna, Philippines. Mycogroe tablet contains spores of ECM fungi *Scleroderma* and *Pisolithus* mixed in a carrier while Mykovam consists of spores, infected roots and other infective propagules of *Gigaspora margarita, Glomus etunicatum, Gl. macrocarpum, Gl. fasciculatum* and other unidentified species of *Glomus.*

Inoculation and Transplanting

Seeds of mangium and falcata were provided by DENR-PENRO-CENRO, Surigao City while mangkono was provided by Taganito Mining Corporation, Surigao, Philippines. Seeds were pre-germinated in seed boxes filled with garden soil. One-month old mangium and falcata and six-month old mangkono seedlings were transplanted into pots filled with either marginal soil or mine waste soil. Inoculation was done during transplanting. Mycorrhizal inoculants used and rate of inoculation were: indigenous mycorrhizal fungi coded as MMC soil inoculant at 10g/plant, Mycogroe at one tablet (0.34g weight/pc) plant, Mykovam at 5g/plant, and combination of Mykovam+Mycogroe. MMC soil inoculant and Mykovam contained 50-100 spores per plant. Inoculants were placed in a one to two inches depth hole where the seedlings were inserted. The roots were in contact with the mycorrhizal inoculants. The uninoculated or control seedlings were planted in a similar manner but without mycorrhizal fungi. Nothing was added in the control treatment. The holes were carefully filled with soil and watered using a wash bottle.

Parameters Measured

Height and Diameter

Plant height was measured one inch above the soil surface to the apical tip and was done once a month for three months. Height increment was calculated as the difference between the monthly height measurement and the initial height obtained during mycorrhizal inoculation. Initial diameter was measured during inoculation time while the final diameter was measured after three months. Diameter increment was calculated similar to that of height increment.

Plant Biomass

Plants were harvested after three months in the nursery. The shoot was cut one inch above the root collar and washed in running tap water in order to remove soil and dirt that may have been splashed over during watering. The stem and leaves were partitioned and wrapped separately with tissue paper. The roots were also washed carefully with running tap water. The fine (<0.2mm diameter) and the coarse roots were separated. Two grams root samples were soaked in 50% ethyl alcohol for the assessment of roots colonized by mycorrhizal fungi. Fine roots, coarse roots, stem and leaves were oven-dried at 70^0C for three days. Dry weights were obtained using an analytical balance after the samples were cooled down for one hour in desiccators containing silica gel.

Mycorrhiza Infection

Fine root samples (0.2g, blot dried in paper towels) were chopped into 2-3 mm length, cleared with potassium hydroxide and stained with trypan blue (Philips and Hayman, 1970). Stained roots were evenly spread in Petri dishes and examined under a dissecting microscope. Mycorrhiza infected roots were counted following the gridline intersect method (Giovannetti and Mosse, 1980). For AM infection, roots with vesicles, hyphae or arbuscules were scored as mycorrhizal while dark blue root tips were considered as ECM colonized root tips. ECM fungi colonized the root tips while AM fungi colonized the older part of the fine roots. Clear root tips or root fragments were scored as non-mycorrhizal. Percent mycorrhizal infection was calculated as the proportion of infected roots over the total roots counted (infected plus uninfected) in ten field views under a stereo microscope.

Plant Phosphorus and Cu Concentration Analyses

Phosphorus and Cu concentrations were analyzed from 0.5mg ground fine roots or shoots after digestion with concentrated H_2SO_4 and 30% H_2O_2 and dilution with distilled water to 100 ml (PCARRD 1991). Sample preparation was done at the Bureau of Soil based at UPLB while sample reading using an Atomic Absorption Spectrophotometer was done at the Bureau of Soils and Water Management, Diliman, Quezon City.

Data Analysis

All data gathered were analyzed by Analysis of Variance (ANOVA) of a Randomized Complete Block Design using a MSTAT-C Statistical Computer Program (MSU 1989). Treatment means were compared using the Duncan Multiple Range test at $p<0.05$ (Duncan 1955).

Results

Mycorrhizal Diversity in Manila Mining Mine Site

Fruit bodies of ECM fungi *Scleroderma* spp. and *Thelepora* sp. (coral like) were found associated with mangium growing at the MMC dumpsite at Barangay Suyoc. The number of sporocarps of *Scleroderma* was three times more than *Thelepora* per mangium tree. Three genera of AM fungi were also isolated from the rhizosphere soil samples namely: *Glomus, Gigaspora* and *Acaulospora* (Table 1). *Glomus* spp. (44 spores per ten grams dry soil) out numbered *Gigaspora* (11 spores per ten grams soil) and the least was *Acaulospora* (4 spores per ten grams). *Entrophospora* sp. was not found in plants other than mangium. Mangium harbored the most diverse and highest spore count of mycorrhizal species and the least are the ferns.

Effect of Mycorrhizal Inoculation on Height and Diameter

Seedlings grown in marginal soil attained greater height increment than those grown in mine waste soil. MMC soil inoculant increased height increment of mangium in marginal soil by 86% and by 63% in mine waste soil relative to the uninoculated counterpart (5.53 cm and 3.81 cm, respectively) (Table 2). Height increment of the uninoculated plants grown in mine waste soil was significantly lower than those control plants grown in marginal soil.

Mangium inoculated with MMC soil inoculant obtained higher height and diameter increments than those inoculated with Mycogroe, Mykovam, and those not inoculated with mycorrhizal fungi (Table 2). Similarly, falcata inoculated with MMC soil inoculant were taller and with bigger stem diameter than the control (data not shown).

The beneficial effect of mycorrhizal inoculation was clearly seen three weeks after transplanting and this continued until harvest (12 weeks) (data not shown). Mangium grown in marginal soil gave higher height and diameter increments than those grown in mine waste soil throughout the duration of the experiment (Table 2). Likewise, mangkono seedlings inoculated with MMC inoculant were healthier and more robust as compared with the uninoculated counterpart (Table 3).

Table 1. Genus of mycorrhizal fungi collected from top four most common host plants thriving in Manila Mining Co., Suyoc dumpsite, Placer, Surigao del Norte, Philippines

Prevalent host plant	Spore count per 10g dry soil sample/description
Grass - Talahib (*Saccharum spontaneum*)	2 *Glomus* sp.1, light yellow 2 *Glomus* sp.2, dark brown 6 *Gigaspora margarita*
Reforestation species- Falcata (*Paraserianthes falcata*)	1 *Glomus* sp.1, light yellow 12 *Glomus* sp.2, dark brown 1 *Glomus* sp.3, light brown 1 *Glomus* sp. 4, black
Fern (different species)	4 *Gigaspora margarita*
Reforestation species- Mangium (*Acacia mangium*)	15 *Glomus* sp.2, dark brown 4 *Glomus* sp.3, light brown 2 *Glomus* sp. 5, dark red 1 *Gigaspora* sp.1, white 4 *Acaulospora* sp.1, light brown 8 *Entrophospora* sp., light brown

Interaction Effects of Soil and Mycorrhizal Inoculation on Plant Dry Weight

Generally, growth of all plant species studied was better in marginal soil than in mine waste soil. MMC soil inoculant significantly increased root, nodule and total dry weights of mangium grown in marginal soil (Table 2). Likewise, in mine waste soil, MMC soil inoculant significantly increased stem and nodule dry weight by 83% and 111%, respectively, relative to the control. Other parameters observed were highest with inoculation with MMC soil inoculant but the differences were not statistically significant as compared with the other treatments. In mine waste soil, the heaviest root, stem, leaf and total dry weight were obtained from mangium seedlings inoculated with MMC soil inoculant. Mangkono seedlings grown in mine waste soil also obtained the highest plant dry weight (root, shoot, stem, and leaf) with inoculation with MMC soil inoculant (Table 3).

Table 2. Interaction effects of soil and mycorrhizal inoculation on growth parameters measured after harvest of non-mycorrhizal and mycorrhizal *A. mangium* seedlings and grown in marginal and mine waste soil of the MMC, Placer, Surigao del Norte

Soil	Treatment	Height increment (cm)	Diameter increment (cm)	Root length (cm)	Leaf area (cm^2)	Root dry weight (g/plant)	Stem dry weight (g/plant)	Leaf dry weight (g/plant)	Nodule dry weight (g/plant)	Total dry weight (g/plant)
Marginal	Control	5.53 b*	0.183 ab*	31.16 a^{ns}	149 a^{ns}	0.192 cd*	0.263 ab*	0.602 ab**	27.60 bcd*	0.746 cd*
	Mykovam	7.73 b	0.155 a-d	36.80 a	152 a	0.244 bc	0.222 abc	0.551 ab	52.50 a	1.059 bc
	Mycogroe	6.52 bc	0.113 cd	34.98 a	166 a	0.285 ab	0.167 cd	0.368 cd	39.8 abc	0.962 bc
	Myk+Myc	7.68 b	0.155 a-d	36.96 a	142 a	0.269 bc	0.212 bcd	0.447 bc	45.3 ab	1.170 ab
	MMC soil inoculant	10.30 a	0.191 a	40.86 a	192 a	0.363 a	0.292 a	0.714 a	55.30 a	1.424 a
Mine tailings	Control	3.81 e	0.174 abc	24.64a	51 a	0.094 e	0.115 e	0.182 de	18.00 d	0.407 de
	Mykovam	4.14 de	0.091 d	27.16a	51 a	0.088 e	0.102 e	0.169 e	17.00 d	0.363 e
	Mycogroe	5.99 c	0.121 bcd	35.79 a	105 a	0.124 de	0.138 de	0.272 cde	26.50 bcd	0.552 de
	Myk+Myc	5.32 cd	0.116 cd	33.50 a	70 a	0.117 de	0.116 e	0.281 de	21.70 cd	0.48 de
	MMC soil inoculant	6.20 c	0.132 a-d	28.93 a	72 a	0.113 de	0.211 bcd	0.247 de	37.90 abc	0.575 de

NS = Not significant; * = Significant at 5% confidence level; ** = Highly significant at 1% confidence level.
Treatment means with the same letter(s) are not significantly different from each other at 5% confidence level using DMRT.

Table 3. Effect of mycorrhizal inoculation on the growth parameters measured after harvest of non-mycorrhizal and mycorrhizal (with MMC soil inoculant) mangkono seedlings and grown in mine waste soil of the MMC, Placer, Surigao del Norte Philippines

Inoculation treatment	Final height increment (cm)	Final diameter increment (cm)	Root length (cm)	Leaf area (cm^2)	Root dry weight (g/plant)	Stem dry weight (g/plant)	Leaf dry weight (g/plant)	Total dry weight (g/plant)	Mycorrhizal infection %
Uninoc	3.79 a^{ns}	0.176 b*	22.90 a^{ns}	184 b*	0.725 a^{ns}	0.717 a^{ns}	1.588 a^{ns}	3.03 a^{ns}	0
MMC soil inoculant	4.07 a	0.241 a	25.03 a	241 a	0.876 a	0.826 a	1.831 a	3.53 a	5.5

NS = Not significant; * = Significant at 5% confidence level.

Effect of Mycorrhizal Inoculation on P Concentration in Mine Waste Soil

Mycorrhizal plants contained higher (0.51%) P than the uninoculated (0.22%) counterpart (Table 4). Phosphorus absorption was increased by almost three times over the non-mycorrhizal plants. Among the tree species studied, falcata inoculated with MMC soil inoculant contained the highest (0.39%) concentration of P and the least was mangkono (0.34%). Phosphorus concentration in falcata and mangium was not significantly different from each other.

Furthermore, absorbed P was more than twice greater (0.50%) in the leaves compared with that in the roots (0.23%). The highest leaf and root P concentration were observed in mycorrhizal falcata. Leaf P concentration was 0.80% in the leaves and 0.40% in the roots. On the other hand, the lowest P concentration was observed in the roots (0.12%) and leaves (0.24%) of non-mycorrhizal or uninoculated falcata. All the uninoculated plants gave lower P concentration in the roots and P concentrations in the leaves were two times greater than in the roots (Table 5).

Effect of Mycorrhizal Inoculation on Cu Concentration in Mine Waste Soil

In general, Cu concentration was highest in mangium (50.62 µg/g) which is significantly higher than that in mangkono (29.38 µg/g) and the lowest was observed in falcata seedlings (19.38 µg/g) (Table 5). Falcata accumulated the highest P concentration but accumulated the lowest Cu concentration. Cu concentration was higher in the roots than in the leaves unlike P concentration which was higher in the leaves than in the roots. Cu accumulation was twice higher in the roots (45.33 µg/g) than in the leaves (20.92 µg/g). Inoculation with MMC soil inoculant increased Cu accumulation by three times in relation to the control or non-mycorrhizal counterpart (17.67 µg/g).

Cu concentration in the roots was highest in mycorrhizal mangium (106.5 µg/g) followed by the roots of mycorrhizal mangkono (64.00 µg/g) and the least were in the roots of mycorrhizal falcata (41.00 µg/g). On the other hand, the leaf Cu concentration was highest in mycorrhizal mangium (38.50 µg/g) and the least was in mycorrhizal mangkono (21.00 µg/g). All the uninoculated counterpart had lower Cu concentration both in the roots and in the leaves (Table 5).

Mycorrhizal Root Colonization

Mycorrhizal roots were higher (6-14%) in mangium grown in marginal soil than in mine soil (3-9%) (Table 6). Falcata seedlings grown in marginal soil and were inoculated with MMC soil inoculant gave the highest percent of root colonization (29%) and plants grown in either marginal or mine waste soil that was not treated with mycorrhizal fungi had the lowest percent of mycorrhizal infection (0%) except for the falcata seedlings grown in marginal and mine waste soil that had 3% and 4% mycorrhizal infection.

Table 4. Phosphorus and Cu concentrations of mangium, falcata and mangkono, grown in mine waste soil of MMC as influenced by mycorrhizal fungi native in the site

Source of Variation			P conc (%)	Cu conc (µg/g)
Factor A (tree)	Mangium		0.373 a[***]	50.62 a[***]
	Falcata		0.392 a	19.38 c
	Mangkono		0.334 b	29.38 b
Factor B (Inoculation)	Uninoculated		0.218 b[***]	17.67 b[***]
	MMC soil inoculant		0.514 a	46.58 a
A x B (Tree x Inoculation)	Mangium	Uninoculated	0.245 b[***]	28.75 c[***]
		MMC soil inoculant	0.500 b	72.50 a
	Falcata	Uninoculated	0.182 e	8.00 e
		MMC soil inoculant	0.602 a	30.75 c
	Mangkono	Uninoculated	0.227 d	16.25 d
		MMC soil inoculant	0.441 c	42.50 b
Factor C (Plant part)	Root		0.230 b[***]	45.33 a[***]
	Leaf		0.502 a	20.92 b

[***] = Significant at 0.1% confidence level.
Treatment means with the same letter(s) are not significantly different from each other at 5% confidence level using DMRT.

Table 5. Phosphorus and Cu concentrations in the roots and leaves of mangium, falcata and mangkono, grown in mine waste soil from MMC mine site as influenced by mycorrhizal fungi native in the site

Plant species	Inoculation treatment	Plant part	P conc (%)	Cu conc (µg/g)
Mangium	Uninoculated	Root	0.190 ef*	29.00 d***
		Leaf	0.300 d	28.50 d
	MMC soil inoculant	Root	0.300 d	106.5 a
		Leaf	0.720 b	38.50 c
Falcata	Uninoculated	Root	0.125 g	10.00 f
		Leaf	0.240 e	6.00 g
	MMC soil inoculant	Root	0.405 c	41.00 c
		Leaf	0.800 a	20.50 e
Mangkono	Uninoculated	Root	0.150 fg	21.50 e
		Leaf	0.303 d	11.00 f
	MMC soil inoculant	Root	0.210 e	64.00 b
		Leaf	0.672 b	21.00 e

*, *** = Significant at 5% and 0.1% confidence level, respectively.
Treatment means with the same letter(s) are not significantly different from each other at 5% confidence level using DMRT.

Table 6. Mycorrhiza infected roots of mangium, and falcata grown in marginal and mine waste soil and mangkono grown in mine waste soil of Manila Mining Co., Placer, Surigao del Norte

Host species	Type of soil	Inoculation treatment	Mycorrhizal infection (%)
Manguim (*Acacia mangium*)	Marginal	Control	0
		Mykovam	5.75
		Mycogroe	11.25
		Mycovam+Mycogroe	13.67
		MMC soil inoculant	5
	Mine waste soil	Control	0
		Mykovam	5
		Mycogroe	8.5
		Mycovam+Mycogroe	3
		MMC soil inoculant	4.67
Falcata (*Paraserianthes falcataria*)	Marginal	Control	3
		Mykovam	4
		Mycogroe	10.41
		Mycogroe + MMC soil inoculant	29
	Mine waste soil	Control	4
		Mykovam	15
		Mycogroe	16.67
		Mycogroe + MMC soil inoculant	17.67
Mangkono (*Xanthostemon verdugonianus*)	Mine waste soil	Control	0
		MMC soil inoculant	5.5

Discussion

A. mangium and *P. falcataria* are the dominant plants thriving in the Manila Mining Corporation, Placer, Surigao del Norte, in Mindanao island, Philippines and the roots were dominantly colonized with *Glomus*. These exotic fast growing trees are nitrogen fixers. There were numerous big nodules with pinkish sap when cut. This might be the reason why these legumes can thrive in mine tailing sites besides being colonized by a quite diverse species of mycorrhizal fungi. Aside from *Glomus*, which is the dominant AM fungi, species under the genera of *Gigaspora*, *Acaulospora* and *Entrophospora* were also observed. This sort of diverse AM population indicates that these mycorrhizal species can be tapped in rehabilitating such environmentally stressed areas.

Mycorrhizal fungi isolated from abandoned mine sites were reported to be more tolerant and more effective in promoting growth of plants planted in mine sites than the commercial mycorrhizal inoculant Mykovam (Kasahara et al. , 2007, Manalo et al. , 2005, Mercado et al. , 2006, Naupal et al. , 2007). In this study, the mycorrhizal fungi isolated in the mine waste soil were also effective in promoting growth of plants grown in marginal soil and that the mycorrhizal species present in Mykovam and Mycogroe which were collected in marginal grasslands and reported effective in promoting growth and yield of a variety of crops and trees grown in infertile soil. Their effectiveness as plant growth promoter was better than the non-mycorrhizal counterpart.

MMC soil inoculant has shown its potential in promoting growth and survival of mangium and mangkono both in marginal and in mine waste soils. Plants inoculated with MMC soil inoculant grew better than those inoculated with the commercial Mykovam and/or Mycogroe. The beneficial effect of inoculation with MMC soil inoculant was clearly seen as early as three weeks after transplanting and this was consistent until harvest. Mangkono seedlings treated with MMC soil inoculant gave greater height increment (4.07cm) than the control counterpart (3.79cm). The better growth of MMC inoculated seedlings was due to the greater amount of P in the plant tissue. Mycorrhizal plants contained higher (0.51%) P than the uninoculated (0.22%) counterpart. Phosphorus absorption was increased by almost three times over the non-mycorrhizal plants. Among the tree species studied, falcata contained the highest (0.39%) concentration of P followed by mangium (0.37%) and the least was mangkono (0.34%). Phosphorus concentration in falcata and mangium was not significantly different from each other. Inoculation with MMC inoculum increased P concentration in mangkono by 94%, mangium by 104% and falcata by 231%, relative to the control. Absorbed P was more than twice greater (0.50%) in the leaves compared with that in the roots (0.23%).

Mycorrhizal plants accumulated more Cu in their tissues but did not show any toxicity symptoms. This could be accounted for by the greater biomass plant produced as a consequence of being mycorrhizal that lead to dilution of Cu within the plant. Similar observations were reported by Aggangan et al. (2007). Aggangan et al. (2007) compared the effectiveness of three isolates of *Pisolithus* collected under eucalypts growing in Western Australia, Philippines and in a mining residue in New Caledonia on *Eucalyptus urophylla* seedlings in the presence or absence of Ni. Plant analysis revealed that inoculation increased plant growth through improved P uptake but did not prevent Ni uptake. However, toxicity was minimized by the dilution effect due to an increase in plant biomass. Moreover, the *Pisolithus* isolate from mine site was more effective in promoting growth of *E. urophylla* grown in soil amended with high Ni than the other isolates not native in mine sites.

In conclusion, *A. mangium* is a candidate reforestation plant in Suyoc mine waste dumpsite, at the Manila Mining Corporation since this tree species can tolerate up to 1,600 µg Cu/g with mycorrhizal inoculation. MMC soil inoculant increased Cu concentration by 152% in *A. mangium*. In mangkono and in falcate seedlings, Cu concentration was increased by 162% and 74% if inoculated with MMC soil inoculant relative to their respective control. The highest Cu concentration was observed in mycorrhizal mangium (72.50 µg/g) and the least was in uninoculated falcata seedlings (8.00 µg/g).

It is recommended that the performance of MMC soil inoculant (comprising *Scleroderma* sp, *Glomus* spp*., and Entrophospora* sp.) in promoting growth of fast growing reforestation species in mine waste soil should be further studied under field conditions. *Acacia mangium* is a good candidate for reforestation or rehabilitation of mine-out areas specifically Manila Mining Corporation, Placer, Surigao del Norte. However, other fast growing tree species should be evaluated for their tolerance in Cu, Ag, and Au mine sites for diversification. It is important to plant mycorrhizal trees in such areas in order to prevent the leaching of Cu and other hazardous heavy metals from the highland to the lowland which may eventually, enter into the food chain. As shown in this study, mycorrhizal infection had greater beneficial effect on plants grown in marginal soil particularly if inoculated with MMC soil inoculant. It is therefore important to include mycorrhizal fungi in the rehabilitation of Cu mine sites not only for Manila Mining Corporation, Placer, Surigao del Norte but to other Cu and Ni mine out sites in the Philippines.

References

Aggangan, N. S., B. Dell and N. Malajczuk. 2007. Nickel tolerance of *Pisolithus-Eucalytus urophylla* under nursery and field conditions. Paper presented during the IUFRO WP 2.08.03 held in Durban, South Africa on October 2007.

Aggangan, N. S., B. Dell, N. Malajczuk and R. E. De la Cruz. 1997. Field performance of *Eucalyptus urophylla* inoculated with an introduced and indigenous strains of *Pisolithus* at three sites in the Philippines. BIO-REFOR, Proceedings of Brisbane Workshop (Kikkawa, J. P. Dart, D. Doley, K. Ishii, D. Lamb and K. Suzuki, eds.). pp. 145-148.

Aggangan, N. S. 1996. Soil factors affecting the formation and function of *Pisolithus-Eucalyptus urophylla* ectomycorrhizas in acid soils in the Philippines. *PhD Thesis*. Murdoch University, Perth, Western Australia. pp.207

Aggangan, N. S., B. Dell and N. Malajczuk. 1996. Effect of soil pH on the ectomycorrhizal response of *Eucalyptus urophylla* S. T. Blake seedlings. New Phytologist 134(4): 539-546.

Brundrett, M., Boughcr, N. L., Dell, B., Grove, T. & Malajczuk, N. (1996). Working with Mycorrhizas in Forestry and Agriculture. ACIAR Monograph 32: 374 pp.

Duncan, D. B. 1995. Multiple Range and Multiple F test. Geometrics 11: 1-24.

Dupre de Boulois, H. (2007). Role of arbuscular mycorrhizal fungi on the accumulation of radiocaesium by plants. Doctoral dissertation, Universite catholique de Louvain (UCL). Retrieved from http://mycorrhiza.ag.utk.eru/theses/the2007.htm on January 22, 2010.

Duncan, D. B. 1955. Multiple range and multiple F tests. Biometrics 11: 1-24

Giovannetti, M. and B. Mosse. 1980. An evaluation of techniques for measuring vesicular-arbuscular infection in roots. New Phytologist 84: 489-500.

Harley, J. L. 1969. The Biology of Mycorrhiza. Leonard Hill, London.

Kasahara, E. S., A. B. Albano, N. S. Aggangan, E. M. Ragragio, N. M. Pampolina and N. K. Torreta. 2007. Contribution of mycorrhizal inoculation on heavy metal accumulation by *Jatropha curcas* L. in marginal and mine soils for better health. Paper presented during the 8th Annual Scientific Meeting and Symposium, Mycological Society of the Philippines, CLSU, Munoz, Nueva Ecija. April 16, 2007.

Landim, M. F. 2003. Brazilian Atlantic Rainforest Remnants and Mycorrhizal Symbiosis-Implications for reforestation. A case study in Sergipe, Northeast Brazil. Doctoral dissertation, University of Bremen (UFT), Plant Physiology and Plant anatomy, Leobener Strasse, D-28539 Bremen Germany. Retrieved from http://mycorrhiza.ag.utk.edu/theses/thes2003.htm on January 23, 2010.

Mercado, G. A., S. M. M. Reyes and N. S. Aggangan. 2006. Growth and heavy metal accumulation of narra (*Pterocarpus indicus*) for bioremediation of abandoned mine sites. Paper presented during the Scientific Meeting of the Mycological Society of the

Philippines held at the Ecosystems Development Bureau, UPLB College, Laguna. April 2006.

[MSU] Michigan State University. 1989. User's guide to MSTAT-C. Design, Management and Statistical Reseach Tool. East Lansing, Michigan: Michigan State University.

Naupal, R. T., N. S. Aggangan and N. M. Pampolina. 2007. Effects of mycorrhizal inoculation and other amendments to copper-rich Mogpog soil to growth and copper-accumulation of *Jatropha curcas* L. Paper presented during the 8th Annual Scientific Meeting and Symposium, Mycological Society of the Philippine*s*, CLSU, Munoz, Nueva Ecija. April 16, 2007.

PCARRD. (1991). Standard Methods of Analysis for Soil, Plant Tissue, Water and Fertilizer. PCARRD Book Series No. 120, Manila, Philippines.

Species Index

Subject Index

S